Springer Proceedings in Business and Economics

Springer Proceedings in Business and Economics brings the most current research presented at conferences and workshops to a global readership. The series features volumes (in electronic and print formats) of selected contributions from conferences in all areas of economics, business, management, and finance. In addition to an overall evaluation by the publisher of the topical interest, scientific quality, and timeliness of each volume, each contribution is refereed to standards comparable to those of leading journals, resulting in authoritative contributions to the respective fields. Springer's production and distribution infrastructure ensures rapid publication and wide circulation of the latest developments in the most compelling and promising areas of research today.

The editorial development of volumes may be managed using Springer Nature's innovative EquinOCS, a proven online conference proceedings submission, management and review system. This system is designed to ensure an efficient timeline for your publication, making Springer Proceedings in Business and Economics the premier series to publish your workshop or conference volume.

This book series is indexed in SCOPUS.

Francesca Pagliara
Editor

Socioeconomic Impacts of High-Speed Rail Systems

Proceedings of the 2nd International Workshop on High-Speed Rail Socioeconomic Impacts, University of Naples Federco II, Italy, 13–14 September 2022

Springer

Editor
Francesca Pagliara
Department of Civil, Architectural
and Environmental Engineering
University of Naples Federico II
Naples, Italy

ISSN 2198-7246 ISSN 2198-7254 (electronic)
Springer Proceedings in Business and Economics
ISBN 978-3-031-26339-2 ISBN 978-3-031-26340-8 (eBook)
https://doi.org/10.1007/978-3-031-26340-8

This Springer imprint is published by the registered company Springer Nature Switzerland AG
The registered company address is: Gewerbestrasse 11, 6330 Cham, Switzerland

Preface

Since 1964 with the Shinkansen in Japan, High-Speed Rail (HSR) deployment has played an important role in reshaping travel patterns and activities of people, consequently changing the way cities develop.

In the last decades, significant investments in HSR systems have been carried out in the world. Spending public money on the construction of HSR lines produces several benefits such as time saving, increase in comfort, induced demand, reduction in congestion and wider economic benefits including the development of the less developed regions. Motivations to develop HSR systems for many countries are not only related to the increasing transport infrastructure capacity or to the "green" transport alternative they represent, but also deal with the promotion of economic growth and regional development to which these systems can contribute.

HSRs have the potential for the formation of a megalopolis. The latter is an integrated economic urban complex, created by the fusion of multiple cities connected by high-speed transportation of 200–300 km/h. It is a geographical area that shares a common labour market and a common market for household and business services. The positive impacts of a megalopolis are larger labour markets and commercial markets, thus greater productivity; megalopolises are better and more effective than cities alone in meeting the economic and social challenges. Between two cities served by HSR, some phenomena can be observed: significant increases in one-day round trips between a pair or group of cities; increase of generated induced demand; induced demand for business trips; increase in the number of daily commuters. Given the power of HSRs in shrinking spaces and shaping places, the interest at the international level among researchers, academics and professionals has increased focusing their studies on the socioeconomic impacts of HSR systems.

This book collects most of the contributions presented at the 2nd edition of the International Workshop on High-Speed Rail (HSR) Socioeconomic Impacts, which was held online on 13 and 14 September organised by me. With over 100 participants and 39 speakers from all over the world, this large-scale workshop aimed to analyse and quantify the socioeconomic impacts of HSR systems.

The workshop was supported by the International Union of Railways (UIC), which in 2015 promoted the UIC Alliance of Universities for the development of

HSR. The Alliance is a global academic network including universities, institutes of technology, polytechnics, engineering, architecture and business schools that have proven substantial involvement in the deployment of HSR. The objective of the Alliance is to promote and implement research programs on these systems; to improve education in HSR; to exchange and disseminate know-how between Universities and UIC members; to develop outstanding talents with HSR expertise. Following the success of both editions, UIC has decided to support this event every year with the objective of building up a community around this topic, interested to meet every year to discuss the current trends on the models and methods to quantify these effects.

The contributions presented in this volume are divided into 6 parts. Part "HSR and Productivity" is dedicated to HSR and productivity. Part "HSR: Competition Versus Integration Versus Inclusion" deals with competition versus integration versus inclusion of HSR systems. Parts "HSR and Land Use Impacts 1" and "HSR and Land Use Impacts 2" present contributions dealing with HSR impacts on the land use system. Parts "HSR: Innovation and Environment" and "HSR: Accessibility and Tourism" report studies on innovation and the environment and on accessibility and the tourism market, respectively. Several are the case studies presented from Europe, China, Japan, India and the USA.

Naples, Italy Francesca Pagliara
fpagliar@unina.it

Contents

The Socio-economic Impacts of HS1: Towards a Re-assessment

John Preston

Abstract With a route length of around 108 km, High Speed 1 (HS1) is the only high speed rail line in operation in the UK. This chapter attempts to provide a socio-economic assessment of HS1, building on the earlier work of the author and of others. International services failed to attract the demand levels that were forecast and instead domestic passenger traffic has filled the capacity gap. However, the user benefits to these domestic passengers were less than that forecast for international passengers and as a result HS1 has failed to achieve a Benefit Cost Ratio in excess of unity, based on conventional transportation costs and benefits. This shortfall might be made up by regeneration benefits, as HS1 was specifically designed to stimulate economic activity in central and east London and in Kent. Some impacts in terms of land values can be detected, initially in central and east London, but subsequently spreading outwards to Kent. There is also some evidence of population gains but, like the changes in land values, these could be abstractive rather than generative. Moreover, evidence of increased productivity in terms of Gross Value Added per capita is difficult to find. The policy issues in terms of future high speed rail development in the UK are discussed.

Keywords High speed rail · Cost benefit analysis · Regeneration · Land value · Productivity

1 Introduction

This chapter builds on two papers published by La Revue d'Histoire des Chemins der Fer in 2018, which in turn have their origins in earlier conference presentations [1, 2]. It also builds on work by others, most notably Atkins et al. [3]. The aim is to provide a socio-economic assessment of High Speed 1 (HS1), previously known as the Channel Tunnel Rail Link (CTRL). This will be done as follows. In the next

J. Preston (✉)
Transportation Group, School of Engineering, University of Southampton, Southampton SO16 7QF, UK
e-mail: J.M.Preston@soton.ac.uk

F. Pagliara (ed.), *Socioeconomic Impacts of High-Speed Rail Systems*,
Springer Proceedings in Business and Economics,
https://doi.org/10.1007/978-3-031-26340-8_1

section, the evolution of HS1 will be briefly reviewed. In Sect. 3, the usage of HS1 and related rail services will be assessed. The previous assessments of HS1, particularly by the National Audit Office, will be reviewed in Sect. 4, whilst Sect. 5 will examine regeneration evidence at stations such as Ashford, London St. Pancras and Stratford International. Section 6 will present evaluation evidence at the corridor, station and Local Authority District (LAD) levels. Finally, Sect. 7 will draw some tentative conclusions.

2 A Brief Chronology

The Channel Tunnel opened in May 1994 and, although there was a High Speed rail link on the French side of the Channel, there was no such link on the UK side [4]. Initially called the Channel Tunnel Rail Link (CTRL), the UK High Speed rail link was opened in two phases. Phase 1, which ran broadly from the Chunnel portal to the M25 (see Fig. 1) opened in 2003, whilst Phase 2 from the M25 to central London opened in 2007 but with Stratford station and domestic services not entering operation until 2009. From 2007, the line was re-branded High Speed 1 (HS1). Initially, a south easterly route was proposed serving London Waterloo, the original terminus for the Channel Tunnel services. However, as a result of lobbying by consultants Arup and the then Deputy Prime Minister, Michael Heseltine, a more easterly route with a terminus at London St. Pancras was confirmed in 1996. The objective was to regenerate London Docklands and the Thames Gateway with stations at Stratford and Ebbsfleet [5]. The re-routing of HS1 has permitted the subsequent repurposing of Waterloo International to provided additional capacity for commuter services—the evaluation of which is beyond the scope of this chapter.

The link was originally developed by the London and Continental Railways Consortium that consisted of National Express, Virgin Group, SG Warburg & Co, Bechtel and London Electric. However, the consortium quickly encountered financial difficulties and despite restructurings in 1998 and 2002 was taken into public ownership in 2009. Moreover, in 2010, HS1 was acquired by Borealis Infrastructure and Ontario Teachers' Pension Plan who, in 2017, sold on to InfraRed Capital Partners Limited and Equitix Investment Management Limited.

The operation of the Channel Tunnel and HS1 has been fraught with difficulties, including issues related to fire, border control and industrial disputes but two years should be particularly highlighted, namely the economic recession of 2008 and Covid and Brexit in 2020.

Fig. 1 Sketch of the HS1 route and stations

3 Usage

Transport Statistics Great Britain [6] provides some aggregate data on the usage of the Channel Tunnel by both passengers and freight. Figure 2 indicates that demand in terms of passengers on both Eurostar and Le Shuttle built-up rapidly over the first four years, peaking at over 18 million in 1998. This ramp-up of demand is in line with other studies (see, for example, [7]). There then followed a period of decline that was reversed in 2003 with the opening of CTRL Phase 1—but the 18 million figure was not exceeded again until 2013. Peak usage of 21.6 million was achieved in 2018, which coincided with the introduction of through services to Amsterdam. However, Covid in 2020 led to a 62% reduction in passenger demand. It should be noted that the Channel Tunnel Pathfinder Prospectus (November 1987) forecast usage by 2003 of 39.5 million passengers. This was revised down in the Rights Issue document (May 1994) to 35.77 million. Even if this forecast is used, at most 60% of the forecasted number of passengers has been achieved, with the short-fall linked to persistent competition from short-sea ferry operators and low-cost airlines and the lack of regional services beyond London and Paris.

As Fig. 3 indicates, rail freight movements through the Channel Tunnel are even more disappointing than passenger movements. They peaked at just over 3 million tonnes lifted in 1998 but in the last decade have fluctuated between 1 and 1.5 million tonnes. The Channel Tunnel Pathfinder Prospectus (November 1987) forecast rail

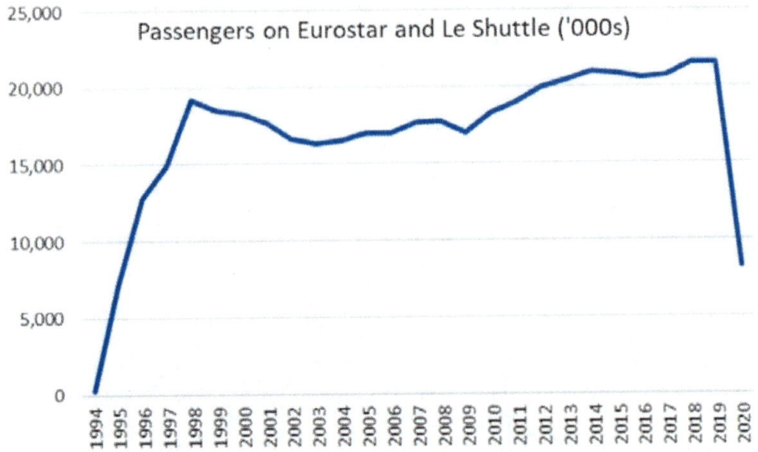

Fig. 2 Passenger usage of the channel tunnel (*Source* [6])

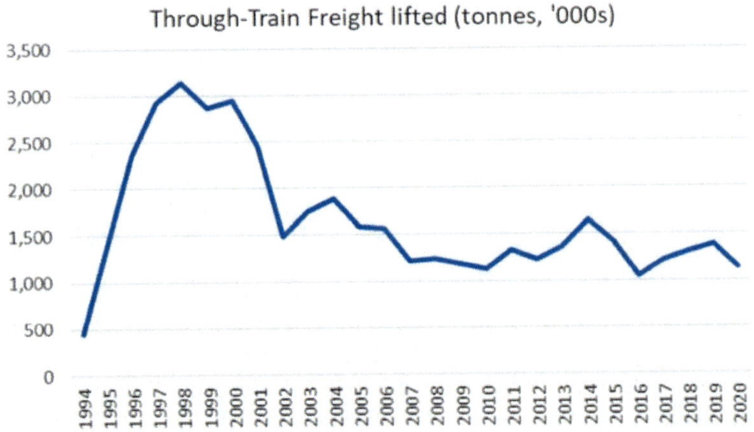

Fig. 3 Rail freight usage of the channel tunnel (*Source* [6])

freight usage by 2003 of 10.6 million tonnes. This was revised down in the Rights Issue document (May 1994) to 10.45 million. Hence, at the very most only 30% of the envisaged rail freight demand has materialised. This may be linked to competition from roll-on, roll-off and lift-on, lift-off ferries as well as Le Shuttle and interoperability issues with multi-national rail freight movements.

Figures for Eurostar usage can be obtained from the trade press and from company accounts. Figure 4 shows that Eurostar demand initially peaked at just over 7 million in 2000. It did not achieve this figure again until 2004 but CTRL and HS1 seem to have presaged a period of sustained growth with demand in 2019 exceeding 11 million. However, in 2020 Covid saw a 77% reduction in demand. The Channel Tunnel Pathfinder Prospectus (November 1987) forecast Eurostar usage by 2003 of

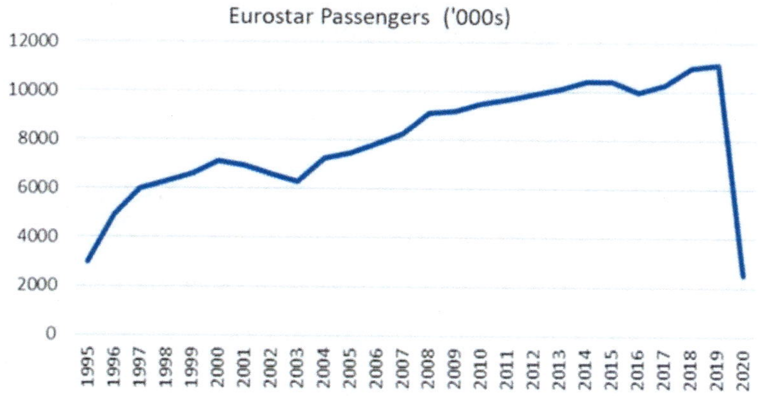

Fig. 4 Eurostar usage (*Sources* Trade press and Eurostar annual reports and accounts)

21.4 million passengers. This was revised down in the Rights Issue document (May 1994) to 17.12 million. This indicates that out-turn demand was only at most a little over 50% of that forecast. Faith (op cit., page 49) notes that British Rail in 1986 had forecast usage of between 13.4 and 15.6 million passengers a year, with nearly half coming from outside London and the South East. Although Eurostar has a reportedly high share of centre to centre (capital to capital) traffic, it has a negligible share of region to region traffic. The regional through services (Trans European Express), including night services, promised for the UK never materialised, even though the trains were ordered. Neither Waterloo nor St. Pancras are particularly well connected to the national network. Much the same might be said of Gare du Nord. Both central London and central Paris represent significant barriers to onward travel, with at least one interchange and sometimes two required.

Domestic High Speed services began full operation in the UK on 13 December 2009, with preview services running from 29 June 2009. Figure 5 shows that demand grew rapidly in the first year and that by mid-2013/14 annual demand was approaching 10 million [8]. However, Fig. 6 indicates that prior to the introduction of High Speed services to St. Pancras, South East Trains classic mainline services were carrying around 55 million passengers per annum, which by mid-2013/14 had reduced to 50 million. Thus, in very broad terms, 50% of High Speed rail usage was abstracted from classic rail, a figure consistent with studies of other High Speed rail lines [9]. It also suggests High Speed services had increased total rail usage by around 9%.

We investigated this further by using station count data collected by the Office of Rail Regulation (ORR - since 2015 the Office of Rail and Road). The first part of this analysis involved carrying out a corridor by corridor assessment of demand change on routes served by domestic HS1 services and on a control route. Domestic services on HS1 began operation in June 2009, so this analysis compares demand in the year 2008–9 (as recorded in the ORR station usage spreadsheets) with demand in 2013–4, 2019/20 and 2020/21. The analysis covered 51 stations, which included 24 stations

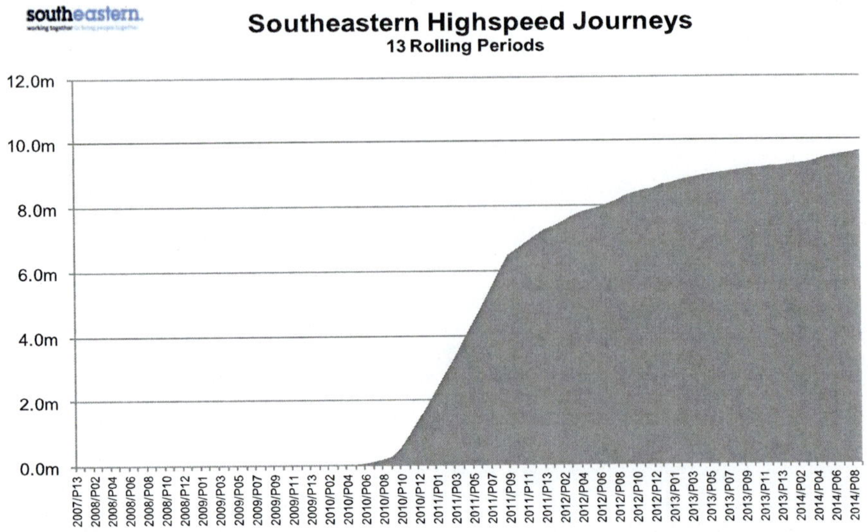

Fig. 5 Usage of domestic high speed services (millions, moving annual average) (*Source* [8])

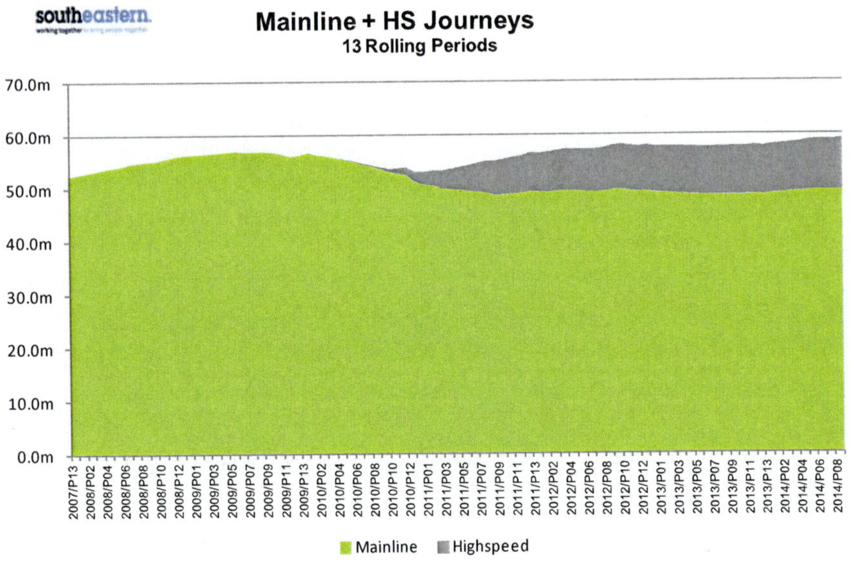

Fig. 6 Usage of South East trains longer distance services (millions, moving annual average) (*Source* [8])

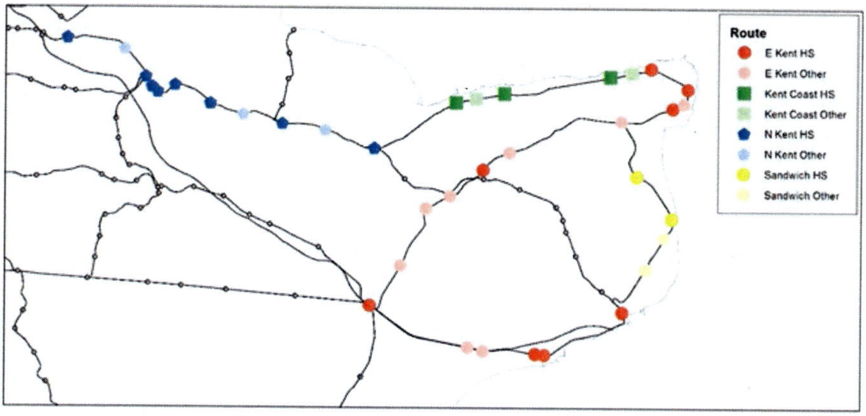

Fig. 7 Corridors served by HS1 domestic services (*Source* [1])

served by HS1 domestic services (including Stratford and Ebbsfleet), 15 stations on the routes served by HS1 domestic services but where these services did not stop, and 12 stations on a control route not served by HS1 services (from Tonbridge to Hastings—although this service did benefit from the re-routing of international services). The routes served by HS1 domestic services were divided into four corridors for the purposes of this analysis, and these are illustrated by Fig. 7 which also shows whether or not stations were served by the High Speed services.

Excluding Stratford and Ebbsfleet (neither open to domestic services in 2008/9), the 22 stations that were served by domestic HS1 services were used by 27.5 million passengers in 2008/9, 28.9 million in 2013/14 (up 5%), 33.8 million in 2019/20 (up 23% compared to 2008/9) and 10 million in 2020/21 (down 64% compared to 2008/9). Stratford and Ebbsfleet stations were used by 2.2 million passengers in 2013/14, 4.9 million passengers in 2019/20 and 1.2 million passengers in 2020/21. Many of these stations are served by both classic and High Speed services, with the former serving termini at the southern edge of the London central business district, whilst the latter serves St. Pancras which is at the northern edge of the central business district.

The impact on the different lines in Kent and on stations served by High Speed Rail services is shown in Table 1. The Hastings line, mainly in adjoining East Sussex, is used as a control (12 stations). This Table indicates a high degree of switching between stations and between lines, with the picture complicated by the fact that the re-routing of international services permitted a recast of the timetables for classic rail services. Additionally, the usage of some stations was affected by external factors, such as the scaling down of activities in the Sandwich area by the Pfizer pharmaceutical company in 2011.

Overall, compared to 2008/9, usage at a sample of 37 Kent stations in 2013/14 increased by 3.1% (excluding Ebbsfleet and Stratford). By contrast usage of stations on the Hastings line declined by 4.7%, suggesting a Difference in Difference growth

Table 1 Change in annual usage of rail stations compared to 2008/9

Year	2013/14			2019/20		
Location	High speed (%)	Other (%)	Overall (%)	High speed (%)	Other (%)	Overall (%)
North Kent	1.4	−27.3	−6.4	20.3	−18.3	9.7
Kent Coast	0.7	−22.4	−8.5	27.0	−25.5	6.0
East Kent	19.9	−14.4	2.8	67.0	8.2	37.6
Sandwich	−19.9	−27.5	−23.7	5.7	32.9	19.3
Hastings	N/A	−4.7	−4.7	N/A	12.6	12.6

N/A = Not Applicable
Source ORR

of 7.8%. By 2019/20, usage of the 37 Kent stations had increased by 20.7% but usage of stations on the Hastings line had increased by 12.6%, so that the Difference in Difference growth was 8.1%. As with the intercontinental traffic, this suggests that the main growth in demand occurred relatively quickly, with subsequent growth related to external factors such as the post-2008 economic recovery.

4 Previous Socio-economic Assessments

There have been attempts to undertake socio-economic assessments of both the Channel Tunnel and HS1 using cost–benefit analysis. Anguera [10, 11] has undertaken ex-post evaluations of the Channel Tunnel, with the first study (based on a ten year horizon) suggesting a Benefit–Cost Ratio (BCR) of 0.41 and the second study (based on a twenty year horizon) giving a BCR of 0.72. By contrast, Boeuf and Goldsmith [12] argue that if a longer-term perspective is taken, Eurotunnel might be expected to make a return of between 3 and 6%.

With respect to the CTRL/HS1, ex-ante, the National Audit Office [13, p. 36] indicated that on transport grounds alone, the case for the CTRL was quite weak, with a BCR of only 1.1 compared to the Department for Transport's May 1998 estimate of 1.5, including regeneration benefits, or 1.26 if regeneration benefits were excluded. If regeneration benefits were included by the NAO (estimated at the time as around £500 million), the BCR was estimated to increase to 1.35. The regeneration benefits were based on the Government's willingness to pay for the extra 50,000 jobs (net) that the CTRL was forecast to create, adjusted for double counting of benefits. Later estimates of the regeneration benefits of CTRL were put as high as £8 billion [14], although this figure is strikingly similar to the £8 billion of regeneration investments later identified at St. Pancras/King's Cross and the Thames Gateway by the NAO [15]. Subsequently, the NAO produced an ex-post BCR for CTRL of 0.8 [16]. They estimated that if the BCR was to be increased to 1.5 (which was the BCR used in the business case in 1998) regeneration benefits of £8.3 billion would be required.

A more recent evaluation by Atkins et al. [3] estimates a BCR of 0.53 for HS1 without wider economic benefits and 0.64 with wider economic benefits, although regeneration benefits are not included. The Present Value of Costs is estimated as £12.6 billion (2010 prices) and the Present Value of Benefits is estimated at £8.0 billion (including wider economic impacts). Hence for a BCR of 1.5 to be achieved, additional regeneration benefits of £10.9 billion would be required. These additional benefits might come from a number of sources. HS1 might reduce imperfections in transport-using sectors, particularly by reducing imperfections in land and labour markets. It might promote agglomeration benefits by increasing productivity, in particular by increasing accessibility to jobs in central London. These wider economic impacts were estimated by Atkins et al. but were only estimated to be around £1.3 billion. In addition to these wider economic impacts, there might also be spatial equity gains by promoting development in deprived locations, including the Thames Gateway and east Kent that would not otherwise attract economic development. In addition, some economic activity might be generated, for example in the cultural and tourist sectors that might not be detected by conventional methods of estimating wider economic impacts. It is evidence of benefits of this type that will be reviewed in the rest of this chapter.

5 Regeneration Benefits: Initial Studies

This section reports on earlier work on regeneration benefits at the four station sites on HS1, before attempting to update these findings. Initial research on Ashford is reported in [17, 18]. The International station opened in 1996 but international services were always modest, peaking at around 30 trains per day and drastically reducing in 2007 to only 8 trains a day. In 2022, Ashford lost its international service completely. For domestic services, the big change came in 2009 (after the initial research) when High Speed domestic services were introduced to Ashford, with around 85 trains per day, reducing the journey time to London from around 90 min (with a fastest time of 84 min) to around 45 min (with a fastest time of 37 min). Unsurprisingly, the opening of Ashford International station had only modest effects but did coincide with an 11% increase in population, a 6% increase in employment and a 3% increase in house prices over that of the South East as a whole (1995–2005). Furthermore, Table 2 shows that the development impacts in Ashford between 1991 and 2001 were relatively modest. Only 25% of the planned commercial developments, 76% of the planned housing and 50% of the planned jobs materialised between 1991 and 2001. But development rates over the next 30 years were anticipated to be substantially greater. Large developments were also expected at the three other sites on HS1 involving over 5 million m^2 of commercial development and almost 20,000 new homes and 90,000 new jobs. Between 2001 and 2031 some 32,000 new dwellings were planned for Ashford, or over 1,000 per year. However, Atkins et al.

Table 2 Comparison of development at four sites

	Commercial development (m^2)	Dwellings	Jobs
Ashford 1991-01 P	430,000	6800	10,000
Ashford 1991-01 A	107,000 (Net)	5200	5000
Ashford 2001-31 P	1,416,000*	32,000	28,000
Ebbsfleet	2,390,000	10,000	27,000
Stratford	770,000	4500	30,000
St. Pancras	550,000	2500	30,000
Total	5,126,000	51,000	115,000

N.B. P = Planned; A = Actual; * = Total m^2 by 2031
Source Based on [17]

[3] find that between 2009 and 2012 housing stock for the entire HS1 corridor only increased by around 23,000, although if pro-rated by population Ashford would be just meeting its targets. The population of the Ashford District in 2001 was around 103,000. By 2011, it had increased to around 118,000 (up 14.5%) and by 2021 it was 132,700 (up 29%).

Work on property prices has been undertaken around St. Pancras and Stratford stations. Cascetta et al. [19] found that comparing 2008 with 2006—the year before the opening of St. Pancras International Station—an increase of 28% in private accommodation prices in the borough of Camden could be observed. This data can be compared with the increase of 15% which took place in London in the same period. However, the boundaries of the London Borough of Camden are by no means contiguous with the catchment area of St. Pancras station. Instead, Cascetta et al. compared changes in property prices within 5 miles of St. Pancreas with properties within 5–10 miles over the period 2006–2008. House prices within 5 miles increased by 28% compared to 11% within 5–10 miles (17% net), whilst office prices increased 20% within 5 miles compared to no change within 5–10 mile (20% net). More spatially detailed work by Pagliara et al. [20] found that the main property price movements were between 2007 and 2009 for St. Pancras, with prices increasing 32% between 250 and 500 m and by 28% between 500 and 1,000 m from the station. For Stratford, the main impact detected was a 64% increase between 2009 and 2010 for properties within 250 m.

6 Corridor, Catchment Area and Local Authority Assessments

The HS1 evaluation by Atkins et al. (op cit., Chap. 4) [3] attempted to assess (but not monetise) regeneration benefits. This was done at two levels: Corridor level and Station level, which is supplemented here by additional work at the Local Authority level.

At the Corridor level, the HS1 corridor in Kent was compared with three other corridors: the Milton Keynes corridor along the M1, the Cambridge corridor along the M11 and the Colchester corridor along the A12. In terms of the increase in the rateable value of business properties between 2005 and 2010 this was found to be greatest in the HS1 corridor at 24%. However, by contrast the increase in the number of business premises over the period was lowest in the HS1 corridor at 0.23%. In terms of changes in average house prices between Quarter 3 2005 and Quarter 3 2012, the increase was 36.5%, the second highest of the four corridors after Milton Keynes (49.2%). The growth in housing stock in the HS1 corridor between 2009 and 2012, at 2.4%, was behind that of the Cambridge (2.9%) and Milton Keynes (2.7%) corridors. Between 2007 and 2011, total workplace employment in the HS1 corridor decreased by 2.3%, the worst performance of the four corridors. By contrast, for the Cambridge corridor, it increased by 1.4%. The analysis of GVA (Gross Value Added) is conflated by the 2008 recession and hence the focus is on the period between 2009 and 2011. Here, there has been a growth for the HS1 corridor of £187 million (in 2010 prices), which is a better performance than either the Milton Keynes or Colchester corridors where the economies continued to contract, but is inferior to that of the Cambridge corridor, which exhibited an increase of £3,108 million. Overall, there is little evidence that the HS1 corridor outperformed the other corridors in terms of economic performance. It is understood that the Department for Transport is currently updating this evaluation, the results of which will be incorporated into later analysis.

At the Station catchment area level, Atkins et al. [3] made some comparisons based on difference-in-differences (DiD), as illustrated by Tables 3, 4 and 5. Table 3 shows that all stations have outperformed the England average over the period 2005–2010 in terms of increases in average rateable values, with the biggest relative growth being at St. Pancras.

Table 4 shows that all stations have outperformed the England average over the period 2005–2010 in terms of the percentage increase in the number of commercial premises, although only just so in the case of Stratford International. The largest relative increase is for Ashford, but this was from a relatively low base compared to some of the other stations. There is also a large relative increase for the St. Pancras 500 m buffer—but from a much higher base.

Table 5 shows that house price increases have been above the national average for St. Pancras and for Stratford International for the period 2009–2013, although in

Table 3 Percentage increase in average rateable values (2005–2010)

	500 m buffer	2 km buffer	DiD 500 m	DiD 2 km
Ashford	8.7	16.0	4.0	11.3
St. Pancras	53.1	46.0	48.4	41.3
Ebbsfleet		28.7		24.0
Stratford Int.		25.7		21.0

England average increase 4.7%
Source [3]

Table 4 Percentage increase in number of commercial premises (2005–10)

	500 m buffer	2 km buffer	DiD 500 m	DiD 2 km
Ashford	4.8	6.1	4.3	5.6
St. Pancras	5.1	2.0	4.6	1.5
Ebbsfleet		4.7		4.2
Stratford Int.		0.6		0.1

England average increase 0.5%
Source [3]

Table 5 Percentage increase in average house prices (2009–13)

	2 km buffer (%)	DiD 2 km (%)
Ashford	7	−10
St. Pancras	50	33
Ebbsfleet	0	−17
Stratford Int.	21	4

England and Wales average increase 17%
Source [3]

the latter case only just. By contrast, they have been below the national average at Ashford and, particularly, Ebbsfleet—but this may reflect new supply outstripping demand.

Some of this data can be updated using information for Local Authority Districts, with Stratford in the London Borough of Newham and Ebbsfleet in the Borough of Dartford in Kent. Table 6 suggests that between 2014 and 2021 there were relative declines compared to the national average in property price in Camden but increases in Newham, Dartford and Ashford. This might indicate possible impacts emerging further out from central London, a pattern that also is apparent with respect to population changes between 2011 and 2021 (Table 7).

By contrast, Table 8 suggests that there has not been any impact in Ashford or Dartford in terms of productivity in terms of GVA per capita compared to England (or indeed Kent and Medway) as a whole. However, the analysis here is complicated by the fact that these two locations have become more attractive for commuting, with the gains in GVA likely to accrue in central London and not in Kent.

Table 6 Percentage increase in average house prices (2014–21). England and Wales average increase 53%

LAD	% increase (%)	DiD (%)
Ashford	62	+9
Camden	27	−26
Dartford	64	+11
Newham	66	+14

Source https://www.gov.uk/check-house-price-trends

Table 7 Population changes 2011–2021. England average increase 7%

LAD	2011	2021	Change (%)	DiD (%)
Ashford	118,000	132,700	+13	+6
Camden	220,300	210,000	−5	−12
Dartford	97,400	116,800	+20	+13
Newham	308,000	351,000	+14	+7

Source ONS, Census

Table 8 GVA per capita (£) and percentage change 2009–2019

	2009	2014	2019	% change (%)
Ashford	19,595	22,199	24,971	+27
Dartford	29,717	33,095	35,327	+19
Kent and Medway	18,564	20,558	24,306	+31
England	22,951	26,166	30,239	+32

Source Kent County Council

7 Conclusions

Although there is a need for a more systematic and robust approach, there is some evidence of regeneration benefits of various sorts associated with CTRL/HS1. Such benefits were initially greatest around St. Pancras and Stratford (see also [21]) where there were complementary factors, not least the 2012 London Olympics and preceding transport investments (such as the Jubilee Line and Docklands Light Rail Extensions). However, there were also adverse macroeconomic factors from 2008 and again from 2020, the latter possibly compounded by Brexit. More recently, there is some evidence that population is beginning to shift outwards along the HS1 corridor, with a knock-on effect on property prices. However, to date, it is unlikely that the regeneration benefits are of the scale required to judge HS1 a socio-economic success. The total HS1 corridor population in 2021 was 2,416,900 (Camden, Kent, Medway and Newham). Suppose we have an interest rate of 3.5% (the current UK Government test discount rate) and a project life of 30 years. From [3] it was estimated that a present value of £10.9 billion in terms of regeneration benefits would be needed to achieve a BCR of 1.5. Using a capital recovery factor, this would equate to annual benefits of £592.6 million or around £245 per person per year. There is no sign that there have been productivity increases of this (or indeed any) magnitude, at least in the parts of Kent served by HS1. Alternatively, given a 2019 GVA per capita of £24,306 (Kent and Medway average, 2019), this would require 24,380 more people. This is equivalent to a 1.3% increase of the Kent and Medway population—but this would need to be both generative and attributable to HS1. Between 2011 and 2021, Kent and Medway's population increased by 7.4% compared to England's 6.6%. The difference in differences is 0.8% so even here there is a shortfall. These benefits may come in the long-run. However, previous policy studies have suggested that up to 15 years is sufficient for long-run effects to manifest themselves [22, 23] which would suggest 2022 is the cut-off for HS1 (2024 for domestic services), although this might be extended for transformative infrastructure projects.

Another yardstick that HS1 can be measured against is traffic. In 2019/20, HS1 traffic peaked at an estimated 23.5 million passengers (of which 47% was international). By 2021/21, this had reduced to an estimated 8.9 million (28% international). Overall, demand was down 62% due to Covid, with international travel particularly badly affected. Out of the 39 HSR services compiled by Preston [9], the 2019/20 usage would rank HS1 16th, whilst the 2020/21 usage would rank HS1 31st. Rapid recovery of both domestic and international usage is vital if HS1 is to be judged a socio-economic success. Due to high costs, required level of demand are likely to be above the threshold of 9 million passengers per annum determined by Nash et al. [24]. That paper reported construction costs in the range of £11 to £79 million per km. If we suppose Nash et al.'s break-even demand calculations are based on the mid-point (£45 million) but HS1 costs are at the high mark, then the social break-even threshold would be around 16 million. Prior to Covid, this was being achieved, although some of the journeys would be relatively short, with commensurately low benefits.

Overall, it is evident that HS1 has not yet met its objectives in terms of transport or socio-economic criteria. In terms of transport, international traffic has been substantially below forecast levels and domestic services have filled the capacity gap. In terms of socio-economics benefits, there is some evidence of regeneration benefits, initially around St. Pancras and Stratford but subsequently spreading to Ebbsfleet and Ashford. However, these benefits are likely to be abstractive rather than generative and do not yet seem sufficient to provide a socio-economic justification of HS1 although other intangible benefits, such as national prestige, may also be important. These findings suggest that other high speed rail lines in Britain, such as High Speed 2 (HS2) currently being built between London, Birmingham and Manchester, may have challenges in attracting passengers and in generating economic development without complementary planning policies.

Acknowledgements An earlier version of this chapter was presented to the 2nd International Workshop on High Speed Rail Socioeconomic Impacts organised by University Federico II (Naples) and UIC on 13th September 2022.

References

1. Blainey, S., Preston, J.: The Channel Tunnel Rail Link in the UK: Is it all that it seems? Twenty Years Under The Channel And Beyond, Lille, March (2015). Also in Rohou, J.L. (ed.): Twenty Years under the Channel and Beyond? La Revue D'Histoire Des Chemins De Fer. 48/49, 183–200. SNCF, Paris (2018)
2. Preston, J.: The Channel Tunnel Rail Link in the UK: What Have Been the Impacts on Economic Development? Twenty Years Under the Channel and Beyond: Accessible Regions and Growing

Regions? Canterbury (2016). Also in Rohou, J.L. (Ed.): Twenty Years Under the Channel and Beyond? La Revue D'Histoire Des Chemins De Fer. 48/49, 367–382. SNCF, Paris (2018)
3. Atkins, AECOM and Frontier Economics: First Interim Evaluation of the Impacts of High Speed 1. Final Report. Department for Transport, London (2015)
4. Gourvish, T.: The Official History of Britain and the Channel Tunnel. Routledge, Abingdon (2006)
5. Faith, N.: The Right Line. The Politics, The Planning and The Against-The-Odds Gamble Behind Britain's First High-Speed Railway. Segrave Foulkes, Kingston (2007)
6. DfT.: Transport Statistics Great Britain (2021)
7. Preston, J., Dargay, J.: The Dynamics of Rail Demand Third Conference on Railroad Industry Structure, Competition and Investment. Stockholm, Sweden (2005)
8. Lucas, V.: Sintropher Project. http://www.sintropher.eu/publications/conferences-proceedings/seminar-debate-catalysing-development-through-transport (2014). Accessed 27 Feb 2015
9. Preston, J.: The Economics of Investment in High-Speed Rail. Round Table 155, International Transport Forum, Paris (2014)
10. Anguera, R.: The channel tunnel—an ex post economic evaluation. Transp. Res. A **40**(4), 291–315 (2006)
11. Anguera, R.: The Channel Tunnel Costs-Benefit Analysis After 20 Years of Operations. Twenty Years Under the Channel and Beyond: Accessible Regions and Growing Regions? Canterbury (2016). Also in Rohou, J.L. (ed.): Twenty Years Under the Channel and Beyond? La Revue D'Histoire Des Chemins De Fer. 48/49, 233–281. SNCF, Paris (2018)
12. Bouef, P., Goldsmith, H.: Une tentative de réhabilitation du tunnel sous la Manche dans une perspective historique. Twenty Years Under The Channel And Beyond, Lille, March (2015). Also in In Rohou, J.L. (ed): Twenty Years Under the Channel and Beyond? La Revue D'Histoire Des Chemins De Fer. 48/49, 79–97. SNCF, Paris (2018)
13. NAO (National Audit Office).: Department of the Environment, Transport and the Regions. The Channel Tunnel Rail Link. HC 302. The Stationery Office, London (2001)
14. DfT (Department for Transport): Facts and Figures on the CTRL. Department for Transport, London (2007)
15. NAO.: Department for Transport. Progress on the Channel Tunnel Rail Link. HC 77. The Stationary Office, London (2005)
16. NAO.: Department for Transport. The Completion and Sale of High Speed 1. HC1834. The Stationery Office, London (2012)
17. Larbie, A.T.: The impacts of high speed trains in north west europe: ashford case study. M.Sc. Dissertation, School of Civil Engineering and the Environment, University of Southampton (2006)
18. Preston, J., Wall, G.: The ex-ante and ex-post economic and social impacts of the introduction of High Speed Trains in South East England. Planning Pract. Res. **23**(3), 403–422 (2008)
19. Cascetta, E., Pagliara, F., Brancaccio, F., Preston, J.: Evaluating regeneration impacts of the channel tunnel rail link.In: 12th World Conference on Transport Research, Lisbon, Portugal, 11–15 July 2010
20. Pagliara, F., Rietveld, P., Preston, J.: London High Speed stations impact on residential property prices: the case of St. Pancras and Stratford station. In: 17th European Colloquium on Quantitative and Theoretical Geography (2011)
21. Ward, J.E., Wright, P.G., Dimitriou, H.T.: Perspectives on "success" from the UK channel tunnel rail link project. Plan. Theory Pract. **15**(3), 423–428 (2014)
22. Preston, J., Robins, D.: Evaluating the long term impacts of transport policy: an assessment of rail privatisation. Res. Transp. Econ. **39**, 134–144 (2013)
23. Preston, J., Almutairi, T.: Evaluating the long term impacts of transport policy: an initial assessment of bus deregulation. Res. Transp. Econ. **39**, 208–214 (2013)
24. Nash, C., Jandova, M., Paleta, T.: When is HSR worthwhile? Lessons from Western Europe and implications for Central and Eastern Europe. European Transport Conference, Milan (2022)

HSR and Productivity

Evaluation of the Economic Effects of High-Speed Rail on the Italian Economy Through a Broader Input–Output Model Approach

Mario Tartaglia, Michele Cerullo, Tommaso Ferraresi, Renato Paniccià, Mara Radicioni, and Enrico Cieri

Abstract High-Speed Rail (HSR) is a very competitive and sustainable mode of transport; however, it requires considerable infrastructure and rolling stock investments. The convenience of realizing this transportation system is discussed in scientific literature and policy-maker debates. As is customary when assessing transport systems, several approaches are used when evaluating the usefulness of building a new HSR, each referring to a specific point of view or a particular methodological context. Among these various approaches, the Economic Impact Analysis focuses mainly the consequences of the investments rather than the broader more general effects on the social and territorial system. Indeed, it focuses on direct, indirect and induced effects of the expense on the economy of the area studied. Several studies have dealt with the impacts of the Italian HSR network, but only a few of them investigated these from the economic point of view. The purpose of this paper is to address the latter perspective. It presents an Input–Output economic model developed combining Supply and I–O tables, so including a wider view compared to some more traditional calculations. This model has been applied to assessing the impact of the complete historical series of investments in the Italian HSR, showing that even if only the economic impact is considered, the balance of this investment can be regarded as unquestionably positive.

Keywords Economic effects · High-Speed Rail · Input–Output · Supply and use tables

M. Tartaglia (✉) · M. Cerullo
Ferrovie dello Stato Italiane, Roma, Italy
e-mail: m.tartaglia@fsitaliane.it

T. Ferraresi · R. Paniccià
IRPET, Tuscany Regional Economic Planning Institute, Firenze, Italy

M. Radicioni · E. Cieri
RFI, Rete Ferroviaria Italiana, Roma, Italy

F. Pagliara (ed.), *Socioeconomic Impacts of High-Speed Rail Systems*,
Springer Proceedings in Business and Economics,
https://doi.org/10.1007/978-3-031-26340-8_2

1 Introduction

The appraisal of social and economic effects of a High-Speed Rail (hereinafter also referred to as HSR) system has been widely debated in the last decades. Several methodological approaches for assessing the effects of HSR have been proposed in the scientific literature, such as Cost-Benefit Analysis, External Cost Analysis, Economic Impact Analysis, and several ad hoc methods. Each of these different approaches takes a specific point of view, so it is crucial to pay attention to the suitability of their joint application. Among the feasible approaches, the Economic Impact Analysis technique mainly focuses on estimating the overall economic impact of a sector, a single activity unit/enterprise or a specific investment project on the national or local economy.

This paper presents the estimate of an Economic Impact Analysis model based on a more comprehensive Input–Output approach, i.e., including the use of the Make tables along with traditional I–O matrices and consequently enables the appraisal process to take into account the entire productive structure of the economy.

The developed commodity driven I–O model is mainly composed by two ex-ante causal relations. The first, is a technical Leontevian production function determining both the intermediate demand and the total demand by the entire study area. The second, is a trade pattern determining the spatial area where this demand will be satisfied through sectoral output. Furthermore, the I–O model is a demand driven model based on fixed prices and supply unconstrained. Final demand is expressed as "products" yielded with various shares (market shares) by sectors (for instance: Agriculture produces 3% of all Electric Power). Finally, the commodity driven model solution combining demand expressed in products and sectoral output is based on the Leontevian inverse i.e., on multipliers.

In particular, a consistent time series of national Supply tables, at current basic prices, have been estimated. Consistency refers to the SNA2008/ESA2010 accounting methodology and Nace Rev. 2 Statistical classification of economic activities. Indeed, since 1995 the Italian National Institute of Statistics (ISTAT) has released yearly SUTs; however, some methodological and classificatory changes have occurred. In particular, in 2013, the new system of national account ESA2010 following the SNA2008 was introduced (replacing the previous SNA95) and a new classification: in 2007, the Nace Rev. 2, replaced the previous Nace Rev. 1.1. Moreover, the ESA2010 was updated in 2019. The previous considerations show the need to make the national Supply, released under various accounting systems and classifications, consistent with the last ESA2010 and Nace Rev. 2 to perform consistent impact assessment. These reconstructed time series include rebalancing the Supply table, released under previous systems and classification, using the available information from national accounts and administrative sources. Once rebalanced and reconstructed, the Make tables extracted from the Supply ones will be used along with ICIO/OECD tables, which has been recently rebuilt completely consistent with SNA2008 and Nace Rev. 2, in our impact exercise.

Although a large number of studies have been dedicated to the analysis of the impacts of the Italian HSR, only a few applications have analysed its specific effects on the Italian national economy. To further investigate this issue, the model presented in this paper has been used to improve the assessment of the economic impact of the investments made in Italian HSR starting from the beginning of the intervention to the current period. In particular, the model has been used for assessing the impact of the investments in HSR over the 1995–2018 period.

2 Literature Review

HSR transport estimating can be viewed as a particular case of the more general evaluation of the effects of a transport system. These estimates are often carried out to provide objective elements in the decision-making processes for assessing investment projects [1] or supporting project proposals or business plans of institutional entities or companies (e.g., [2]). Typically, in most cases, the targets against which these impacts are estimated are identified in the economic, social and environmental subsystems [3]. Three main points of view can be assumed by impact analyses [4]: that of the entity in charge of implementing or operating the HSR system, that of the users of the system itself (or instead, of the entire transport system integrated with the HSR network), and finally that of non-users of the transport system, in which case the impacts are defined as externals with reference to the relevant economic theories.

Although there are numerous applications in the specific literature relating to the estimation of the impacts of a transport system [5] and also of an HSR system, in particular, the methods to be applied are subject to debate and do not yet constitute a well-defined and homogeneous framework [6]. Some of the numerous approaches proposed in the scientific literature for assessing the effects of a general transport system or HRS-type system may be considered more consolidated in terms of their use in practical case studies, also because they are the ones most frequently referred to in the legislative guidelines of many countries. These include Cost-Benefit Analysis, External Cost Analysis, Multi-Criteria Analysis, Environmental Impact Analysis, Wider Economic Impact Analysis, and Economic Impact Analysis. Apart from these most common methods, studies and analyses use others, including many ad hoc methods. Each of these analytical methods is based on specific approaches, assumptions and methods and adopts a specific point of view. For this reason, in general, the various methods mentioned cannot be used the same way, nor can their results be compared.

The Cost-Benefit Analysis is the most frequently used method for assessing the feasibility of interventions in the project planning phase. This method compares the costs and benefits of a specific transport system configuration by expressing them in monetary values, generally for a period corresponding to the useful life of the interventions of which the effects are to be evaluated [7, 8]. Aspects considered in the analysis include not only investment costs and operating costs and revenues but also effects on the social and economic system owing to elements such as travel time and traffic congestion, air quality, climate-changing emissions, noise, energy consumption, and accidents [9]. A community is the point of view adopted by the Cost-Benefit Analysis.

External Cost Analysis determines the economic value of the so-called external transport costs. Such an analysis is based on the economic concept of externality, i.e., the set of positive and negative as well as intentional and unintentional effects that a specific activity produces on economic systems other than those carrying out the activity [10]. Hence, externality refers to the fact that these effects, produced by a specific economic system, are not entirely absorbed by those who produce them but cause consequences on the economic system outside them. For this reason, the External Cost Analysis perspective is not that of the entire community but its subset external to the system producing the effects. The application of the concept of externalities to the world of transport, initially referring only to environmental costs [11], has since evolved to include multiple components of the ecosystem, also based on constantly updated methodological guidelines, such as those of the European Commission [12]. The assessment of external costs is also often used in the context of Cost Benefits Analysis, and here it is crucial to consider that the two techniques' points of view and scope are heterogeneous [13].

The Multi-Criteria Analysis is very similar in its methodological approach to the Cost-Benefit Analysis [14] and, similar to the former, in the transport field, it is mainly used in the decision-making process when evaluating projects [15]. Compared to Cost-Benefit Analysis, it expresses effects in non-monetary terms but by using physical or even qualitative measures, which are usually normalised to unit values using utility functions; it is also used to weigh the various effects to represent the policy-making objectives [16].

The Environmental Impact Assessment can be defined as "the systematic examination of unintended consequences of a development project or program, with the view to reduce or mitigate negative impacts and maximize on positive ones" [17]. The primary purpose of this type of analysis is also to evaluate projects within policy-making processes, and it is performed using a variety of methods and approaches depending on the country, but generally aimed at providing a snapshot of the environmental effects of a transport system's design [18].

Accordingly, the Wider Economic Impact analysis has the objective of extending beyond the Cost Benefits Analysis by taking into account effects that are usually not considered, such as those on economic productivity, economic growth, spatial accessibility, equity and fairness of the availability of transport systems, social and territorial cohesion, etc. [19]. There is extensive debate in scientific literature surrounding the applicability of the Wider Economic Impact analysis, especially regarding the risk of double counting of effects [20]. Despite this debate, the Wider Economic Impact analysis is included in the guidelines for evaluating the effects of transport systems in many countries [21, 22], although some current institutional guidelines explicitly prohibit its use in Cost Benefits Analysis (e.g., [8]).

Compared to all the methods mentioned above, the Economic Impact Analysis is expressly committed to evaluating the overall economic impact of the realisation and operation of a transport system, thus applying an approach focused on assessing the economic development implications of an investment [23]. These economic effects can be classified into direct effects (owing to spending directly on transport demand), indirect effects (or Leontevian, owing to spending on intermediate goods activated by enterprises that produce for direct demand) and induced effects (or Keynesian, owing to the increase in household income produced by the two previous effects, income that is, in turn, spent leading to a further increase in demand for goods and services) [24]. The analysis is generally conducted using Input–Output (IO) models; other types of models are sometimes used for this purpose [25], such as econometric models, economic simulation models, or General Economic Equilibrium (CGE) models, for example in [26, 27] to assess the effects of HSR in China and in [28] on HSR in Singapore. The Input–Output approach is based on the approach initially introduced by economist Wassily Leontief in 1936 [29]. The approach is an extension of the general interdependency theory, which interprets economic systems in a given area in terms of structural interrelationships between their components [30]. While Economic Impact Analysis has not always been part of the institutional approaches of various countries to assess the effects of HSR systems [31], several studies focusing on the economic impact of HSR can be found in literature [32]. Specifically, some exciting applications of Input–Output models have been developed for evaluating portions of HSR networks being designed or in operation in France [33], China [34], Republic of Korea [35, 36], Australia [37], California, USA [2, 38], Florida, USA [39], India [40]. On the contrary, in the Italian HSR, the authors have yet to be made aware of any analysis of the entire network based on the Economic Impact Analysis approach, particularly in Input–Output type models.

3 The Case Study: Italy's High-Speed Rail, A New Way to Travel

In a short time, HSR has changed the habits and customs of millions of people, shortening distances and reducing travelling time. The Ferrovie dello Stato Italiane Group has completed over one thousand kilometres of High-Speed/High-Capacity lines—the most significant infrastructure work in Italy since the end of World War II.

The High-Speed Rail network was the start of a substantial change in the public transport sector, a revolution in Italian mobility that has stepped up the pace of the country's cultural and social life, making the main city districts into one giant metropolis: Italy. The HS trains race throughout Italy—crossing the Apennines and an area which—according to UNESCO—includes half of the world's artistic heritage and where 65% of the national demand for labour and mobility is concentrated. What is known as "Italy's fast underground" connects Milan to Rome in about three hours, Naples in about four hours and Florence in just one hour and 45 min. We travel more often and at speeds of up to 300 kms per hour.

The Italian High-Speed Rail system is the result of the intelligence and work of Italian enterprise. Millions of people have chosen to travel this way: train over air and road traffic. This way, the atmosphere was saved from more than 600 thousand tonnes of CO_2 emission. The new fast lines overcame some tough challenges, mainly due to the topography and hydro-geology of the land and the highly built-up areas it crosses. The HS/HC system has made it possible to dedicate the railway lines to different types of transport—both HS and metropolitan routes—to benefit the whole network. The connections between high-speed and conventional networks offer practical alternatives and benefits for commuters who can pick up the most efficient and regular services. The total capacity of railway traffic has more than doubled on the new lines with the combined HS and conventional systems. Moreover, the new tracks all have a low environmental impact. The new Italian HS network has moved travellers off the roads and planes onto trains, making it a determining factor for the modal rebalancing of the national transport system.

The Ferrovie dello Stato Italiane is the first in Europe to have adopted the level 2 European Railway Traffic Management System/European Train Control System (ERTMS/ETCS) on the new HS/HC lines. This exclusive system controls circulation guaranteeing the highest level of technology and safety, eliminating the possibility of human error and guaranteeing control of the train moment by moment. Therefore, top security at top speed and more frequent trains that can also circulate every few minutes. Fully developed in Italy, the EU adopted the level 2 ERTMS/ETCS as the European standard of reference for the new high-speed transnational networks. The Ferrovie dello Stato Italiane was the first in the world to use it, obtaining the Best Paper Award in 2006 in Montreal, a significant award by the international railway community.

The HS stations in the major metropolitan hubs of Turin, Milan, Bologna, Florence, Rome and Naples have been renovated or built from scratch to designs by renowned architects who have won international competitions. The first new HS station completed was Roma Tiburtina, inaugurated on 28 November 2011. Stations are protagonists of significant urban redevelopment operations and the expression of a new architectural language, conceived as spaces dedicated to railway activities and meeting and communication places. The restyling of Roma Termini and, more recently, Milano Centrale stations are pilot projects of the new way of interpreting stations as city squares.

3.1 The History of High-Speed Rail in Italy

High-Speed Rail in Italy started in 1975 with the construction of the "Direttissima" Rome-Florence line with a design speed of over 250 km/h to connect two of the main Italian cities without intermediate stops. The "Direttisima" is the earliest High-Speed Rail line built in Europe, after about ten years compared to Japan but before France, which is the leading European HS network with about 60 billion people-km

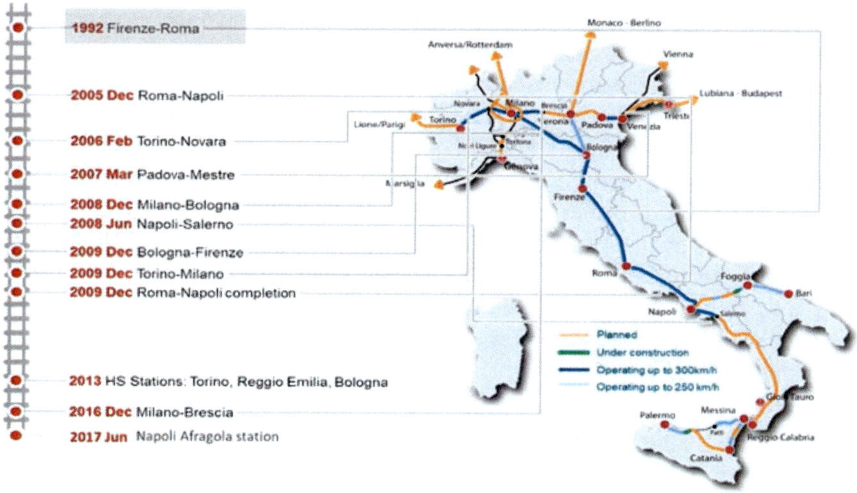

Fig. 1 Italy's HSR system evolution

transported in 2019. The construction of the "Direttissima" line was completed only in 1992 when the service began (Fig. 1).

The construction of the HSR in Italy went through a regulatory framework, technological development and transport policy choices that made it possible. The decision-making and implementation process of the Italian HSR network started in 1986 with the National General Transportation Plan, which identified a network of HSR lines as a tool for both relaunching the national rail transport and complying with EU directives.

In 1991, the company TAV SpA was established with the corporate purpose of the executive design and construction of the lines and infrastructures and anything else necessary for the High-Speed Rail system and with a deed of concession, FS transferred to TAV SpA, the design, construction and economic exploitation of the Milan-Naples and Turin-Venice High-Speed lines.

1996 was an important year for the evolution of the High-Speed network in Italy with Article 2, paragraph 15 of Law 662/1996; a revision of the overall project started, moving from the simple High-Speed network to a High-Speed/High-Capacity network, implementing the technical characteristics to allow the transport of goods.

The dividing of Ferrovie dello Stato SpA was completed in the early 2000s, which resulted in the foundation in July 2001 of Rete Ferroviaria Italiana, the infrastructure operator. At that time, the role and concession to TAV was taken into consideration again. The TAV was fully purchased by Ferrovie dello Stato in 1998, which meant that the Government was responsible for construction costs. In 2010, TAV SpA was incorporated into Rete Ferroviaria Italiana. During the years in which construction took place, various financial laws provided resources to the AV system, allowing, still today, the completion of the entire system: for example, the construction of the third Pass of the Giovi, of the section Brescia–Verona–Padua that is part of the Torino–Milano–Venezia Axis.

On the construction side, the main stages that marked the beginning of the mobility change began on 19 December 2005 when the Rome–Naples High-Speed/High-Capacity line was inaugurated. The route runs for 204 kms through the territory of 61 municipalities. The trains are moved by the Level 2 ERTMS system, the first application at the European level on a line in commercial service.

Construction to build the first station dedicated to High-Speed Rail in Italy, Torino Porta Susa, began in 2006. Destined to become the main Turin terminal for regional, national and international rail traffic, Torino Porta Susa is part of a larger High-Speed station project planned by Rete Ferroviaria Italiana to be developed in Bologna, Roma Tiburtina and Napoli Afragola. The projects related to the Nodes and the construction of the High-Speed stations that complete the entire system are carried out in succession to the lines, making the High-Speed services expanded and improved and fully usable by the passengers.

In 2006, the High-Speed route between Turin and Novara was inaugurated on the occasion of the XX Winter Olympic Games, with a trip on the first High-Speed train from Turin Porta Nuova to Malpensa Airport. Open for operation for about 85 kms, the line can be travelled from the Turin Stura interconnection to the Novara interconnection and rejoin the "old" Turin–Milan route.

In 2007, the Roma Tiburtina station project was launched, and the station was inaugurated in 2011. The new station represents the arrival and departure point in the capital of Italian and foreign High-Speed trains. It became the most critical exchange node of the Roman urban and metropolitan mobility network.

The Naples–Salerno High-Speed/High-Capacity line "Via Monte del Vesuvio", opened for commercial use in April 2008, extended the High-Speed to Salerno and allocated the Naples–Torre Annunziata–Salerno coastline to the exclusive metropolitan service. The line runs for about 29 kms and is connected to the Rome–Naples High-Speed/High-Capacity line and the interconnection with the Naples–Salerno coastline.

In December 2009, the commercial service began on the entire Turin-Salerno High-Speed axis with the non-stop Milan-Rome connection in 2 h and 59 min, which significantly brought the two cities closer together with very high standards of punctuality and service. This revolution in the country's mobility and people's lifestyle makes it possible to leave and return on the same day. This creates new ways of commuting, allowing people to keep their roots in their cities. Together with the A1 Milan–Naples motorway, the High-Speed/High-Capacity system is considered the most extensive work built in Italy in the post-war period.

The Reggio Emilia High-Speed Mediopadana station was inaugurated in 2013. It was designed by the Spanish architect Santiago Calatrava, representing the only intermediate stop on the Milan–Bologna High-Speed line.

The Bologna High-Speed station was inaugurated in the same year. It is the keystone of the project to upgrade the railway junction of the Emilian capital, which has always been affected by intense traffic of local, national, and international trains and goods.

The construction of the Treviglio–Brescia section started in 2012 and the Brescia Est–Verona section in 2020.

The new Napoli Afragola High-Speed station, designed by the architect Zaha Hadid was inaugurated in 2017. The sinuous structure abstractly leads to the image of a modern running train. The new Napoli Afragola High-Speed station represents, together with Napoli Centrale station, a hub serving the territory, able to improve the supply, efficiency and regularity of rail transport and, at the same time, to create the basis for the development of the surrounding areas.

4 Methodology and Data

In order to evaluate the impact of the investment expenditure in High-Speed rail on the Italian economy we use an input–output model which combines the global Inter-Country Input–Output Tables (ICIO) produced by OECD [41], the only tables covering a wide time span including the HSR implementation, and the Supply tables provided by Istat, from 1995 to 2018 [42, 43]. This combination has led to set up a commodity driven I–O model (Fig. 2). The reason of the implementation of this type of I–O model relies on the particular feature of the impact. Indeed the HSR investment expenditure is specified by products (goods and services) as the I–O structure (ICIO) is symmetric sector by sector, consequently we need to assign the investments products to the proper producing sector and this information is provided by the market share computed on the Make matrix which is part of the Supply table.

Indeed before passing products to sectors we need to perform two important operations: (i) transforming the investments products flows from purchasing prices to basic prices, the latter ones are consistent with ICIO tables evaluation prices; (ii)

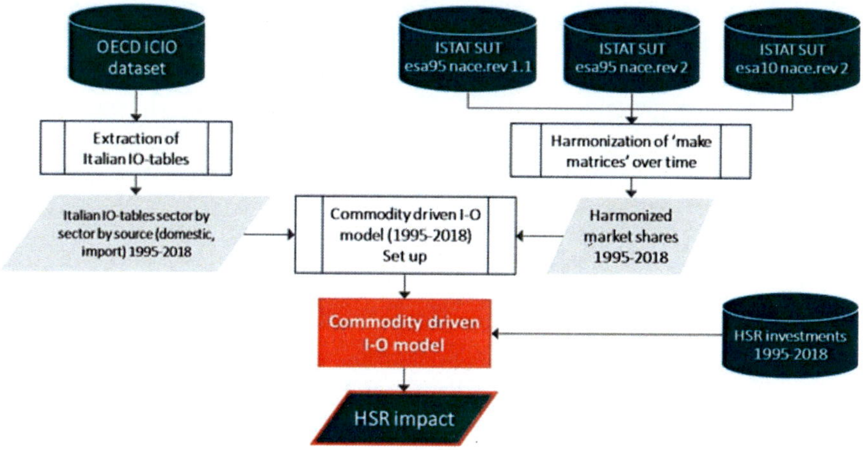

Fig. 2 Methodology and data

subtracting from the amount at basic prices of the investments products the amount of product imported from abroad.

The first operation has been done through proper tax and trade margins matrices as the final import has been estimated through import propensities provided by Istat.

HSR investment basically regards four types of goods and services: (i) construction works, (ii) fabricated metal products, (iii) electrical equipment (iv) engineering services. These products are grouped through the related classification items CPA classification.

As reminded in Chap. 1, over the time span period, two important changes occurred: (i) classification changed from Nace Rev. 1.1 to Nace Rev. 2, (ii) national accounts methodology moved from SNA 1995 to SNA 2008. These changes forced us to implement a transition procedure in order to produce consistent scenarios with the ICIO tables built according SNA 2008 with Nace Rev. 2.

After the preliminary operations and once ensured consistency over time we then estimated market shares from the Make tables. This allowed us to reassign products to sectors of the Italian economy and to set up a proper final demand scenario for the ICIO I–O model for each single year of the time span.

Once the sectoral final demand vector has been reconstructed, we ran the symmetric sector-by-sector input–output model to estimate the impact of HSR investment in terms of output, value-added and employment. A schematic representation of the methodological framework used is shown in Fig. 2.

4.1 The Input–Output Model

The model related to the ICIOit table is based on two main causal relations:

- a Leontevian technical relation, which determines the regional demand of intermediates and, along with the exogenous final demand, the total demand of each area;
- an allocative relation (multiregional trade pattern), which determines the output by distributing between national and foreign KAUs (kind-of-activity units).

In a closed system it is possible to formalize the above relations as follows:

$$d = A \cdot x + f \tag{a}$$

$$x = T \cdot d \tag{b}$$

where d is the total demand of the system (final and intermediate), x is the vector of total output and f the final demand. In Eq. (a), the relation between production and demand for intermediate goods and services is quantified by the input cost coefficient matrix A. Equation (b) shows the spatial allocation pattern of demand, represented by the trade matrix coefficients T. The model assumes competitive imports.

The basic model shown in Eqs. (a) and (b) is the theoretical starting point for specifying a complete Italy-Rest of the World bi-area structural model, in which every single intermediate and final trade cell expresses a proper allocation parameter according to the Isard approach [44].

Below is the structural form underlying the commodity driven model[1]:

$$x + mw_x + D \cdot mw_f = A \cdot x + D \cdot f + D \cdot ew \tag{1}$$

$$mw_x = (M_x \times \widehat{A \cdot x}) \cdot i \tag{2}$$

$$mw_f = [(M_f \times f) \cdot i] \tag{3}$$

where:

- x sectoral output at basic prices.
- mw foreign imports: intermediate (subscript x and by sector) and final (subscript f and by product).
- f product final demand.
- ew product exports to RoW.
- M foreign import share coefficients for intermediate (subscript x and by sector) and final (subscript f and by product) demand.
- D market shares matrix.
- i proper unit vector.

A reduced form is associated with model (1)–(3). In particular, solving the basic price output, we could get:

$$x \equiv \{I - [(1 - M_x) \times A]\}^{-1}\{D \cdot [(1 - M_f) \times f \cdot i + ew]\} \tag{4}$$

Considering that:

$$R_x = [(1 - M_x) \times A] \tag{5}$$

where R is the intermediate flows, net of the RoW imports. Equation (4) could be written as:

$$x = (I - R_x)^{-1}\{D \cdot [(1 - M_f) \times f \cdot i + ew]\} \tag{6}$$

an Isard type reduced form of model for intermediate and final goods.

Recursively, it is possible to determine value added as:

$$y = \widehat{V} \cdot x \tag{7}$$

[1] Important notation: symbol " × " stand for *element by element* matrix product, as symbol "." represesents the genuine matrix multiplication.

where $\widehat{V}$ = diagonal matrix of value added share per unit of output, or:

$$y = \widehat{V} \cdot B_I \cdot \{D \cdot [R_f \times f \cdot i + ew]\} \tag{8}$$

where $B_I = (I - R)^{-1}$.

The Leontevian inverse B_I, will be the array containing the direct and indirect impact multipliers of output also called Type I multipliers. Type II multipliers are based on the partial endogenization of household consumption, in our case, the household expenditure related to wages. The hypothesis is that wages generated by the production process will constitute additional disposable income which will be, in turn, transformed in additional consumption according to an average propensity to consume. If so, the structural model Eq. (3) will be changed as follows:

$$mw_f = [(M_f \times f + M_c \times c) \cdot \mathrm{i}] \tag{3b}$$

And the endogenous component of household consumption by product will be:

$$c = Hx \tag{4a}$$

where:

c endogenous component of household consumption.
H consumption coefficients.
M foreign import coefficients for consumption product (subscript c).

The reduced form derived from introducing (3b) and (4a) will be as follows:

$$x \equiv \{I - [(1 - M_x) \times A + [D \cdot ((1 - M_c) \times H)]]\}^{-1}\{D \cdot [(1 - M_f) \times f \cdot i + ew]\} \tag{9}$$

or using the Isard model notation as in (6)

$$x = [I - (R_x + R_c)]]^{-1}\{D \cdot [(1 - M_f) \times f \cdot i + ew]\} \tag{10}$$

The Type II multipliers used in our impact analysis will be defined as:

$$B_{II} = [I - (R_x + R_c)]^{-1} \tag{11}$$

5 Results and Discussion

As discussed in Sect. 3 above, the Italian HSR project was formally designed in the mid-1980s when it was included in the National General Transportation Plan. However, some work on the country's railway network that would later be an integral part of the new system had already been undertaken in the previous decade. Specifically, the construction of a new High-Speed Florence–Rome line was begun in the 1970s and completed in 1992, but with technical characteristics mainly similar to the pre-existing network, such as the power supply and signaling system. For these reasons, although the Florence–Rome line is now, to all intents and purposes, an entire section of the Italian HSR network, we will not consider it in the investment package of the HSR project published in 1986. In doing so, it can be assumed with an acceptable approximation that the first works for the construction of the HSR network started in the middle of the decade following the formal adoption of the project, and we will conventionally assume the beginning of these works in 1995. This assumption, which is in any case acceptable considering the number of annual investments made, which in the first five years of the 1990s were modest, also makes it possible to assume the Supply and Use Tables that the Istat began to produce and publish in 1995 as the fundamental input for the calculation model. Moreover, as can be seen in Fig. 1, a large part of the new T-shaped HSR network, from Milan in the north to Salerno in the south and from Turin in the west to Brescia (with a planned extension to Venice) in the east, was completed and commissioned by 2018 [45].

The foregoing is supported by the graph in Fig. 3, in which we can see the trend of investments (expressed at current prices) that Italy has made, through its own company Ferrovie dello Stato, which became an industrial group in the early 2000s and was subsequently renamed Ferrovie dello Stato Italiane, between 1995 and 2018. As can easily be seen, from 1995 onwards, investments grew steadily until 2004, starting from around € 200 million in 1995 to the remarkable figure of over € 4.2 billion in 2004. From that year onwards, there was a gradual decrease in the amount of annual resources invested until 2010, the year when the substantial part of the network currently in operation (Turin–Milan and Milan–Naples–Salerno) was activated. From 2010 to 2013, the annual investments, though very modest compared to the previous decade, increased again, mainly due to the completion of the last major new stations in the station network (Turin, Reggio Emilia and Bologna). From 2014 onwards, the amount of investments dropped further and settled around an average of € 150 million per year, with no major works to be completed.

With regard to the composition of investments, as can be expected, it showed a significant share related to design activities only in the first years, from 1995 to 2002, with an average of € 17 million per year, a peak of more than € 380 million in 1998 and a total cumulative value over the period considered of about € 1 billion. In any case, the share of investments relating to civil works immediately began to appear preponderant, assuming very high values that abundantly exceeded € 1 billion per year in the 2001–2005 period, peaking at more than € 2.7 billion euro in 2003, and totaling about € 17 billion over the period considered from 1995 to 2018. As work on civil works progressed, investments in signaling systems began to become evident, and these reached considerable proportions from 2003 to 2009 in particular, years in which they exceeded the figure of half a billion euro per year. On the other hand, investments in track and fitting-out only began to be significant after about five years of preparatory civil and technological works, becoming considerable only between 2003 and 2008, and in any case amounting in total, in the time period to which we

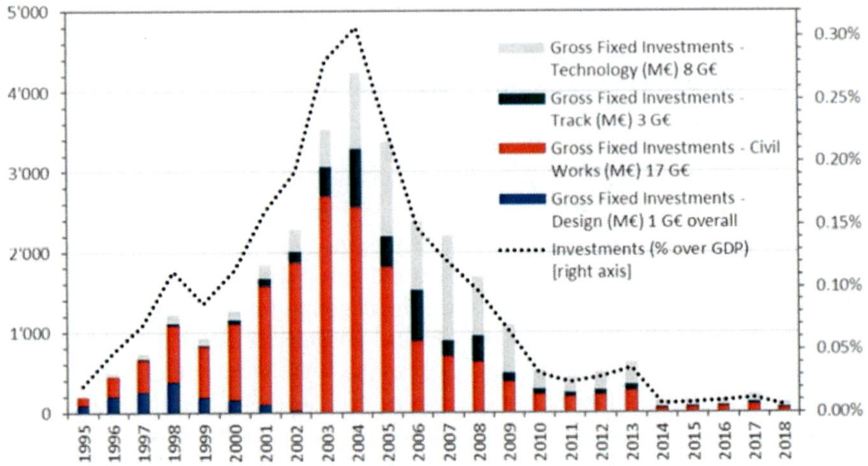

Fig. 3 Italy's HSR investments history, current prices

are referring, to only about € 3 billion, three times the amount invested in planning but only 10% of the total expenditure.

Overall, the total investment generated for the implementation of the Italian HSR network over the 1995–2018 period amounted to approximately € 30 billion, with an annual average of more than € 1.2 billion and a peak, in 2004, of approximately € 4.25 billion.

Based on the historical investment series outlined in Fig. 3, mapped in the detailed classifications of products in Sect. 4, the model has been applied according to the following steps:

1. Quantification of sectoral impact scenario:
    a. Quantification of HSR investments by product 1995–2018 at purchasing prices.
    b. Estimate of HSR inv. by product at basic prices.
    c. Estimate of HSR inv. by product at basic prices net of final import using final import coefficients.
    d. Using matrices D (market share) estimate of sectoral domestic HSR investments at basic prices.
2. Simulation for every year of the sectoral impact scenario in the corresponding I–O model based in the ICIO tables.

The results obtained in the simulations are summarised in Figs. 4, 5 and 6 and in Table 1. Figure 4 shows the trends in total economic output and value added generated by investments, the sums of the respective direct, indirect and induced components, compared with the trends in investments. As expected the value added and output generated follow the investment pattern both picking in 2004. Important

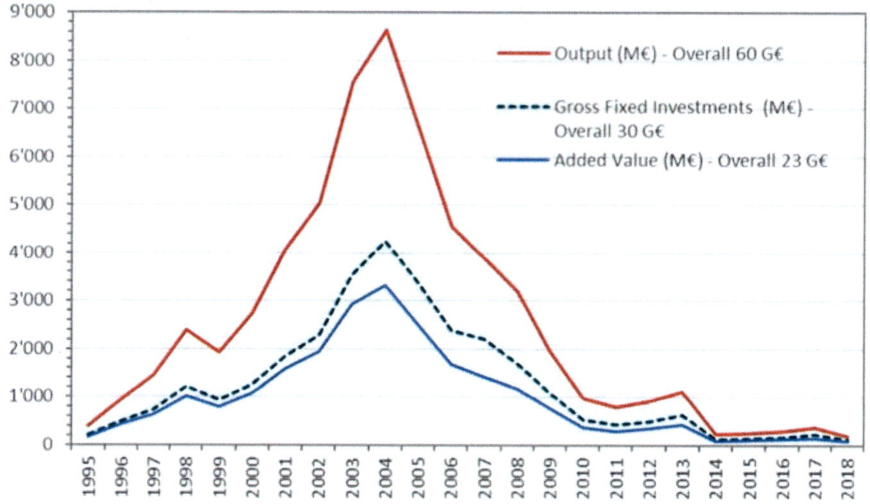

Fig. 4 Impact of Italy's HSR investments: output and value added

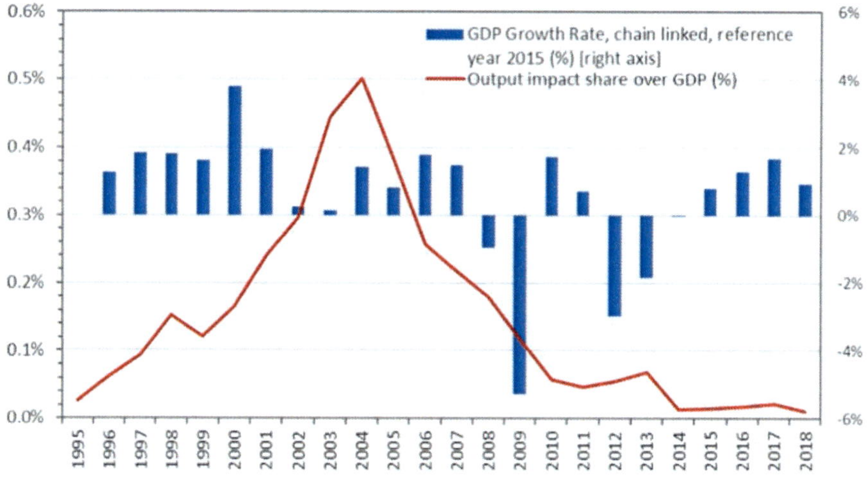

Fig. 5 Impact of Italy's HSR investments on the country's GDP

to note that the impact of output almost double the amount of investment showing a good performances of the impact in terms of indirect diffusion amongst sectors.

The value added generated is also significant and was almost proxing the investment flows.

The economic effects calculated using the I–O model can also be shown in terms incidence on Italy's national GDP. In Fig. 5 the GDP generated by the HSR investments is expressed as percentage of the actual GDP, in the same Figure the economic cycle reported by the GDP at constant prices is shown for contextualizing the impact

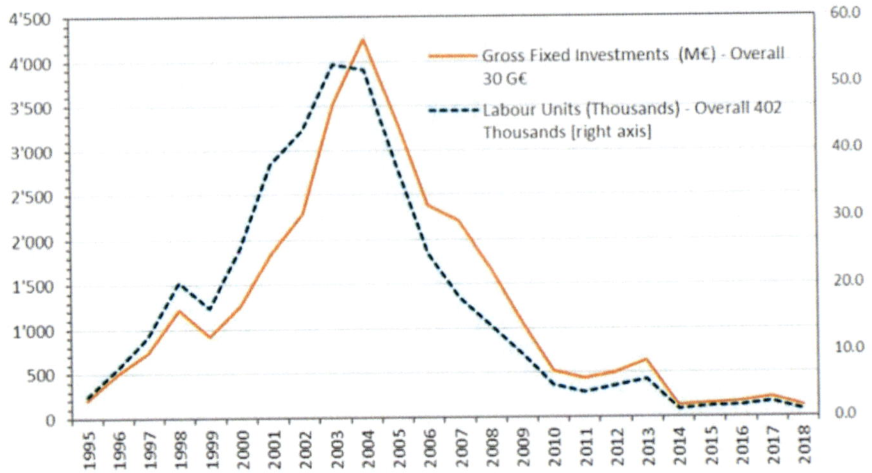

Fig. 6 Impact of Italy's HSR investments on employment

Table 1 Minimum and maximum annual impacts of Italy's HSR investments

	Minimum (Year)	Maximum (Year)
Value added (M€)	74.7 (2018)	3324 (2004)
Value added multiplier	0.8 (2018)	0.9 (2004)
Output (M€)	186 (2018)	8623 (2004)
Output multiplier	1.61 (2018)	2.03 (2004)
Labour units (Thousands)	1.3 (2018)	52.9 (2003)

generated. Of course, it takes on lower values when investments (and economic output) are lower but to be noted that in the peak year (2004) a maximum value of 0.5% has been reached. On the other hand, there is no apparent correlation between the GDP contribution of the impact and the GDP Growth Rate, as seen from the two respective series in Fig. 5.

In a recursive way impact generated on output should result in increasing labour demand and Fig. 6 shows the trend in labour input demand, expressed in units of labour, yearly sustained by the investments. As expected, the two sets of values are strictly linked. The number of units of labour generated by the HSR investments averaged almost 17,000 per year.

To conclude, Table 1 shows indicators that well summarise the results of the model. The economic impacts of the investment programme for the implementation of the Italian HSR system between 1995 and 2018 peaked in 2004,[2] at the peak of investment, and a minimum fourteen years later, in 2018. During the two years with the lowest and highest impact, the added value created ranged from a minimum of almost € 75 million to a maximum of over € 3.3 billion, with investment multipliers

[2] One exception is the number of units of labour supported, which peaked in 2003.

of about 0.8 and 0.9 respectively. The total economic output, on the other hand, ranged from a minimum of almost € 190 million to a maximum of over € 8.6 billion, with investment multipliers of about 1.61 and 2.03 respectively. Accordingly, the demanded units of labour ranged from a minimum of 1,000 to a maximum of almost 53,000.

More interesting are the multipliers, in particular in the peak year the value added multiplier was 0.9 as the for output 2.03. These indicators summarise the overall results obtained by the model developed by considering the direct, indirect and induced impacts produced by the HSR investments expenditure on the Italian economy, showing a positive balance, regardless of the other types of effects such as transport, social, territorial and environmental effects.

6 Conclusions and Further Perspectives

This study produced an EIA model commodity driven by combining estimated time series of national Supply tables, consistent with the changes in classification and methodologies of national Supply/make tables for the 24-year period between 1995 and 2018, with the ICIO I–O tables.

The model has been applied to evaluate the year-by-year economic impact of the investments made in Italian High-Speed Railway in that period, which includes all substantial investments made for the conception and development of the Italian High-Speed Rail system. As shown above, the model allows to perform a highly accurate evaluation of the economic impact.

The model application results show that at least the economic balance of the expenditure investment, considering the direct, indirect and induced effects on the country's economy, can be considered positive. Although there is a strong debate in Italy about the cost-benefit of the HSR project, the effects resulting from the expenditure for investment show that the HSR project in Italy are undisputed.

Further developments of the model are in-depth spatial analysis, spatial extension, and epistemological scope extension.

Spatial in-depth analysis allows us to allocate the economic impact to each elementary area included in the study region. These analyses can be based on the development of multi-regional modelling to estimate better the relationship between the supply and the use sector at a very detailed spatial division. This will also allow us to show how various economic actors contribute to the economy of a territory, even if they are based in areas different from those impacted.

The spatial extension will allow us to model the economic impact in other countries where HSR systems have been developed.

Finally, the epistemological scope extension aims to estimate the actual and more structural impact on the supply system, taking into account the effect of the HSR on the overall price system, productivity, competitiveness, among others. This last issue could be challenging as it is on the frontiers of the filed-specific literature.

CRediT authorship contribution statement Tartaglia, Mario: conceptualization, supervision, review, editing, writing (Sects. 1, 2, 5, 6); Cerullo, Michele: coordination; Ferraresi, Tommaso, and Paniccià, Renato: writing (Sect. 4); Cerullo, Michele, Radicioni, Mara, and Cieri, Enrico: writing (Sect. 3).

References

1. Henke, I., Cartenì, A., Di Francesco, L.: A sustainable evaluation processes for investments in the transport sector: a combined multi-criteria and cost-benefit analysis for a new highway in Italy. Sustainability **12**(23), 9854 (2020). https://doi.org/10.3390/su12239854
2. California High-Speed Rail Authority: The economic impact of california high-speed rail. https://www.hssr.ca.gov (2020). Retrieved 1 Nov 2022
3. ITF: Quantifying the socio-economic benefits of transport, ITF Roundtable Reports, OECD Publishing, Paris. https://doi.org/10.1787/9789282108093-en (2017). Retrieved 1 Nov 2022
4. Almujibah, H., Preston. J.: The total social costs of constructing and operating a high-speed rail line using a case study of the Riyadh-Dammam Corridor, Saudi Arabia. Front. Built Environ. **5**, 79. https://doi.org/10.3389/fbuil.2019.00079 (2019)
5. Wang, L., Xue, X., Zhao, Z., Wang, Z.: The impacts of transportation infrastructure on sustainable development: emerging trends and challenges. Int. J. Environ. Res. Public Health **15**(6), 1172. https://doi.org/10.3390/ijerph15061172 (2018)
6. Mouter, N. (ed.):New Methods, reflections and application domains in transport appraisal. In: Advances in Transport Policy and Planning, vol. 7, pp. 1–7 (2021)
7. de Rus, G., Socorro, M.P., Valido, J., Campos, J.: Cost-benefit analysis of transport projects: Theoretical framework and practical rules. Transp. Policy **123**, 25–39. ISSN 0967-070X. https://doi.org/10.1016/j.tranpol.2022.04.008 (2022)
8. European Commission: Guide to cost-benefit analysis of investment projects: economic appraisal tool for cohesion policy 2014–2020. https://doi.org/10.2776/97516 (2014)
9. Donais, F.M., Abi-Zeid, I., Owen, E., Waygood, D., Lavoie, R.: A review of cost-benefit analysis and multicriteria decision analysis from the perspective of sustainable transport in project evaluation. EURO J. Decis. Process. **7**(3–4), 327–358. ISSN 2193-9438. https://doi.org/10.1007/s40070-019-00098-1 (2019)
10. Rodrigue, J.P.: The Geography of Transport Systems, Fifth Edition. Routledge, New York, 456 p. ISBN 978-0-367-36463-2 (2020)
11. Rainer, F., Bickel, P.: Environmental External Costs of Transport. Springer (2001)
12. European Commission, Directorate-General for Mobility and Transport, Essen, H., Fiorello, D., El Beyrouty, K., et al.: Handbook on the external costs of transport: version 2019—1.1, Publications Office. https://data.europa.eu/doi/10.2832/51388 (2020)
13. Harvey, J.: Externalities and cost-benefit analysis. In: Economics Revision Guide. Palgrave, London. https://doi.org/10.1007/978-1-349-13313-0_17 (1994)
14. De Brucker, K., Macharis, C., Verbeke, A.: Multi-Criteria Analysis in Transport Project Evaluation: An Institutional Approach. European Transport, No. 47, pp. 3–24 (2011)
15. Broniewicz, E., Ogrodnik, K.: Multi-criteria analysis of transport infrastructure projects. Transp. Res. Part D Transp. Environ. **83**, 102351. ISSN 1361-9209. https://doi.org/10.1016/j.trd.2020.102351 (2020)
16. Ward, E.J., Dimitriou, H.T., Dean, M.: Theory and background of multi-criteria analysis: toward a policy-led approach to mega transport infrastructure project appraisal. Res. Transp. Econ. **58**, 21–45. ISSN 0739-8859. https://doi.org/10.1016/j.retrec.2016.08.003 (2016)
17. Agardy, F.J., Nemerow, N.L.: Environmental solutions. Elsevier, Amsterdam. https://doi.org/10.1016/B978-0-12-088441-4.X5000-X (2005)

18. Nita, A., Fineran, S., Rozylowicz, L.: Researchers' perspective on the main strengths and weaknesses of Environmental Impact Assessment (EIA) procedures. Environ. Impact Assess. Rev. 92, 106690. ISSN 0195-9255. https://doi.org/10.1016/j.eiar.2021.106690 (2022)
19. Veryard, D.: Quantifying the socio-economic benefits of transport. Discussion paper 2016-06 from the roundtable summary and conclusions, 9–10 November 2015. International Energy Agency, International Transort Forum, Paris, France (2016)
20. Chen, C., Vickerman, R.: Quantifying the economic and social impacts of high-speed rail: some evidence from Europe and the People's Republic of China. In: ADBI Working Paper 962. Asian Development Bank Institute, Tokyo. https://www.adb.org/publications/quantifying-economic-and-social-impacts-hsr-evidence-europe-prc (2019)
21. Wang, W., Zhong, M., Hunt, J.D.: Analysis of the wider economic impact of a transport infrastructure project using an integrated land use transport model. Sustainability. **11**(2), 364 (2019). https://doi.org/10.3390/su11020364
22. Wangsness, P.B., Rødseth, K.L., Hansen, W.: A review of guidelines for including wider economic impacts in transport appraisal. Transp. Rev. **37**(1), 94–115 (2017). https://doi.org/10.1080/01441647.2016.1217283
23. ten Raa, T.: The Economics of Input-Output Analysis. Cambridge University Press, New York (2005)
24. Adept Economics: Economic impact assessment. An overview. Briefing note (2019)
25. Queensland treasury overview of some alternative methodologies for economic impact analysis. Queensland Treasury, Office of Economic and Statistical Research, Brisbane (2006)
26. Chen, Z., Xue, J., Rose, A.Z., Haynes, K.E.: The impact of high-speed rail investment on economic and environmental change in China: a dynamic CGE analysis. Transp. Res. Part A Policy Pract. **92**, 232–245. ISSN 0965-8564. https://doi.org/10.1016/j.tra.2016.08.006 (2016)
27. Chen, Z.: Measuring the regional economic impacts of high-speed rail using a dynamic SCGE model: the case of China. Eur. Plan. Stud. **27**(3), 483–512 (2019). https://doi.org/10.1080/09654313.2018.1562655
28. Kumagai, S., Isono, I., Hayakawa, K.: Economic impacts of high-speed rail between Kuala Lumpur and Singapore: an application of IDE-GSM. IDE Discussion Paper No. 700, Institute of Developing Economies, Japan External Trade Organization. http://doi.org/10.20561/00050312 (2018)
29. Leontief, W.: Input-Output Economics. 2nd edn. Oxford University Press, New York, 1986 (1966)
30. Leontief, W.: Input–output analysis. In: The New Palgrave Dictionary of Economics. Palgrave Macmillan, London. https://doi.org/10.1057/978-1-349-95121-5_1072-2 (2008)
31. Gleave, S.D.: High Speed Rail: International Comparisons. Commission for Integrated Transport, London (2004)
32. Blanquart, C., Koning, M.: The local economic impacts of high-speed railways: theories and facts. Eur. Transp. Res. Rev. **9**, 12 (2017). https://doi.org/10.1007/s12544-017-0233-0
33. Fouqueray, E.: Impact économique de la construction de la LGV SEA Tours-Bordeaux sur les régions traversées. Revue d'Economie Régionale et Urbaine **2**(2016), 385–416 (2016)
34. Zhou, X., Lin, X., Ji, X., Liang, J.: Effects of high-speed railway construction and operation on related industries in China. Sustainability **13**(11), 6119 (2021). https://doi.org/10.3390/su13116119
35. Li, Z.: Research on the benefit of high-speed rail's investment promotion. Inf. Technol. J. **13**, 1371–1375 (2014). https://doi.org/10.3923/itj.2014.1371.1375
36. Lee, C., Ma, J., Oh, K.K.: Economic impact analysis on regional industries by high-speed rail investments: application of an input-output model. Transp. Res. Rec. **2672**(10), 247–259 (2018). https://doi.org/10.1177/0361198118786809
37. Wang, J., Charles, M.B.: The potential impacts of High Speed Rail on regional economic development in Australia: towards a multi-regional input-output approach (2010)
38. Nixon, H., Holian, M., Niles, J., Pogodzinski, J.M.: Measuring the economic impact of high-speed rail construction for California and the central valley region. Report 18-08, Mineta Transportation Institute, San José State University, San José, CA 95192-0219 (2018)

39. Lynch, T.A., Sipe, N., Dirac Drive, E.P., Polzin, S.E., Chu, X.: An Analysis of the Economic Impacts of Florida High Speed Rail. Florida Department of Transportation, Tampa, FL 33620-5375 (1997)
40. Sugimori, S., Hayashi, Y., Takeshita, H., Isobe, T.: Evaluating the regional economic impacts of high-speed rail and interregional disparity: a combined model of I/O and spatial interaction. Sustainability. **14**(18), 11545 (2022). https://doi.org/10.3390/su141811545
41. OECD: OECD inter-country input-output database. http://oe.cd/icio (2021). Retrieved 1 Nov 2022
42. ISTAT, Annual National Accounts. Years 1995–2018. https://www.istat.it/it/archivio/233518 and https://www.istat.it/en/archivio/233525 (English version) (2019). Retrieved 1 Nov 2022
43. ISTAT, i.Stat Datawarehouse. Gross domestic product and main components, http://dati.istat.it/?lang=en&SubSessionId=f34c402a-8d5e-46fc-9ae8-596fb4432544 (2022). Retrieved 1 Nov 2022
44. Isard, W.: Introduction to Regional Science, Englewood Cliffs, N.J, Prentice-Hall (1975)
45. FS, Ferrovie dello Stato Italiane: FS Italiane Group: ten years of Italian High Speed. https://www.fsitaliane.it/content/fsitaliane/en/media/press-releases/2019/12/5/fs-italiane-group--ten-years-of-italian-high-speed.html (2019). Retrieved 1 Nov 2022

Comparison of Preliminary, Initial, and Final Construction Costs of Italian High-Speed Railways

Francesco Bruzzone, Federico Cavallaro, and Silvio Nocera

Abstract The planning of high-speed railways (HSRs) in any megaproject faces challenges in providing reliable forecasts about financial aspects. This study presents the condition of Italian HSRs by focusing on the lines built or under construction in the last 30 years. The primary technical characteristics are illustrated for each, together with an analysis of the preliminary, initial, and final costs (when the line is completed). Our research reveals an interesting aspect: the final costs are always higher than the preliminary costs, with the cost variation presenting peaks of more than 360% and an average value of 165%. However, except for Turin–Milan HSR, they are often lower than the funds available at the beginning of construction work. Hence, Italian HSRs seem to demonstrate a specific capacity for project management and tendering processes to correctly estimate costs and timing. Meanwhile, Italian HSRs' unitary costs are generally higher than the European Union average, an aspect that is only partially explainable by Italy's geomorphological specificities, which might hide some marginal buffer that could be included in the initial phase of project construction.

Keywords High-speed railway · Construction cost · Preliminary cost · Final cost · Italy

1 Introduction

In the second half of the twentieth century, high-speed railways (HSRs) saw broad diffusion, particularly in Europe and Asia. The operational rationale behind their adoption is multiple; HSRs can replace conventional railways, support them along busy corridors, open new connections, and erode significant shares of air and road

F. Bruzzone · F. Cavallaro
Politecnico di Torino, Viale Mattioli 39, 10125 Turin, Italy

F. Bruzzone · S. Nocera (✉)
Università IUAV di Venezia, Santa Croce 191 Tolentini, 30135 Venice, Italy
e-mail: nocera@iuav.it

F. Pagliara (ed.), *Socioeconomic Impacts of High-Speed Rail Systems*,
Springer Proceedings in Business and Economics,
https://doi.org/10.1007/978-3-031-26340-8_3

traffic on certain routes, thus reducing environmental externalities [1, 2]. Italy was the first European country to open an HSR: the *Direttissima* line between Florence and Rome in 1977 [3]. The need for an integrated network rather than opening single lines is highlighted. Since 1986, the construction of a common, interoperable, continental HSR network has been a priority of the European Union (EU) within the wider TEN-T and TEN-R (trans-European network) frameworks. In the last decades of the past century and renewed vigour during the first years of the new millennium, national and transnational efforts to build HSRs with common characteristics and upgrade conventional lines to those standards are registered in the EU [4].

The TEN-T framework and HSRs are part of the European Commission's transport strategy, envisioning doubling the number of HSR passengers by 2030 and tripling by 2050, overcoming most of the current obstacles to long-distance and cross-border rail travel [5]. The last closed European funds call selected 135 proposals related to HSR networks for a total co-funding of approximately 5.4 billion Euros [6]. HSR projects are developed due to a complex multilevel decision, which goes from the European scale to the local one [7]. In the former case, the primary connections must be agreed upon among the member states. In the latter case, executive projects must include local adaptations to territorial contexts that specify initial schemes. Thus, relevant differences may occur between the initial and final layouts (and costs).

This aspect is pivotal in territorial development and drives relevant structural changes. Projects include significant expenditures, notable discrete policy, political choices, and profound impacts and transformations on crossed regions, with relevant consequences on local economic competitiveness and attractiveness, rail services provision, and regional accessibility and equity [8]. The complexity of HSR projects, the multifaceted and broad impacts they generate on existing and future natural and anthropic environments, and the fact that they respond to multiple governance and economic priorities, being considered strategic and essential for transnational and national development, contribute to their promotion [9]. By funding new HSR connections, including cross-border links, the European Commission aims to avoid the obstacles of commercial, economic, social, and environmental viability as obstacles to infrastructure capacity and capacity allocation. A report published in 2021 [10] synthesises the critical elements to ensure the effectiveness of HSR investments, showing how European institutions are working to reach the ambitious targets expressed by EU documents, including the Next-Generation EU plan, while overcoming issues, such as infrastructural deficiencies and lack of capacity.

A relevant issue is to understand whether their planning and management processes are effective and if they can assess the benefits and burdens for each involved and/or affected stakeholder and all those elements involved in the management process, including those of a more stochastic nature, to correctly forecast the costs and benefits of the works. The significant investments allocated to HSR projects suggest including them under the category of "megaprojects", a term proposed by Flyvbjerg to denote primary transport interventions (among the most relevant, see Flyvbjerg et al. [11] and Flyvbjerg [12, 13]). Megaprojects are characterized by some specificities: first, decision making, planning, and management are often weak and subject to numerous changes during the development and construction of projects;

second, the over commitment to certain concepts (sometimes due to political or financial reasons) makes the analysis of alternatives weak or absent; third, the complexity of the projects and the inability to correctly estimate projects benefits, as well as incorporate the stochastic element in the project's process, often make time and budget contingencies inadequate. Consequently, misestimations and misinformation regarding costs, schedules, benefits, and risks are considered the norm throughout project development, decision making, and communication. Cost overrunning is observed consistently throughout megaproject management and building, with HSRs being no exception [13].

This study looks at the Italian condition, discussing if and how the trend observed by Flyvbjerg and that mentioned above for megaprojects is valid for HSRs considering the renewed transnational policy framework discussed above. Section 2 individually analyses all HSR lines opened or under construction within the last 30 years, comparing the cost figures provided in initial, intermediate, and final official reports. Section 3 provides a more aggregate vision that sheds light on the evolution of Italian conditions related to HSR planning and financing. Finally, Sect. 4 draws conclusions that can be helpful for policymakers in evaluating the risk of overrunning costs derived from the construction of such infrastructure.

2 Recent Developments of the Italian HSRs

As mentioned in the introduction, Italy was the first European country to introduce a pioneering HSR shortly after Japan, which was the absolute precursor in the field. The first section of the Rome–Florence line opened in 1977, allowing a maximum speed of 250 km/h and an average speed above 200 km/h [14]. The lines designed afterwards followed a different concept, named "Alta Velocità/Alta Capacità" (high speed/high capacity, HS/HC), based on high-speed operations (at least 300 km/h) and on the circulation of passenger and freight trains. Apart from the Rome–Florence line, the Italian HSR was developed in the new millennium, with funding ensured to start by 1991. Between 2006 and 2008, the HSR between Turin, Milan, and Bologna opened, shortly followed by the Bologna–Florence stretch, marking a turning point in the history of Italian HSR. Since then, the main line (*Dorsale*) between Turin and Salerno via Milan, Bologna, Florence, Rome, and Naples has been completed. Services were reorganised nationwide under the brand *Frecce*, which distinguishes Frecciarossa, Frecciargento, and Frecciabianca according to the level of service and number of stops. For the first time, Italy was connected from north to south by a continuous HSR line and served by a recognisable fast-train brand.

With the push of the TEN-T framework, investments were allocated to cross-border corridors. According to the last version of the TEN-T layout [15], Italy is crossed by four main corridors: the Baltic–Adriatic, Scandinavian–Mediterranean, Rhine–Alpine, and Mediterranean (Fig. 1). Within this framework, significant upgrades to the transnational rail infrastructure have been made since the early 2000s, aiming to improve the HSR connections with France (through the new Turin–Lyon line), Switzerland (through the Gotthard and Ceneri tunnels), and Austria (through the Brenner base tunnel and the new Pontebbana and Semmering lines). Other connections are still under discussion (e.g., the upgrade of the Venice–Trieste line and the connection with Slovenia; [16]).

As of 2022, the network is still under strong expansion, pushed by the next-generation EU plan. Figure 1 shows several sections that are under construction. Technical solutions have changed at times to allow intercity and local trains on certain sections of HSR as well [16, 17], thus partially compromising the complete interoperability of the network according to the TEN-T standard. Once completed, the Italian HSR network will remind of a "T", with a West–East corridor from the French border to Venice (extended to the Slovenian and Austrian borders on upgraded

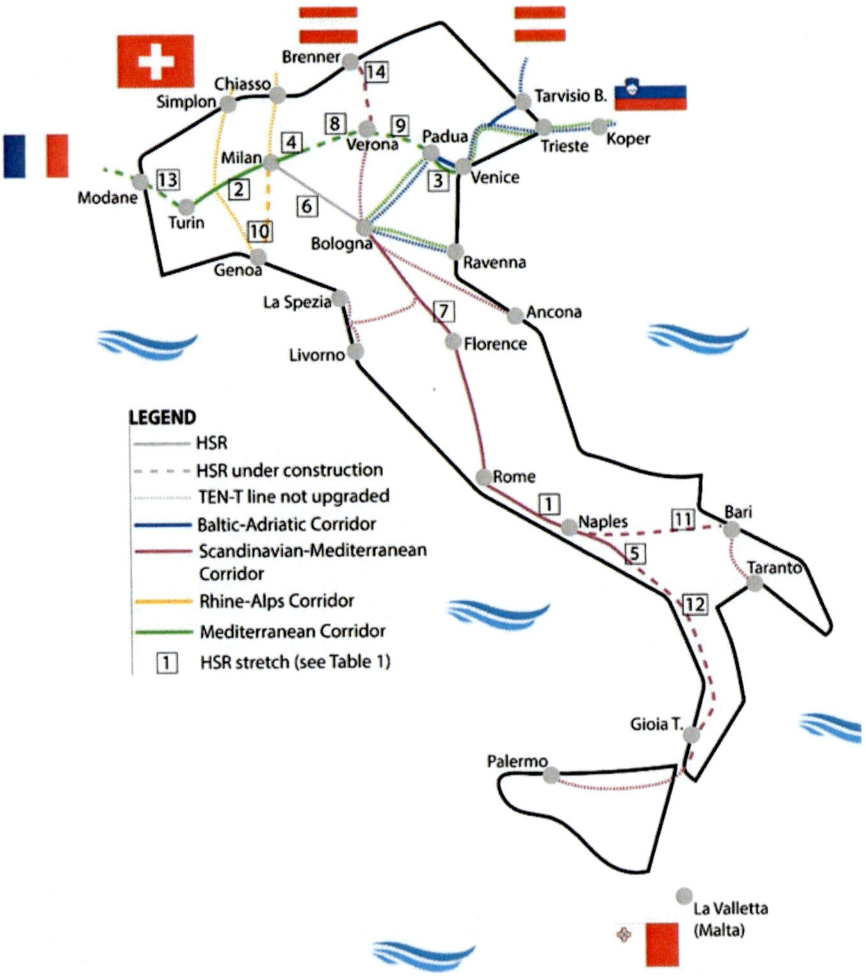

Fig. 1 HSR and TEN-T corridors in Italy (own elaboration)

conventional lines) and with a North–South line from Turin/Milan to Bari along the *Dorsale* line previously mentioned. This will be complemented by the upgraded lines of the TEN-T network within Italy and their connection with neighbouring countries [18].

The Italian HSR system is considered a commercial success, eroding a significant share of air traffic and changing the mobility habits of the Italian people [4]. The full-scale operation of the HSR network has also allowed the coexistence of two operators (a unique case worldwide until late 2021): the publicly-owned Trenitalia and the private Italo [19]. Since the *Frecce* brand of Trenitalia was launched and the long-haul service was reorganised nationwide, starting in 2012, services and passengers along the "T" have seen constant growth, a success beyond initial expectations. According to Legambiente [20], around 20 million passengers travelled on the HSR in Italy in 2010. In 2019 (the year before the COVID-19 pandemic), the number was close to 50 million passengers on the *Frecce* network and approximately 13 million passengers on Italo trains. The number of trains running each day has likewise grown consistently from 108 Eurostar trains circulating daily in 2008 to over 370 daily services operational in 2022.

The rest of this section presents the milestones of the development of the Italian network in chronological order, highlighting any financial or technical peculiarity which might have impacted the development of specific stretches. The temporal dimension is a critical aspect valid for all economic and financial figures provided in this paper. We have reported all currencies to $€_{2022}$ using values made available by the Italian Institute of Statistics [21], allowing a coherent comparison of costs that consider the temporal discount.

2.1 *Rome–Naples (2005)*

The Rome–Naples HSR (Fig. 1, number 1) was opened to revenue service with a timetable change on 9 December 2005, 11 years after construction and 14 years after the completion of the competitive tender. The line is approximately 222 km long, comprising 54 tunnels. It is the first Italian railway to comply with HS/HC standards fully, being electrified at 25 kV AC (rather than the Italian standard, 3 kV CC) and using the European Rail Traffic Management System/European Train Control System signalling [22]. The line is designed for 300 km/h operations and allows a theoretical one-hour travel time between the two termini of Roma Termini and Napoli Centrale. As of 2022, scheduled services take between 65 and 75 min depending on the type of rolling stock and the time of the day. In 1991, when a tender was awarded, the entire line was expected to cost 4.04 billion €. At the time of the inauguration, this figure had risen to an estimated value of 8.1 billion €, then reduced to 6.35 billion € in a more accurate estimate conducted later by Beria et al. [23]. This last value indicates an increase in costs of 57% compared to the initial forecasts (Table 1).

Table 1 Preliminary, initial, and final costs for Italian HSRs. Own elaboration on multiple sources

No	Line and opening year	Type of line	Length (km)	Cost type[a]	Total costs (B€)[b]	Unitary costs (M€/km)[b]	Cost overrunning[c] (%)
1	Rome–Naples (2005)	Segregated HS/HC	222	(a) Preliminary (1991)	4.04	18.20	–
				(b) Initial (2006)	8.10	36.49	200
				(c) Final (2010)	6.35	28.60	157
2	Turin–Milan (2006)	Segregated HS/HC	122	(a) Preliminary (1991)	2.17	17.79	–
				(b) Initial (2002)	3.87	31.72	179
				(c) Final (2010)	10.04	82.30	463
3	Padua–Venice Mestre (2007)	Upgraded line	24	(a) Final (2009)	0.60	25.00	–
4a + 4b	Milan–Treviglio (2007) Treviglio–Brescia (2016)	Mix of segregated HS/HC and upgraded line	30 + 52[d]	(a) Preliminary (2018)	1.66	20.24	–
				(b) Initial (2014)	3.24	39.51	195
				(c) Final (2016)	2.57	4a: 31.34 4b: 42.30	154
5	Naples–Salerno (2008)	Upgraded line	32	(a) Preliminary (1986)	3.08	96.00	–
6	Milan–Bologna (2008)	Segregated HS/HC	205	(a) Preliminary (1991)	3.01	14.68	–
				(b) Initial (2014)	9.30	45.36	309
				(c) Final (2016)	7.78	37.95	258
7	Bologna–Florence (2009)	Segregated HS/HC	78	(a) Preliminary (1991)	2.17	27.82	–

(continued)

Table 1 (continued)

No	Line and opening year	Type of line	Length (km)	Cost type[a]	Total costs (B€)[b]	Unitary costs (M€/km)[b]	Cost overrunning[c] (%)
				(b) Initial (2006)	7.74	99.23	357
				(c) Final (2009)	6.38	81.79	294
8	Brescia–Verona	Segregated line at 3 kvCC	48	(a) Preliminary (2014)	4.41	91.87	–
				(b) Initial (2019)	2.74 (after a project review)	57.08	62
9	Verona–Padua	Segregated HS/HC	77	(a) Preliminary (1991)	1.83	23.77	–
				(b) Initial (2018)	4.47	58.05	244
10	Terzo Valico	Segregated HS/HC	53	(a) Preliminary (1991)	3.21	60.57	–
				(b) Initial (2009)	6.87	129.62	214
				(c) Final (in progress; 2022)	7.46	140.75	232
11	Naples–Bari	Segregated HS/HC	150	(a) Preliminary (2016)	6.17	41.13	–
				(b) Initial (2022)	6.22	41.47	101
12	Salerno–Reggio C	Segregated HS/HC	450	(a) Preliminary (2022)	13	28.89	–
				(b) Initial (2021)	32.61	72.47	251
13	Turin–Frejus	Segregated HS/HC	85	(a) Preliminary (2003)	18.44	216.94	–

(continued)

Table 1 (continued)

No	Line and opening year	Type of line	Length (km)	Cost type[a]	Total costs (B€)[b]	Unitary costs (M€/km)[b]	Cost overrunning[c] (%)
				(b) Initial (2012)	8.50 (after a project review)	100.00	46
14	Verona–Fortezza	Segregated HS/HC	180	(a) Preliminary (2009)	3.07	17.06	–
				(b) Initial (2022)	5.00	27.78	163

[a] The year refers to the date of publication of the source used and not the date of financing
[b] Values are expressed as €[2022]
[c] Total expenditure compared with preliminary costs (expressed in %)
[d] Including all links with existing network (not counted in cost/km)

2.2 *Turin–Milan (2006)*

The HSR between Turin and Milan (Fig. 1, number 2) was inaugurated at two different moments: the stretch between Turin and Novara opened on 7 February 2006, after less than four years of construction and coinciding with the opening of the XX Winter Olympics in Turin. The line was immediately used for HS shuttle services at the Milan Malpensa Airport. The rest of the railway to Milan was opened later, between September and November 2009, completing a 122-km section running parallel to the already existing freeway on primarily flat terrain; the line runs at ground level for 80% of its extension [24]. The rush for opening in time for the Olympic Games likely contributed to the high cost of the railway; the line costed 10.04 billion €, 4.63 times more than the initial estimate of 2.17 billion €, and 159% more than the 3.87 billion € estimated at the start of construction [25, 26]. In terms of cost-kilometre (with all accessory road and rail infrastructures included), despite the easier terrain, the Turin–Milan line costed 102.7 million €/km, compared to 36.7 million €/km of the Rome–Naples line and just 21.3 million €/km of the LGV Sud-Est, a contemporary French HSR linking Paris to Lyon [27]. When looking at the costs/km of the HSR solely, the Turin–Lyon line still costed 2.87 times more than the Rome–Naples line, which includes many tunnels and crosses a mountainous terrain.

2.3 *First Sections of the Milan–Venice Line (2007)*

The railway between Milan and Venice is a fundamental tile of the corridor between France, Austria, and Slovenia. It has been updated to HSR standards in different phases and because of some changes in political priorities, according to different technical standards depending on the subsection [28]. The first two sections opened in 2007 between Padua and Venice Mestre (Fig. 1, number 3; 24 km) and between Milan and Treviglio (Fig. 1, number 4; 30 km) were electrified at 3 kV DC and allowed maximum speeds of 200–220 km/h. Both sections are used by all types of trains in regular service, including local and freight traffic. Moreover, both sections were built along the former conventional line and operated according to a "slow–slow–fast–fast" pattern (Fig. 2), meaning that the two "high-speed" tracks were managed independently from the two older tracks. This goes against the common practice throughout Europe, which suggests a "fast–slow–slow–fast" combination, where fast trains run on the outside tracks and slow trains run on the inside tracks, with island-type platforms at stations and apparent advantages for connecting switches between the two lines.

Not much data is available regarding the costs of the line between Venice Mestre and Padua; the line was financed as part of large-scale infrastructural projects in the Veneto area and received funding from multiple sources and in numerous different batches. Moreover, the project and planning process has been internal to the FS Group, the national rail operator, thus limiting the role of the ministry and

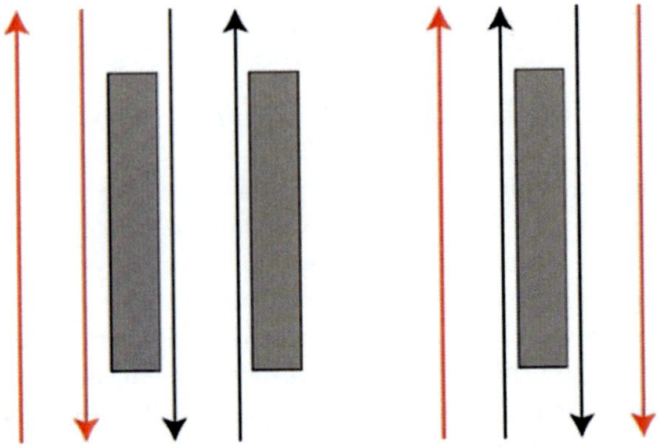

Fig. 2 Fast (red) and slow (black) tracks on the Milan–Treviglio and Padua–Venice Mestre railways (left) and in more conventional European examples (right), with the location of platforms. Own elaboration

public authorities and the quality of public information. According to Transpadana [29], a public–private organization promoting the development of the rail connection between Lyon and Ljubljana via northern Italy, the 24 km of the line costed approximately 0.60 billion €, with a unitary cost equal to 25 million €/km.

Similar issues arise when looking at the costs of the Milan–Treviglio section. According to the infrastructure development plan of the Lombardia region [30], the 30 km long line costed 351.8 million € (11.72 million €/km). The line is not a proper HSR but more of a doubling of the existing line; 11.72 million €/km is an extremely low value, and some doubts can be brought forward regarding the accuracy of the esteem considering the overall costs of the Treviglio–Brescia section (see paragraph 2.7) and of the entire Milan–Verona stretch, recently estimated at approximately 5 billion €, setting a cost per kilometre of around 28 million € [31].

2.4 Naples–Salerno (2008)

The HSR line between Naples and Salerno (Fig. 1, number 5), opened in April 2008, was built according to conventional standards and electrified at 3 kV DC. Used by all types of trains, including HS services, it is 32 km long, has a maximum speed of 250 km/h, and was first proposed and partially financed by law 17/1981, which also attributed to the line its name "*Linea a monte del Vesuvio*" [32], which is still used. The line's funding was confirmed in 1986 by the Italian Parliament [33] for a total of 3.08 billion €. No further information is available on the actual final costs of the line, which opened almost 40 years later to revenue services. In addition, research in press archives did not provide any beneficial results regarding delays, cost overruns, financing, or other potential issues which might have affected the construction period.

2.5 *Milan–Bologna (2008)*

The Milan–Bologna line (Fig. 1, number 6) is among the first to be built according to HC standards. The line allows a maximum speed of 300 km/h for the majority of its extension [34], except for a section around Modena that is limited to 240 km/h owing to tight curves, on which various tests are being conducted to overcome the issue and reach a uniform speed limit of 300 km/h [35]. The line is 205 km long and terminates at a dedicated underground station in Bologna, allowing fast trains to cover the distance from the Milan Central station in approximately 1 h. Moreover, the line hosts the only in-line intermediate station on an Italian HSR, namely Reggio Emilia AV Mediopadana, which opened in 2013 and is considered a local landmark. According to the Italian Parliament's record on the implementation of the "Legge Obiettivo" [36], the line overall costed 7.78 billion €, approximately 38 M€/km, including the entrance tunnel to Bologna station and 32 km of viaducts and bridges (Table 1). This compares with a preliminary estimate of only 3 billion € and funding of 9.3 billion €, indicating savings in the construction phase.

2.6 *Bologna–Florence (2009)*

The HSR line between Bologna and Florence opened in 2009 (Fig. 1, number 7) and constituted an engineering challenge as the Appennini mountain belt had to be crossed. The HSR substitutes the Bologna–Prato line, which opened in 1934, and allowed the connection between Bologna and Florence in an approximately 1 h non-stop travel, thanks to the 18-km long Great Appennine tunnel, the longest double-track tunnel in the world at that time [37]. The HSR allows for a travel time between 34 and 37 min at 300 km/h, which is approximately 78 km long (of which 74 km are in a tunnel), including Bologna station and only excluding the urban section in Florence. The HSR has been heavily criticised for its impact on the ground, freshwater, and mountain ecosystems, with numerous controversies and restoration projects following construction works [38]. Overall, the challenging line costed 6.38 billion € [36], more than 80 M€/km and approximately 300% more compared to the original estimations dated 1991.

2.7 *Treviglio–Brescia (2016)*

After the completion of the HSR between Turin and Salerno, the effort shifted to the transversal axis between the French border at the Frejus tunnel and the Austrian/Slovenian border northeast of Trieste. Excluding international connections, which are now under construction or in the planning phase, the missing HSR sections between Treviglio, Brescia, Verona, and Padova started a lengthy planning process in the early 2000s and are now in an advanced phase of construction or already in service. The Treviglio–Brescia section (Fig. 1, number 8) opened for revenue services on 11 December 2016 when the 39.6 km-long 300 km/h line (joined by 11.7 km of interconnecting tracks with existing railways) entered service, allowing for a scheduled travel time of just 31 min between Milan Central Station and Brescia.

Construction works on the line started in late 2011 after the approval of the Ministry Committee for Economic Programming and were completed in less than five years for 2.2 billion € [39]. Including all connecting tracks, this corresponds to about 42 million €/km, a slightly higher value compared to other Italian HSR sections built on flatland; this can be partially explained by the presence of 4.1 km of bridges, 1 km of tunnels, and 1.1 km in the trench and numerous environmental mitigation projects, including approximately 10 km of soundproof barriers and several rainwater collection basins [39].

2.8 *Lines Under Construction*

As mentioned, a paradigm change has recently been introduced in Italian HSR projects; formerly, HSRs were designed as fully segregated HS lines designed for 300 km/h operations and for mixed passenger-freight traffic (freight trains, however, have never been routed on HSR due to excessive maintenance costs, timing issues, and lack of suitable locomotives and rolling stock, except a single adapted ETR500, branded "mercitalia fast", and launched in 2018; [40]). More recently, HSR has been designed to develop a conventional railway with upgraded standards allowing 200–250 km/h operations, a broad loading gauge, and interoperability between passengers and freight operators. This new paradigm, known as *Alta Velocità di Rete*, is characterised by lower construction and operation costs [41] and can provide better performance in terms of territorial accessibility [16], although penalising longer-distance movements [2]. Excluding cross-border connections with neighbouring countries, Italy is now promoting three principal HSR axes: the completion of the Milan to Venice corridor, split into the Brescia–Verona (Fig. 1, number 9) and Verona–Padua sections (Fig. 1, number 10); the so-called "Terzo Valico" across the Ligurian Apennines between Milan/Turin and Genoa (Fig. 1, number 11); and the two lines south of Naples, towards Bari and Reggio Calabria (Fig. 1, numbers 12 and 13, respectively).

The *Brescia–Verona–Padua* line is being developed as a new independent HSR powered at 3 kV CC to increase interoperability with existing non-HSR rolling stock, given the frequent stops located along the line, which will serve all mid- and large cities along the corridor (the Garda area, Verona, Vicenza, and Padua). Overall, the line is 122 km long and mostly runs parallel to the existing railway. It is financed for approximately 8.9 billion € (73 million €/km), a sum that includes several side-projects, among which are the renovation and relocation of several kilometres of the parallel A4 freeway and the conventional railway and electric Bus Rapid Transit systems in Verona and Vicenza [42, 43]. The cost of HSR is only slightly lower, at approximately 58 M€/km (Table 1).

Terzo Valico consists of a 53 km HSR across the mountains and offers a third access to the city and port of Genoa, part of the Rhine–Alps TEN-T corridor. The HSR will have a maximum steep grade of 12.5‰ compared with 17.5 and 35‰ of the existing railways. Of the total length, approximately 70% will be in natural and artificial tunnels, including a significant 27 km tunnel. The line will allow 250 km/h operations and will cut the travel time between Turin, Milan, and Genoa to approximately 1 h. As of late 2022, the line has been funded for a total of 7.46 billion € and is due to be completed by late 2024 [44, 45].

The *Napoli–Bari* line was built according to the HS/HC standards. The line is funded for 6.22 billion € (41.3 million €/km) and is expected to open in 2027, allowing reaching Bari from Naples in less than 2 h and from Rome in approximately 3 h [46]. Being 150 km long, of which 80% is in tunnels, the line will be the first HSR crossing the country west to east south of the Padana Plain in the north. As of 2022, the construction is tendered or ongoing on the totality of the line [47].

This is not the case for the line between Salerno and Reggio Calabria, which is currently undergoing an advanced project phase that evaluates different project alternatives, their benefits, and their overall impact on the territory [48, 49].

The line will likely be built partially as a fully segregated HS/HC 300 km/h line and partially as an improvement of the existing railway to ensure services to the numerous relevant settlements located along the coastline. Overall, the line is expected to be approximately 450 km long, allowing the connection between Rome and Reggio Calabria in around 3.5 h [50]. As of 2022, the costs for the whole line are expected to be between 22 and 40 billion € [51].

Aside from national connections, Italy is also heavily investing in primary HSR connections to France (across the Frejus tunnel towards Saint-Jean-de-Maurienne, Chambery, and Lyon) and Austria (across the Brenner pass and future Brenner base tunnel). The eastern connection to Austria and Slovenia is left to the Pontebbana railway, which was upgraded in the late 1990s but not to HSR standards. In contrast, the short section between Milan and the Swiss border is equipped with European interoperable signalling but does not require further upgrades, as the most challenging section is in Switzerland (and is being upgraded within the AlpTransit context, with the new Ceneri and Gotthard Base Tunnels [52]).

The Turin–Lyon HSR (Fig. 1, number 14) has been under construction since the early 2000s. After the last project review, the common Italian/French section of the line is expected to cost 8.5 billion €, including the access line from Turin and the Frejus base tunnel (57.5 km), for 85 km (100 million €/km). For further insights into the criticism and benefits expected from the railway, readers are reminded of Bruzzone et al. [2].

The Brenner line (Fig. 1, number 15) is being tracked between Verona and Fortezza/Franzensfeste in South Tyrol, from where the Brenner Base tunnel heads towards Innsbruck. Excluding the independently managed tunnel, the 250 km/h line is expected to cost approximately 5 billion € over 180 km of tracks (27.8 million €/km), including an underground section in Trento and another 6 km tunnel under the Isarco River [53]. Although works on the Brenner Base Tunnel began in 2008, the Italian access line remains in the final project phase. Plans are in place to simultaneously open the two infrastructures.

Section 3 systemises the current situation of HSR costs in Italy and further discusses the causes and consequences of various technical and political choices, as well as historical and actual management issues that contribute to shaping the development and operations of Italian long-distance railways.

3 HSR Costs in Italy: Overview and Discussion

This section attempts to systematise the main financial aspects related to Italian HSR lines that are completed or under construction and seek explanations for the observable characteristics and phenomena. Table 1 shows a synthetic overview of the costs estimated during the preliminary discussion (column "preliminary costs"), at the start of construction works (column "initial costs"), and at the opening date (column "final costs") for Italian HSR lines opened or under construction during the last 30 years. For the values presented in Sect. 2, all values are expressed in €$_{2022}$.

In short, Table 1 shows a unique but consistent trend; while final costs are generally higher than planned expenditures, they are often lower than the funds available at the beginning of the construction works. For instance, the Milan–Bologna–Florence lines cost approximately 333% more than that initially forecasted but 17% less than the estimates included in the executive project. The only notable exception to this trend is the Turin–Milan HSR, which shows a 159% cost increase even in comparing final and initial costs. As suggested elsewhere, the attentiveness and reliability of project management seem to be inversely proportional to the rush with which the project is carried out. In the case of the Turin–Milan line, the HSR became part of an even wider megaproject, the XX Winter Olympics (held in February 2006), and was thus a subject to amplified issues. In less exceptional situations, when the planning and tendering processes follow regular phases and timing, it seems that executive projects and institutional documents approving funding and launching tenders can correctly estimate overall costs, which sometimes even see a final reduction, possibly due to competitive tendering.

In relative terms, it is interesting to see how the unitary values range from 28 M€/km to over 100 M€/km. Lower values are in line with the international reference values (e.g., [54]), whereas the highest values registered for the Turin–Lyon and Terzo Valico lines (under construction) and the Bologna–Florence line (operational) are up to four times higher than the average. In these cases, the primary causes of cost increase are the types of infrastructural work required to make the line operational (tunnels, bridges, and viaducts) and the reduced time for realizing the work.

According to an independent audit commissioned by the EU, the European HSR takes an average of 16 years to be constructed and opened for revenue services [54]. The political and civil debate on such complex megaprojects begins well ahead, with the first cost estimates often entering the technical-political discourse up to

40–50 years before the start of construction [55]. Even when updating these values to the current costs, it is intuitively visible that these are low and are likely to be inaccurately determined. As shown in Table 1, the initial underestimation of costs is consistent; in most cases, they exceed the preliminary values by 200%, and the Milan–Bologna and Bologna–Firenze lines even exceeded by more than three times (309% and 357%, respectively). However, unlike what was debated by Flyvbjerg and confirmed by other scholars [2, 3, 4, 13, 14, 23, 25, 27, 54], Italian HSRs, which are already operational (except for the Turin–Milan line, see above), have seen final costs comparable with the estimations conducted at the beginning of the construction works. In this comparison, the norm is a cost decrease of approximately 20%. This phenomenon characterises the Italian panorama and seems different from other countries investing in HSR.

The relevant cost increase between the preliminary and final figures (for instance, the Milan–Brescia and Bologna–Florence lines) can be explained in two ways. On the one side, the initial estimates are often low for political reasons to ensure that the project can raise interest and gain political and public support, even at the expense of transparency and correctness in communication and political debate [11–13]. On the other side, most HSR projects were first designed and brought into the parliamentary debate in the late 1980s or early 1990s, when the concept of HSR itself was very different and required much less infrastructural works and efforts, as no voltage and signalling innovations were required and in particular, the high-speed concept in Italy was tailored around "Pendolino" tilting trains, powerful passenger electric multiple units able to climb steep grades and take tight curves at up to 250 km/h [56]. Thus, the required infrastructural works were much less invasive than those related to the HS/HC concept of the early XXI century, demanding long tunnels, wide curves, and fully segregated catenary and signalling.

Meanwhile, the fact that the initial costs of Italian HSR project construction are more accurately determined is partially due to the competitive tendering process, as mentioned above, and in part due to the long but meticulous bureaucratic apparatus that Italian megaprojects need to go through, considering anti-criminal laws [57]. The outcomes of our research suggest that the process is time-consuming but allows for the establishment of precision costs during construction.

Recently, a new change in the priorities for HSR has occurred with governments throughout Europe, Italy being no exception, orienting funding towards strongly upgraded conventional lines rather than fully segregated HSR [2, 16]. It will be interesting to see if in the next decade, the observed trends will continue or if project management and cost definition will further improve or instead degrade with the de-escalation from "megaproject" management to a multitude of smaller-scale interventions scattered throughout the rail network, often managed and tendered separately.

4 Conclusions

The issues of megaproject management and funding are mostly related to transparency and consistent cost overruns. This work discussed the case of the Italian HSR to verify whether such issues are also valid in the context of railways. We first attempted to systemise the costs of HSR that have been opened for revenue service or are under construction in the last 30 years by distinguishing preliminary costs from initial and final costs. Our main finding is that the final costs significantly overrun the preliminary project costs, as proposed during the very early stages of the project. However, in most cases, they do not overrun the funds available at the beginning of construction, despite the length of construction works being often more than a decade.

The case of Italian HSRs demonstrates a certain capacity for project management and the tendering process to correctly estimate the costs and timing. However, more elements should be added to the discourse when widening the perspective. The literature often stresses that Italian HSR's construction costs are generally higher than those observed in other European countries, even when looking at comparable geomorphological conditions of land interested in the project. This might hide some "buffer" investments that could otherwise be saved.

In the future, it will be interesting to study the evolution of the trends discussed here. Within the new "reduced" HSR paradigm, by upgrading conventional lines to HSR standards rather than providing new fully segregated lines with smaller tenders, lower costs, and smaller infrastructural projects in general, a change in project management accuracy can be expected, the magnitude and characteristics of which will surely be an interesting field of study. Regarding this, it is important to assess how the costs of the Italian HSR will be compared to similar infrastructures in neighbouring countries. Finally, it would be interesting to see if upcoming projects will be able to sustain the same traffic and induce the same benefits in terms of travel time, convenience, and modal shift from non-rail modes as HSR, which opened at the beginning of the century, a fundamental tile to approach the validity of HSR investments.

References

1. Dalla Chiara, B., De Franco, D., Coviello, N., Pastrone, D.: Comparative specific energy consumption between air transport and high-speed rail transport: a practical assessment. Transp. Res. Part D Transp. Environ. **52**, 227–243 (2017)
2. Bruzzone, F., Cavallaro, F., Nocera, S.: The effects of high-speed rail on accessibility and equity: evidence from the Turin-Lyon case-study. Socio-Econ. Plan. Sci. 101379 (2022). https://doi.org/10.1016/j.seps.2022.101379
3. Feigenbaum, B.: High-speed rail in Europe and Asia: lessons for the United States. Policy Study 418. Reason Foundation (2013)

4. Cascetta, E., Cartenì, A., Henke, I., Pagliara, F.: Economic growth, transport accessibility and regional equity impacts of high-speed railways in Italy: ten years ex post evaluation and future perspectives. Transp. Res. Part A Policy Pract. **139**, 412–428 (2020)
5. European Commission. Action plan to boost passenger rail. https://transport.ec.europa.eu/news/action-plan-boost-passenger-rail-2021-12-14_en. Accessed 21 Oct 2022
6. European Commission. Transport infrastructure: EU invests EUR 5.4 billion to support key projects across the continent. https://cinea.ec.europa.eu/news-events/news/transport-infrastructure-eu-invests-eur-54-billion-support-key-projects-across-continent-2022-06-29_en. Accessed 10 Oct 2022
7. Cullino, R., Fabrizi, C.: Senza la base non si può cantare. La nuova linea ferroviaria Torino-Lione nell'esperienza italiana e francese. In: Proceedings of the XXXII Italian Conference of Regional Sciences, 15th–17th September 2011, Turin (2011)
8. Cavallaro, F., Bruzzone, F., Nocera, S.: Spatial and social equity implications for high-speed railway lines in Northern Italy. Transp. Res. Part A Policy Pract. **135**, 327–340 (2020)
9. Ravazzoli, E., Streifeneder, T., Cavallaro, F.: The effects of the planned high-speed rail system on travel times and spatial development in the European alps. Mt. Res. Dev. **37**(1), 131–140 (2017). https://doi.org/10.1659/MRD-JOURNALD-15-00051.1
10. UNECE United Nations Economic Commission for Europe. Trans-European Railway High-Speed—Masterplan study—A general background to support further required studies. ISBN 978-92-1-117264-5 (2021)
11. Flyvbjerg, B., Bruzelius, N., Rothengatter, W.: Megaprojects and Risk: An Anatomy of Ambition. Cambridge University Press (2003)
12. Flyvbjerg, B.: Megaprojects and Risk: A Conversation with Bent Flyvbjerg. SSRN Scholarly Paper at https://papers.ssrn.com/abstract=2278321 (2004)
13. Flyvbjerg, B.: What you should know about megaprojects and why: an overview. Proj. Manag. J. **45**, 6–19 (2014)
14. Tomai, F.: Nascita e sviluppo del sistema alta velocita' in Italia. Working Paper, Universitat Politècnica de Catalunya. http://upcommons.upc.edu/pfc/bitstream/2099.1/6457/3/02.pdf. Accessed 21 Nov 2016
15. European Union. Regulation (EU) No 1315/2013 of the European Parliament and of the Council of 11 December 2013 on Union guidelines for the development of the trans-European transport network and repealing Decision No 661/2010/EU Text with EEA relevance (2013)
16. Cavallaro, F., Bruzzone, F., Nocera, S.: Effects of high-speed rail on regional accessibility. Transportation (2022). https://doi.org/10.1007/s11116-022-10291-y
17. RFI. Linea Adriatica. Nuovi interventi di potenziamento e velocizzazione. https://www.mit.gov.it/comunicazione/news/adriatica-definito-il-progetto-per-velocizzazione-e-miglioramento-della-linea. Accessed 22 Oct 2022
18. RFI. Corridoi TEN-T. https://www.rfi.it/it/rete/in-europa/corridoi-ten-t.html. Accessed 21 Oct 2022
19. Cascetta, E., Coppola, P.: High speed rail (HSR) induced demand models. Proc. Soc. Behav. Sci. **111**, 147–156 (2014)
20. Legambiente. Rapporto Pendolaria 2019 (2020)
21. ISTAT. Rivaluta. https://rivaluta.istat.it/Rivaluta/. Accessed 21 Oct 2022
22. Ferrovie on Line. A tutta velocità con l'ERTMS! https://www.ferrovie.it/portale/articoli/1025. Accessed 19 Oct 2022
23. Beria, P., Grimaldi, R., Albalate, D., Bel, G.: Delusions of success: costs and demand of high-speed rail in Italy and Spain. Transp. Policy **68**, 63–79 (2018)
24. WeBuild. L'alta velocità Torino-Milano. https://www.webuildgroup.com/it/progetti/ferrovie-metropolitane/linea-ferroviaria-velocita-capacit-torino-milano. Accessed 21 Oct 2022
25. Bocca, R. Alta voracità. l'Espresso 30/12/2006; In: Wikinotizie https://it.wikinews.org/wiki/I_costi_dell%27Alta_Velocit%C3%A0. Accessed 21 Oct 2022
26. Cicconi, I.: Il libro nero dell'alta velocità ovvero il futuro di tangentopoli diventato storia. Koinè Nuove Edizioni. ISBN 888982817X (2011)

27. Malan, A.: I finanziamenti: in Francia l'alta velocità costa un quarto rispetto all'Italia. https://st.ilsole24ore.com/art/SoleOnLine4/Speciali/2007/tav/speciale-tav-finanziamenti-malan.shtml?uuid=924c3c9c-12a3-11dc-9aa7-00000e251029&DocRulesView=Libero&correlato. Accessed 21 Oct 2022
28. Latorre, M.: Tra Brescia e Verona una ferrovia ad alta capacità dimezzata. TrasportoEuropa, 2020/12/22
29. Transpadana https://web.archive.org/web/20180612142833; http://www.transpadana.org/mediterraneo/padova-veneziamestre.html (2008)
30. Regione Lombardia. D.g.r. 4915/2021
31. OTINord. Ferrovia alta capacità Milano-Verona. https://www.otinord.it/progetti/ferrovia_alta_capacita_milano_verona. Accessed 17 Oct 2022
32. Repubblica Italiana. L. 17/1981. http://legislature.camera.it/_dati/leg10/lavori/stampati/pdf/0132_004002_F010.pdf (1981)
33. Camera dei Deputati. X legislatura. Disegni di legge e Relazioni. http://legislature.camera.it/_dati/leg10/lavori/stampati/pdf/0132_004002_F010.pdf. Accessed 21 Oct 2022
34. Italferr. Linea AV/AC Milano-Bologna. https://www.italferr.it/content/italferr/it/progetti-e-studi/italia/alta-velocita---alta-capacita/milano-bologna.html. Accessed 21 Oct 2022
35. Ferrovie On Line. Più Veloci a Modena, 10/06/2015. https://www.ferrovie.it/portale/articoli/906. Accessed 21 Oct 2022
36. Camera dei Deputati. L'attuazione della "Legge Obiettivo". http://documenti.camera.it/leg16/dossier/testi/Am0075.htm (2009)
37. Lunardi, P.: Design and Construction of Tunnels. Springer, Berlin, Heidelberg (2008). https://doi.org/10.1007/978-3-540-73875-6
38. Rumiz, P.: Quei torrenti inghiottiti dagli scavi della Tav—ambiente—Repubblica.it. https://www.repubblica.it/2009/03/sezioni/ambiente/tav-torrenti/tav-torrenti/tav-torrenti.html (2009)
39. Ferrovie on Line. A 300 km/h da Milano a Brescia, 11/12/2016. https://www.ferrovie.it/portale/articoli/4400. Accessed 22 Oct 2022
40. Ferrovie on Line. Mercitalia Fast—L'ETR 500 per le merci ad alta velocità, 29/10/2018. https://www.ferrovie.it/portale/articoli/7940. Accessed 22 Oct 2022
41. Spirito, P.: Il gioco delle tre carte per l'alta velocità ferroviaria al Sud. Rivista giuridica del Mezzogiorno 1055–1064 (2021). https://doi.org/10.1444/102163
42. Consorzio Cepav Due. AV/AC Brescia Est—Verona. https://www.cepavdue.it/progetto/il-progetto/. Accessed 22 Oct 2022
43. Il Progetto—Alta capacità / Alta velocità Verona Padova. http://veronapadova.it/il-progetto/. Accessed 22 Oct 2022
44. FS Italiane. Terzo valico—Nodo di Genova. https://www.fsitaliane.it/content/fsitaliane/it/opere-strategiche/terzo-valico.html. Accessed 22 Oct 2022
45. MIMS, Commissario Terzo Valico Costi e tempi. https://terzovalico.mit.gov.it/il-terzo-valico/costi-e-tempi/. Accessed 22 Oct 2022
46. FS Italiane. Napoli-Bari. https://www.napolibari.it/. Accessed 22 Oct 2022
47. RFI. https://www.rfi.it/it/rete/i-nostri-progetti/Per-la-lunga-percorrenza.html#2126982277. Accessed 22 Oct 2022
48. RFI. Nuova linea con caratteristiche di Alta Velocità Salerno—Reggio Calabria e interventi sulla linea Battipaglia-Potenza-Metaponto-Taranto (1^ fase). Direzione strategia, pianificazione e sostenibilità (2021)
49. RFI e MIMS. Contratto di programma 2022–2026, Parte Investimenti (2022)
50. RFI. Nuova linea ferroviaria AV Salerno—Reggio Calabria. https://www.rfi.it/it/rete/opere-sottoposte-a-dibattito-pubblico/nuova-linea-ferroviaria-av-salerno---reggio-calabria.html. Accessed 22 Oct 2022
51. Beria, P., Debernardi, A.: Pnrr, perché il mega-investimento per l'alta velocità tra Roma e Reggio Calabria è inutile. Domani, 13/04/2022. https://www.editorialedomani.it/economia/pnrr-perche-il-mega-investimento-per-lalta-velocita-tra-roma-e-reggio-calabria-e-inutile-ljk3vstv. Accessed 20 Sept 2022
52. AlpTransit. https://www.alptransit-portal.ch/it/. Accessed 22 Oct 2022

53. FS Italiane. Brennero e Fortezza-Verona. https://www.fsitaliane.it/content/fsitaliane/it/opere-strategiche/brennero-e-linea-fortezza-verona.html. Accessed 22 Oct 2022
54. European Court of Auditors. La rete ferroviaria ad alta velocità in Europa non è una realtà, bensì un sistema disomogeneo e inefficace. (Publications Office, 2018)
55. Camera dei Deputati. Infrastrutture strategiche e prioritarie. Programmazione e realizzazione (2021)
56. Santanera, O. I treni FIAT: ottant'anni di contributo FIAT alla tecnica ferroviaria. (Automobilia, 1997)
57. Visentin, M.: Infrastrutture: costruirle è un percorso a ostacoli I D. Immordino. *Lavoce.info*. https://www.lavoce.info/archives/67977/infrastrutture-costruirle-e-un-percorso-a-ostacoli/ (2020)

HSR: Competition Versus Integration Versus Inclusion

The Impact of HSR on Same-Day Intercity Mobility: Evidence from the Yangtze River Delta Region

Haixiao Pan, Ya Gao, and Khandker Nurul Habib

Abstract One objective of China's High-speed rail (HSR) development is to promote regional cohesion, which can be reflected by the flow of people between city pairs. As a fast-speed intercity transport mode, same-day intercity mobility has been regarded as an essential measurement for regional cohesion and transport integration. The reduced time by HSR has redefined the business and commute trips which highlights time efficiency. Due to the difficulty in obtaining large samples of data for such trips, we adopted mobile phone data to detect and analyze the spatial distribution and travel behaviour characteristics of same-day return travellers. The efficiency analysis measured by total travel time between city pairs indicates that HSR is less competitive with cars within 300 km for same-day return trips. The variance in HSR passengers' travel time over the same distances could be due to no direct services and the time required for access/egress. Using a 20-week mobile phone data, we adopted a rule-based method for detecting intercity travellers based on their temporal and spatial geographic locations. Results showed that most travellers travelled within 3–3.5 h, and few conducted a same-day return trip regularly. GDP, service frequency, and distance between origin and destination have been examined to explain the mobility of same-day return travel. The findings of our paper are expected to improve our understanding of same-day return travel behaviour and promote HSR travel for efficient round trips.

Keywords High-speed rail · Same-day trips · Intercity mobility · Mobile phone data · China

H. Pan · Y. Gao (✉)
Department of Urban Planning, Tongji University, 1239 Siping Road, Shanghai 200020, China
e-mail: gaoya@tongji.edu.cn

H. Pan
e-mail: panhaixiao@tongji.edu.cn

Y. Gao · K. N. Habib
Department of Civil & Mineral Engineering, University of Toronto, 35 St. George Street, Toronto, ON M5S 1A4, Canada
e-mail: khandker.nurulhabib@utoronto.ca

F. Pagliara (ed.), *Socioeconomic Impacts of High-Speed Rail Systems*,
Springer Proceedings in Business and Economics,
https://doi.org/10.1007/978-3-031-26340-8_4

1 Introduction

Since the first High-speed rail (HSR) launched in China in 2008, the network has noticeably expanded in the past fifteen years. In total, 41,000 km of HSR service length have been operated to connect cities across the country by 2022. In agglomerated regions, such as the Yangtze River Delta (YRD) Region and the Pearl River Delta (PRD) region, the increased density and service frequency of HSR has facilitated the intensity of economic activities. Three types of trips in terms of travel time and travel purposes have been identified in previous research: commuting trips within 1 h, same-day business/tourism trips within 2–3 h, and occasional trips [1, 2]. Among the three types, trips that can be completed within a day are key to regional cohesion and therefore receive tremendous attention from researchers [3–5].

Same-day intercity mobility is generally defined as the ability to complete an intercity around-trip travel within a day. Researchers have adopted location- or schedule-based approaches to measuring same-day intercity accessibility and efficiency [3, 4, 6]. Those approaches mainly evaluate the same-day return travel from the supply side. It is partly due to the lack of revealed data for such trips, either unavailable or conveniently obtainable.

Recently, there has been growing interest in using mobile phone data or other user location data for large-scale mobility behaviour analysis [7, 8]. The mobile phone data can provide the almost real-time anonymized geolocation information to reveal the spatial dynamics of intercity mobility flow. Also, obtaining panel data within a certain period is relatively convenient, from which the periodic travel characteristics of travellers can be detected.

Our objective is threefold in this study: (1) from the supply perspective: we aim to identify the efficiency of possible same-day return city pairs within the YRD region based on the total travel time of both HSR and other competitive modes; (2) from the demand side, using mobile phone data, we try to detect the same-day intercity travellers, and analyze their periodical travel behaviour and spatial distribution characteristics; (3) We attempt to reveal where and to what extent the intercity same-day travel via rail is occurring based on identified rail users passing through the selected railway station areas. The findings of this study are expected to improve our understanding of same-day return travellers and to promote HSR efficiency to support the same-day return travel demand of intercity travellers.

2 HSR in the Yangtze River Delta Region

The Yangtze River Delta (YRD) region consists of Shanghai city and three provincial administration areas (Jiangsu, Zhejiang, and Anhui) (Fig. 1). It is one of the most populous and active regions in China, with 235 million inhabitants in the 2020 census. HSR operation in the YRD region began by connecting major cities, such as Shanghai,

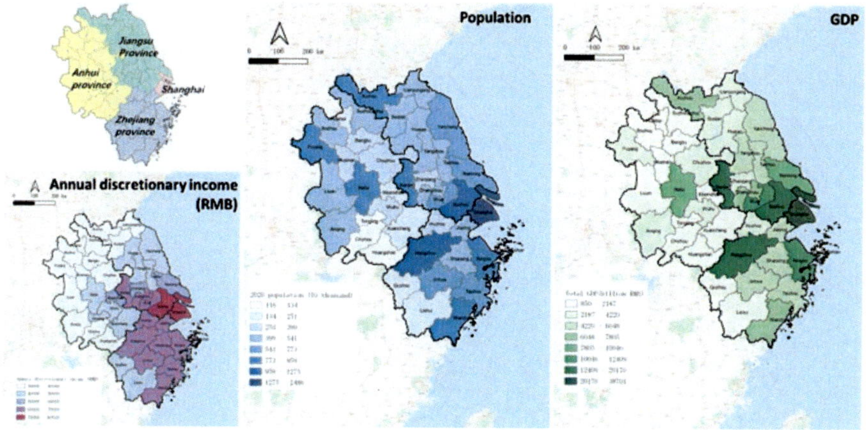

Fig. 1 Population, GDP and discretionary income of 41 cities within the Yangtze River Delta Region by the end of 2020

Nanjing, and Hangzhou, when the passenger volume of existing conventional rail and travel time costs once restricted the regional interaction.

For the past decade, the High-speed rail network has expanded to encompass almost all cities (40 out of 41) within the region. The operated HSR lines have reached 5977 km by the end of 2020 (Fig. 2). The increase in rail mileage for the past decade mainly resulted from the HSR lines. In the next fifteen years, new HSR lines (travelled at 250–350 km/h) and intercity rail lines (travelled at 120–200 km/h) have been planned in the 2020 YRD Transport Integration Development Plan [9]. The new lines aim to provide direct services without transfers, add capacity to congested corridors and connect small and medium-sized cities along the corridors (Fig. 3). The city of Zhoushan, separated by the sea from other cities, is expected to connect to the region's HSR network by 2026.

As a fast-speed intercity transport mode, the time savings by HSR has boosted mobility between cities [10], especially along the Shanghai-Nanjing and Shanghai-Hangzhou HSR corridors. On a typical weekday, around 130 trains (including G and D trains) travel from Shanghai to Nanjing, a 300 km distance with a minimum 59-min in-vehicle time. A similar service frequency operated between Shanghai and Hangzhou, 180 km, with a minimum of 45-min in-vehicle times. New travel patterns, including intercity commuting and same-day return business/leisure trips, have been observed along the highly frequent service corridor [11–13].

3 The Efficiency of Same-Day Return City Pairs

Previous research has proposed different approaches to measure daily mobility via HSR. For example, Liu et al. [3] evaluated the daily accessibility of daytime

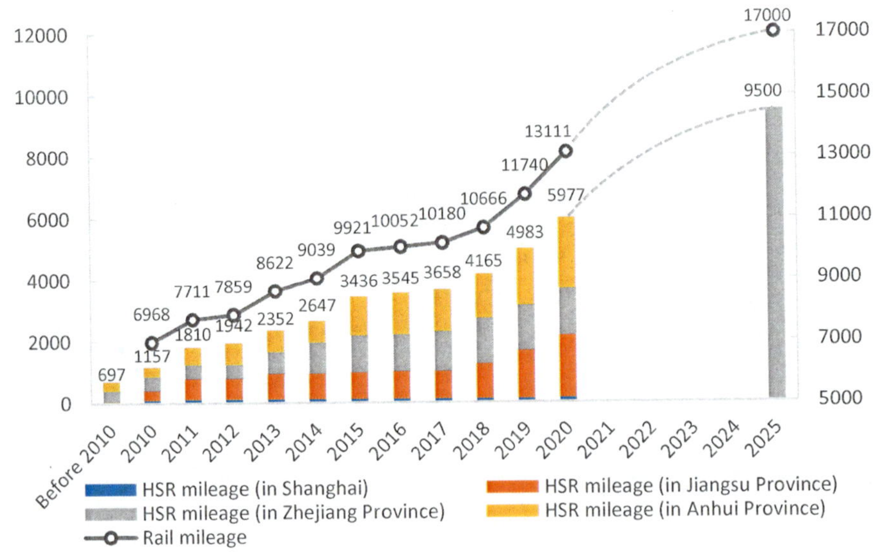

Fig. 2 Shows HSR and total rail mileage within the Yangtze River Delta Region

round-trips in Shandong Province by proposing an improved potential-based daily accessibility index, accounting for GDP, population, area, average travel time, and train frequencies between city pairs. Unlike the previous measurement index, train frequency was added as a weight, and travel time was distinguished into ordinal intervals. They use a threshold of 4 h for one-way business journeys that could be completed within a day. However, the node-based approach (calculating travel time between different HSR stations) did not consider the intra-city travel time, which may account for almost half of the travel time within short travel distances [14]. Thus, the approach proposed may overestimate the cities that could be accessed within a day.

Considering time constraints spent on destination cities and the associated costs for leisure and business same-day return trips, Moyano et al. [4] measured the efficiency of each city-to-city link in the Spanish HSR network. The total population weighted efficiency for tourism travel and the amounts of high-skilled jobs for business travel to obtain the global values of efficiency for comparison. The approach was more precise to capture aggregated efficiency for different travel purposes. However, the approach requires values of time (VOT) parameters obtained by previous research, which is hard to implement without prerequisite information. Also, the above studies only focus on the HSR mode without considering the possible competitive modes for intercity travel, such as cars.

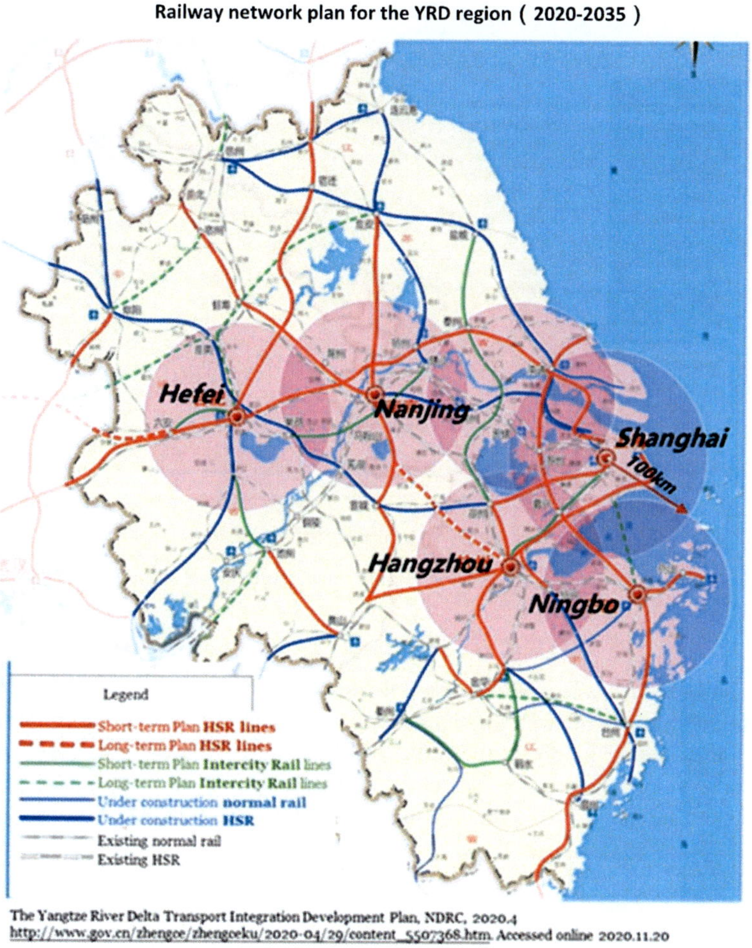

Fig. 3 The railway network plan for the YRD region (2020–2035). (NDRC 2020)

3.1 HSR and Competitive Mode Travel Time for Possible City Pairs

To investigate which travel mode could be more competitive in total travel time, we compare the travel time and distance between different city pairs in the YRD region. Of all 1640 city pairs, the city pairs with the top 25% GDP of origin and destination city were selected for further analysis since GDP was commonly found to yield a robust explanatory power in modelling and forecasting intercity travel demand [15]. We obtained the travel time and distance by retrieving the Gaode map API [16] for the fastest route. Various components of total travel time, such as access/egress, waiting, transfer if possible, and in-vehicle times were included based

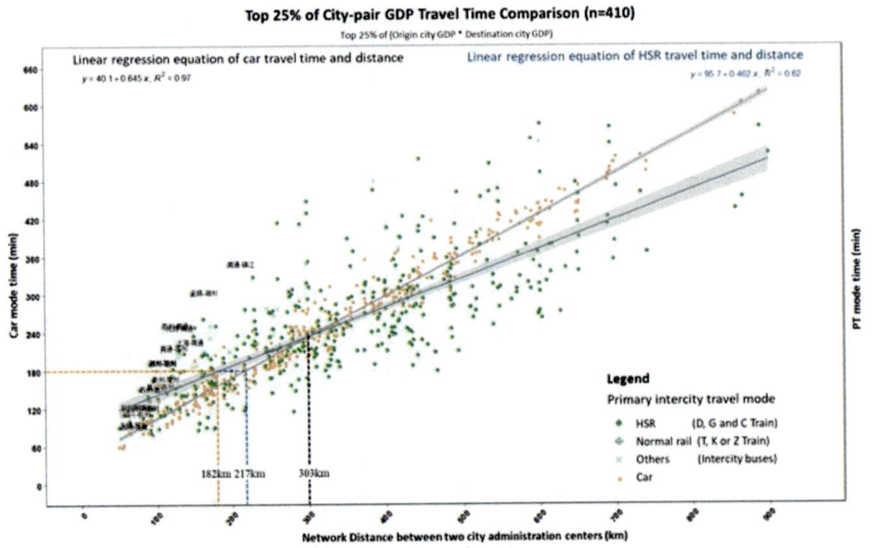

Fig. 4 Travel time and distance comparison for different intercity travel modes for city pairs in the YRD region

on the route planned. Travel distance was measured by network distance between city administration centers. Using a map app to plan an intercity trip is common for intercity travellers. Therefore, the travel time obtained from the map has valuable meaning for intercity travellers. To ensure the data quality, we double-checked the obtained map API time based on travellers' revealed travel time between certain city pairs.

Figure 4 shows the relationship between travel time and distance for cars, HSR, and other public transit modes between city pairs. Intercity railway lines (marked as C trains in China) are categorized into the HSR type since the speed are more similar to the D train.

Results show that regarding total travel time based on the equation of HSR and car travel time and distance, HSR is hard to compete with car mode within 303 km. Despite the same travel distances, HSR travel time varies significantly between city pairs. This variance could result from the following reasons:

(1) the access/egress time in origin and destination cities. For example, from Huzhou-Shaoxing, with a travel distance of 135 km. The total travel time by public transport is 127 min with only 44 min for in-vehicle time. The high speed of HSR was weakened by the access time and egress time to HSR stations.
(2) short-distance city pairs without direct services. For instance, from Nantong to Zhenjiang, passengers have to interchange at the station of Changzhou or Nanjing.
(3) geographic location reasons, such as the city pair of Wuxi-Huzhou, divided by the Tai hu Lake, without straight routes available.

Expanding the HSR network to connect potential city pairs with direct links and better inter-intra-city connections between remote stations to central districts may improve the competitiveness of HSR and further facilitate the interaction between city pairs within the region.

3.2 *Count of Cities for Return Trips Within a day*

We further calculated the cities that could be accessed for a return trip based on a 3-h total travel time threshold of public transport (PT) modes. This threshold is selected based on the aforementioned articles indicating a 2–3 h travel time for a return trip to be completed within a day [1, 2].

In Fig. 5, the size of the ring represents a city that could access more cities or could be arrived at by more cities for a round-trip using public transport. Consistent with the result of the global efficiency analysis in Spain [4], node cities placed in the central position of the network achieve more accessible cities than the peripheral ones. The multi-direction connections link the central position cities, such as Nanjing, Hangzhou, and Bengbu, to more cities. We also observed that cities along the Shanghai-Nanjing corridor have access to more cities than cities along the Shanghai-Hangzhou corridor, thanks to the relatively shorter distances between cities along the Shanghai-Nanjing corridor. The situation of Jiaxing is also remarkable. Jiaxing is located between Shanghai and Hangzhou; however, the city could access fewer cities in a daily round trip due to the longer access/egress time from the administration center to the HSR station.

To identify the potential corridors for HSR promotion, we compared the travel time of public transport and car modes. Total travel time between the range of 3 and 4 h by public transport (PT), and with a longer travel time compared to cars are displayed in the left graph in Fig. 6. The high-volume corridor between the city pairs in the YRD region was identified in previous research [17]. Through the comparison between the two graphs, we could identify corridors of high intercity volume, but the HSR network and services are currently weak to compete with car mode for same-day return trips (marked by dash lines in the right graph). For example, Changzhou, Wuxi, and Suzhou show a high passenger volume to the city of Huzhou; however, the PT travel time for the three city pairs was 1.65, 1.96, and 1.93 times longer than car mode, respectively. The longer travel time by PT modes was caused by no direct services available between those city pairs. The connections between large cities and small-medium cities to large cities have been promoted by the HSR network. Our findings show the necessity to improve the efficiency of medium-to-medium city connections in detected corridors.

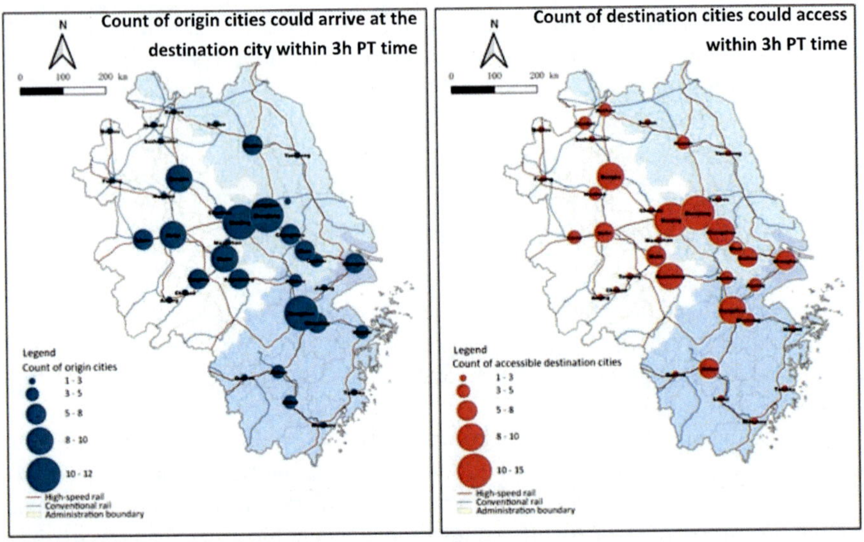

Fig. 5 Count of origin cities that could arrive at the destination city using public transport within 3 h (left) and count of destination cities that the origin city could access using public transport within 3 h (right)

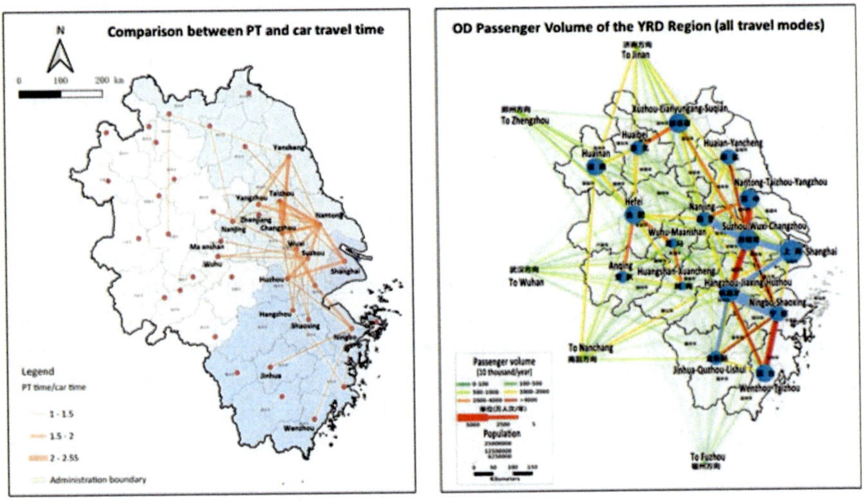

Fig. 6 Comparison between public transit time and car time (left), and OD passenger volume of the Yangtze River Delta Region (Zi 2021) (right)

4 Travel Behaviour and Spatial Distribution Characteristics of Intercity Travellers

Problems of using the traditional survey method for collecting intercity travel data have been discussed in previous research, which usually involves a high response burden, low number of respondents, and potential bias [18–20]. Two alternative data sources, GPS, and mobile phone data have been adopted in recent years [18]. Compared to GPS data, mobile phone data has the advantage of large sample sizes and easily obtain data at any period. This section explains how mobile phone data can be converted into individual trajectories, which are then used to identify intercity same-day return travellers' travel behaviours and to detect rail users.

4.1 Detect the Intercity Same-Day Return Travellers Using Mobile Phone Data

The mobile phone data used in our study was the cellular network-based positioning data from CHINA TELECOM. It is one of the three telecom companies in China, with a market penetration of 28.0% in Shanghai and 29.3% in Zhejiang province. The network-based positioning data utilize one or several base stations to locate a cell phone periodically, usually for 30 min. With the periodical positioning data, individual trajectories could be derived. The spatial resolution of the data varies from a few metres to several hundred metres based on the distribution of base stations [8]. Data resolution in the central area is usually more precise than in the peripheric area resulting from denser base stations inside the outer ring road of Shanghai (less than 1 km inside the outer ring road compared to around 3 km outside). The location estimation accuracy around the selected rail stations in Shanghai and Zhejiang provinces is generally around 150–500 m.

Data was collected between Feb 22 (a Monday) and July 11 (a Sunday) in 2021 for 20 full weeks. Since weekday trips involve more same-day return trips, trips within the 97 weekdays are kept for further analysis. By eliminating the holidays and weekends, the left same-day return trips are more likely to be business trips. We follow three steps to detect intercity travellers. Each stage of this procedure is detailed as follows:

First, an individual's home locations in Shanghai are detected. All night stay locations of a person between 11 p.m. and 5 a.m. (next day) on the 20 weeks are labelled, and the observed night stay location on each day can be counted. In our study, the count of night stay locations for most days and at least 60 days (based on three days per week for the 20 weeks) are labelled as home location.

Second, an individual's trip destinations for the same-day return trip are identified. The individual trajectories that depart from and end up with the home location (0–7 a.m. and 11 p.m. to the next day at 7 a.m. should be at the home location) and be observed in another city in Zhejiang province between 9 a.m. and 5 p.m. within a day.

Then the destination location in Zhejiang province, where a person being charged most inside a city from 9 a.m. to 5 p.m., was identified as the destination location for the individual's same-day return trip. Zhejiang province currently contains 11 cities, including Jiaxing, Huzhou, Hangzhou, Shaoxing, Ningbo, Wenzhou, Quzhou, Jinhua, Lishui, Taizhou, and Zhoushan. Due to the privacy issue in tracing an intercity traveller trajectory individually between subsidiaries of CHINA TELECOM (Shanghai and Zhejiang province are two subsidiaries), only charging reference in terms of call and internet usage records in another subsidiary is available.

Third, after home and destination stays are detected, the detected individuals are aggregated by origin Jiedao or destination city (Jiedao is the basic administrative unit of Chinese cities). For trip mode identification, speed-based methods, buffer zones around rail stations, or other machine learning methods have been applied to GPS data to detect travel modes [21]. In our study, if a person's trajectory passes by the selected railway station area, the individual is labelled as a rail user who used the station. We limit the railway station area to the block of the railway station instead of the buffer zone for higher accuracy. Examples of data collected are shown in Table 1.

In total, 34,567 individuals were detected from their homestay in Shanghai to a destination in Zhejiang Province for a same-day return trip within the 97 weekdays. After expanding the detected individuals to the whole population, the total number of intercity same-day travellers from Shanghai to Zhejiang Province was 1272 intercity travellers per weekday. We also detected 123,919 individuals from their homestay in Zhejiang to Shanghai for a round trip within a day. The total number of intercity same-day travellers from Zhejiang to Shanghai was 4360 individuals per weekday after expanding to the whole population in Zhejiang Province.

Table 1 Examples of data collected from home Jiedao in Shanghai to destination cities in Zhejiang Province

ID	Home stay	Destination city	Count of total trips within 97 weekdays	Detected rail user (1 = yes)	Rail station detected			Detected station for most days
					SHHQ	SHS	SH	
1	Jinze town	Hangzhou	9	0	0	0	0	NA
2	Xietu road	Hangzhou	8	1	0	1	0	SHS
3	Xinhong	Hangzhou	6	1	1	0	0	SHHQ
4	Beicai	Jiaxing	14	1	1	0	0	SHHQ
5	Changbaixincun	Jiaxing	5	0	0	0	0	NA

Note SHHQ-Shanghai Hongqiao Railway Station; SHS-Shanghai South Railway Station; SH-Shanghai Railway Station

4.2 Travel Behaviour and Spatial Distribution of Detected Intercity Same-Day Return Travellers

After detecting the intercity same-day return travellers using mobile phone data, we further explored the travel behaviour and spatial distribution characteristics of travellers.

Figure 7 shows the detected intercity same-day return travellers and the total number of trips stratified by destination cities in Zhejiang Province. In general, Hangzhou was the destination city for most one-day return travellers. Of the 34,567 individuals detected, 33,249 went to Hangzhou as the destination city. Both Hangzhou and Shanghai are core nodes in the regional structure. Our findings show that the same-day return travel demand between core cities was much stronger than between other nodes. Jiaxing, an intermediate city between Shanghai and Hangzhou, ranks as the secondary destination city receiving same-day return travellers from Shanghai. However, the intercity volume is far below Hangzhou. The difference in intercity flow may owe to the lower GDP, less population, and fewer attractions in Jiaxing.

The home stays of detected travellers from Shanghai are clustered in specific Jiedao units (Fig. 8). The adjacent subdistricts between Shanghai and Jiaxing, such as Zhujing and Fengjing Jiedao, showed more same-day return travellers. Also, more detected travellers were found from Jinshan and Fengxian new towns to Zhejiang province. Those subdistricts were connected to Zhejiang Province by highways (shown with brown lines in the bottom right graph). We also identified more same-day return travellers in spatial units near the three main railway stations, Shanghai South Railway Station, Shanghai Hongqiao Railway station and Shanghai Railway Station.

The detected travellers from Zhejiang Province to Shanghai show similar spatial distribution characteristics in Fig. 9, with more travellers distributed along the highway and railway corridors. The results indicate the effect of both highway and

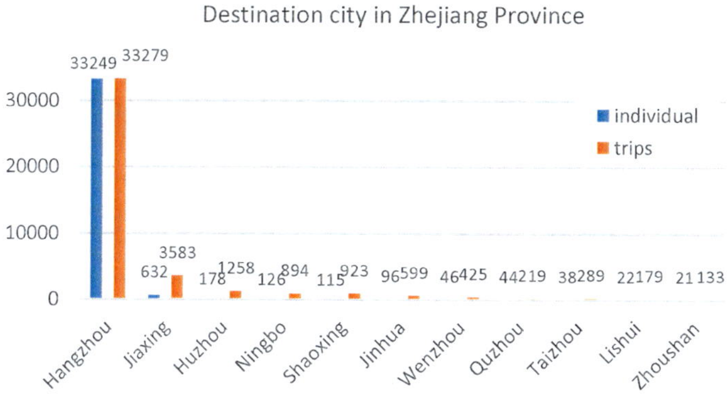

Fig. 7 Detected intercity same-day return travellers stratified by destination city in Zhejiang

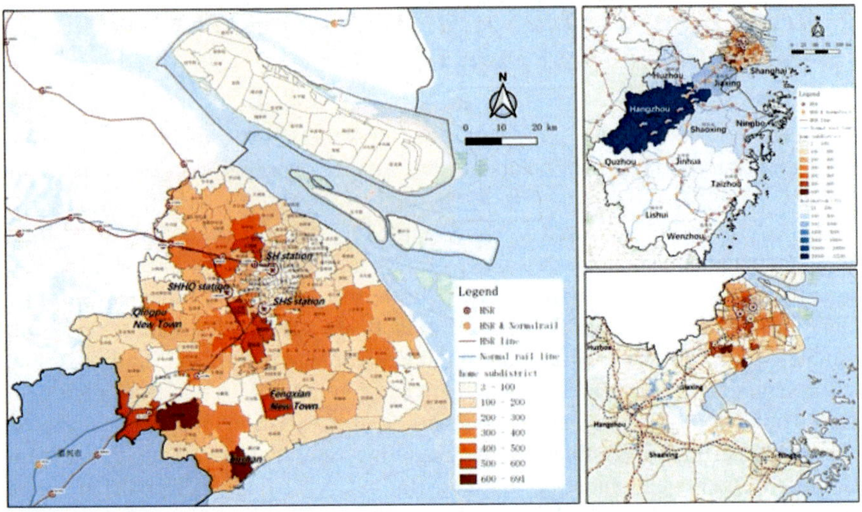

Fig. 8 Spatial distribution of detected traveller's to Zhejiang province aggregated by home Jiedao in Shanghai

rail networks in supporting the same-day return travel demand. The distance decay effects are noticeable along corridors, with most detected round-trip travellers to Shanghai within a 300 km radius. Subdistricts far from Shanghai than Tonglu, Yiwu, and Ningbo have fewer detected same-day return travellers. The total travel time using public transport modes from Shanghai to Tonglu, Yiwu, and Ningbo is between 3 to 3.5 h. Also, the in-vehicle time is about 1.5 h for the three OD pairs, which accounts for less than half of the total travel time. It highlights the need for promoting intra-city efficiency for round-trips within a day.

One advantage of mobile phone data is to capture intercity travellers' periodical travel behaviour. The detected individuals were grouped by their count of days to the destination cities in Zhejiang Province in Fig. 10. We found most travellers only travelled once from Shanghai to the destination city, and the majority of one-time travellers were to the destination city of Hangzhou. For groups that travelled to the destination cities for more than two days, more travellers arrived at the city of Jiaxing. The shorter distance between Shanghai and Jiaxing may lead to more frequent same-day intercity travellers. We also observed more same-day return reverse flow from Zhejiang province to Shanghai. Although Hangzhou is still dominant in the origin cities, we also detected same-day return travellers from other cities, such as Ningbo, Shaoxing and Wenzhou, to Shanghai in the one-time group. It implies an uneven in-and-out flow between Shanghai and other cities in Zhejiang province. However, the volume between Shanghai and Hangzhou is relatively balanced compared to the other city pairs.

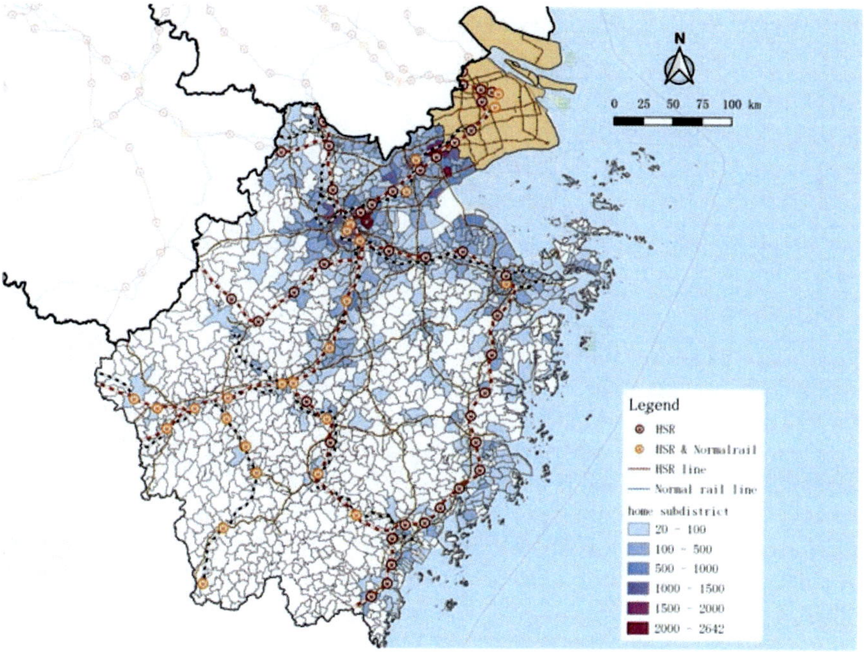

Fig. 9 Spatial distribution of detected traveller's to Shanghai aggregated by home Jiedao in Zhejiang Province

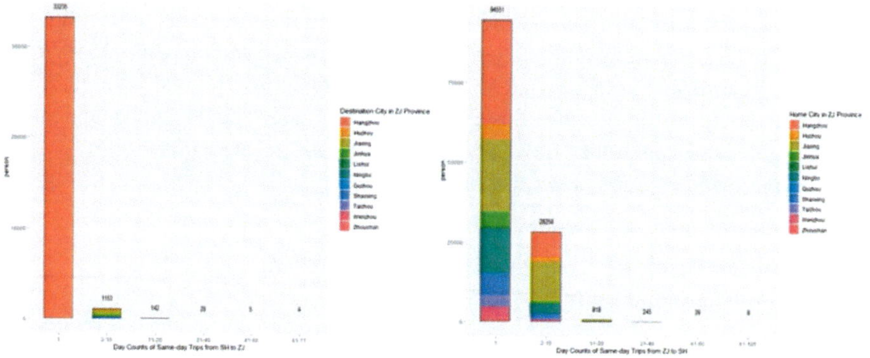

Fig. 10 Counts of detected same-day return travellers grouped by round trip days within the 97 weekdays

4.3 *The Explanatory Variables for Intercity Same-Day Travel Mobility*

The selection of the explanatory variables to model intercity same-day travel mobility is based on previous research and our interest. GDP, population, and distances

Table 2 Estimation results of intercity same-day travel mobility

	Estimate	Std. Error	t value	P value
β_0	−9.35438	3.81479	−2.452	0.0162*
β_1	−0.34092	0.10833	−3.147	0.0023***
β_2	0.859683	0.20951	4.103	0.0001***
β_3	0.005461	0.00243	2.247	0.0272*

Note *, **, and *** indicate statistical significance respectively at 10%, 5%, and 1% level. Adjusted R-squared is 0.64

between city pairs are commonly tested in previous gravity models for intercity mobility [22]. We added the attribute of rail service frequency, including D, G, and K trains. Although K trains are of the conventional rail system, they are included owing to more K trains from Shanghai South Railway Station to Hangzhou compared to the total number of G and D trains.

A gravity model was employed as it can provide effective estimation results and is widely used in previous intercity mobility analyses. The formula of same-day return travel mobility between the origin subdistrict Jiedao *i* and the destination city *j* is as follows:

$$SDR_{ij} = \exp\left(\beta_0 + \left(\beta_1 \ln\left(D_{ij}{}^2\right) + \beta_2 \ln\left(GDP_i * GDP_j\right) + \beta_3 SF_{ij}\right) + \varepsilon\right) \quad (1)$$

where D_{ij} stands for the distance and SF_{ij} represents the service frequency of G, D, and K trains in total. β_0 and ε are the intercept and error, respectively. Model results show that apart from *GDP* and distance, the number of trains to destination cities significantly positively affects same-day return travel mobility. Unlike previous research, which tested HSR service as a dummy variable of with and without HSR services [22], our result indicates that the service frequency should not be neglected in promoting same-day return travel mobility (Table 2).

5 Travel Behaviour and Spatial Distribution Characteristics of Rail Users

The detection of rail users is described in Sect. 4. After identifying the rail users, we analyzed the spatial distribution and catchment area of rail users of same-day return travellers.

5.1 Spatial Distribution of Detected Rail Users

To control the effect of detected numbers of rail users influenced by the different market penetration in Shanghai and Zhejiang, we compute the rail user percentage by dividing the detected rail users in Jiedao units by the total rail users identified in Shanghai and Zhejiang separately.

In Fig. 11, Jiedao units with darker colours represent a high proportion of rail users. We observed that spatial units of Jiedao near railway stations show a higher proportion of rail users. It is understandable since better accessibility in terms of distance to railway stations may positively affect rail users. Also, Jiedao units in big cities near Shanghai, such as Hangzhou Jiaxing and Ningbo, have a higher proportion of rail users. Those cities have higher GDPs and enjoy a shorter distance to Shanghai. However, the city of Huzhou is an exception due to low rail service frequency, as shown by the size of the yellow circle in the bottom left graph.

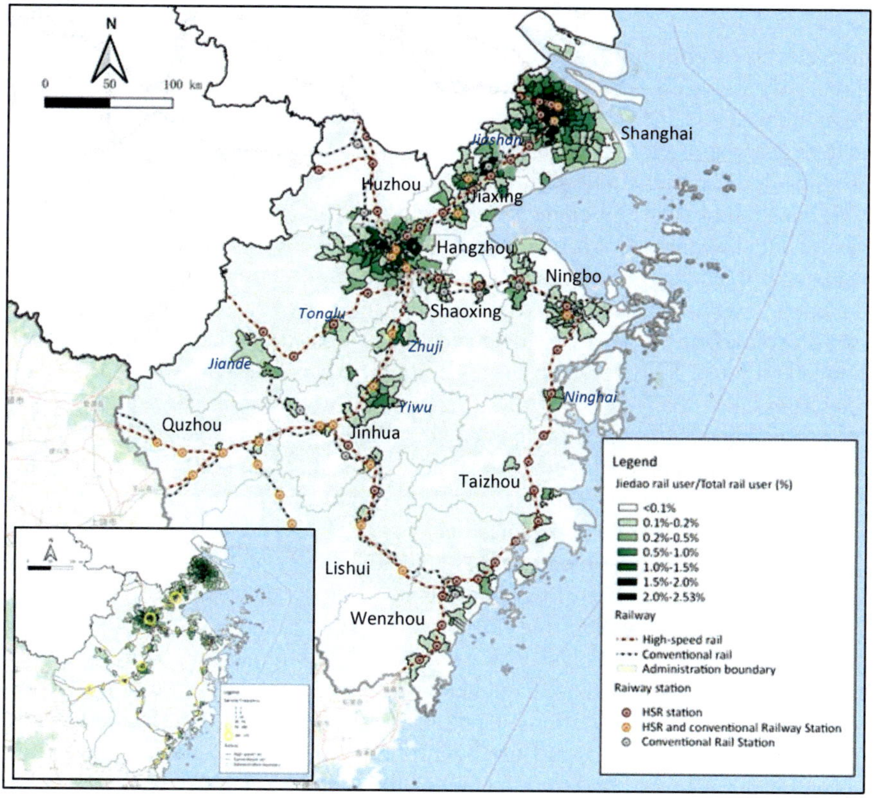

Fig. 11 Distribution of detected rail users in Shanghai and Zhejiang Province. (*Note* Shanghai rail user percentage calculated by detected rail user in Shanghai within a Jiedao divided by total detected Shanghai rail users; Zhejiang rail user percentage calculated by detected rail user in Zhejiang within a Jiedao divided by the sum of detected Zhejiang rail users)

Furthermore, we found relatively more rail users from the Quzhou–Jinhua–Yiwu–Hangzhou corridor to Shanghai than the Wenzhou–Taizhou–Ningbo–Hangzhou corridor or the Jiande–Tonglu–Hangzhou corridor. One possible explanation is that the Quzhou–Jinhua–Yiwu–Hangzhou corridor has a higher service frequency. There are 84 trains from Yiwu to Shanghai, compared to 47 trains from Ningbo to Shanghai and only 10 trains from Jiande to Shanghai per day (the number of services to the three main stations of Shanghai was summed up for efficiency calculation). High service frequency seems to trigger more rail passengers for same-day return trips.

Besides that, the catchment area based on home Jiedao was larger in mega-cities, for example, Shanghai and Hangzhou, than in medium and small cities. We infer that the subway system in mega-cities could induce the potential use of rail for same-day return trips.

5.2 Travel Time and Distances of Rail Users

Since the timestamp information in destination cities is unavailable for our data source, we retrieve the Gaode map [16] to obtain rail users' travel time and distance. Travel time and distances are calculated based on the detected rail users' homestays and their destination city's administrative centers. Due to the privacy issue mentioned in session 4, the administration center is treated as the destination place.

Figure 12 shows that of the detected rail users from Shanghai to Zhejiang Province, most travelled within 200 km, and more than 75% travelled less than three hours. It implies that the domain market share of the same-day return trip from Shanghai to Zhejiang by rail was within relatively shorter distances.

The comparison between the detected rail passengers using different railway stations and its relationship with distance decay explains different identified situations (Fig. 13). Shanghai Hongqiao (SHHQ) Railway Station has a higher HSR service frequency with a larger catchment area. Most detected travellers lived 25 km away from the station, leading to a longer access distance and time. The log-logistic decay function adopted by Martínez et al. [23] and the simple log form decay function were used to capture the distance decay for SHHQ and the other two railway stations. The fewer people living near Hongqiao Railway station could result from less residential land use in its closest hinterland and lower density nearby. Compared to the SHHQ railway station, the other two stations, Shanghai South (SHS) and Shanghai (SH) railway station, have lower HSR service frequencies. However, most detected travellers lived within 20 km network distances, and distance decay effects are apparent in these two stations. Unlike SH railway station, SHS station shows a higher number of detected rail users with more conventional trains. It indicates conventional rail could also be competitive with HSR for same-day return trips.

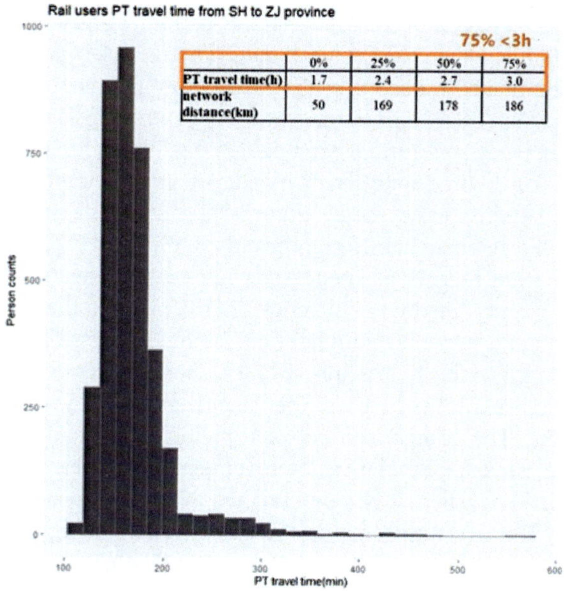

Fig. 12 Travel time and travel distances of detected rail users from Shanghai to Zhejiang Province

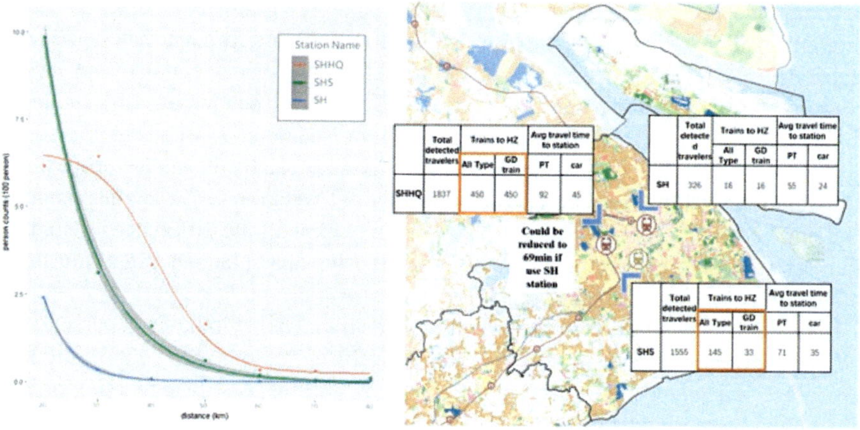

Fig. 13 Distance decay of three main railway stations in Shanghai. (*Note* The number of services to the different stations in Hangzhou was summed up for efficiency calculation)

6 Conclusion and Discussion

This study analyzes the same-day return travel between city pairs in the Yangtze River Delta Region and the efficiency of using High-speed Rail and other public transport modes for round trips within a day. From a demand side, we detected the same-day return trips between Shanghai and cities in Zhejiang Provinces using mobile phone data. The spatial distribution and periodical travel behaviour characteristics were identified with the 20-week data. The rail users were then detected by identifying individuals' trajectory paths passing through the railway stations. A comprehensive analysis of rail users for same-day return trips is expected to boost HSR and its efficiency, and further promote regional interaction and cohesion.

Travel time analysis of city pairs demonstrates that the efficiency of HSR is not as competitive as car mode within 300 km under the current network and services. Also, HSR efficiency varies significantly in different corridors. The supply of direct links, less transfer time and better inter-intra-city connections could be expected to enhance the same-day travel efficiency of HSR. Due to the geographic distribution of cities within a network, periphery cities could access fewer cities within a day than the central nodes. For periphery cities, shorter access/egress time is essential to improve the accessibility of those cities since most one-day return trips were within a 3-h travel time. The comparison between public transit and car mode travel time highlights the need for efficiency improvement to compete with car modes in high-volume corridors, such as Ningbo-Taizhou, Nantong-Suzhou, and Nantong-Changzhou.

The detected same-day return travellers using mobile phone data enable us to explore the spatial distribution and travel behavioural characteristics of intercity travellers, which are difficult to capture using traditional survey methods. We observed less demand for travel time longer than 3–3.5 h, and high service frequency corridors show more detected same-day return travellers. The gravity model results show that the GDP of the origin and destination city and rail service frequency is positively related to same-day return mobility. At the same time, the distance in between exerts a negative effect. Through the detected periodical travel behaviours of travellers, we found only a small amount of travellers from Shanghai to Zhejiang for high-frequency round trips. From Zhejiang to Shanghai, a relatively higher proportion of travellers travelled more than once within the 97 weekdays. The higher income level and better job opportunities in Shanghai may explain the difference. Also, the link between core cities Hangzhou and Shanghai shows a balanced in-and-out flow compared to core-periphery city pairs.

Detecting rail users allows us to investigate the travel behaviour of rail passengers and compare the catchment area of different railway stations. We found a higher percentage of detected rail users living near railway stations. Also, a higher proportion of rail users were found along corridors with high service frequencies from Zhejiang to Shanghai. For the catchment area of different railway stations, big cities show larger catchment areas than medium and small-sized cities for same-day return trips. The heterogeneity in distance decay for different railway stations was identified in Shanghai. The variance in distance decay may be due to the supply of land use near the railway stations and the number of train services to connecting cities. Reasonable conventional rail services can compete with HSR for same-day return trips.

Despite the above findings, there are certain limitations in our paper. Firstly, the data source is from one of the telecommuting companies in China with around 30% market penetration. Although the detected same-day return travellers were large enough to interpret their travel behaviour characteristics, future research using different data resources is needed. Also, for the detection of rail users, the probabilistic method, clustering method, and other advanced machine-learning models were adopted to identify the intercity travel patterns of people based on combined duration, speed, timetable, or route information [24, 25]. The data available only allows us to use the spatial analysis method to detect rail users, which may lead to fewer rail users being detected with irregular records of clients. Furthermore, the destination staying inside a city would be helpful in detecting the trip purpose of people. Future studies using high spatial resolutions for destination cities could potentially improve our understanding of the same-day return trip for different travel purposes.

References

1. Coronado, J.M., Garmendia, M., Moyano, A., Ureña, J.M.: Assessing Spanish HSR network utility for same-day tourism. Recherche Transports Sécurité (RTS), IFSTTAR, 161–175 (2013). https://doi.org/10.4074/S0761898013003026.hal-01670633
2. De Ureña, J.M., Coronado, J.M., Garmendia, M., Romero, V.: Territorial implications at national and regional scales of high-speed rail. In: de Ureña, J.M. (ed.), Territorial Implications of High Speed Rail, 1st edn, pp 129–162. Routledge (2012). https://doi.org/10.4324/9781315612065
3. Liu, Z., Zhu, A.X., Zhang, W., Ren, M.: An improved potential-based approach to measuring the daily accessibility of HSR. Transp. Res. Part A Policy Pract. **150**(October 2020), 271–284 (2021). https://doi.org/10.1016/j.tra.2021.06.009
4. Moyano, A., Rivas, A., Coronado, J.M.: Business and tourism high-speed rail same-day trips: factors influencing the efficiency of high-speed rail links for Spanish cities. Eur. Plan. Stud. **27**(3), 533–554 (2019). https://doi.org/10.1080/09654313.2018.1562657
5. Xu, J., Zhang, M., Zhang, X., Wang, D., Zhang, Y.: How does city-cluster high-speed rail facilitate regional integration? Evidence from the Shanghai-Nanjing corridor. Cities **85**(April 2018), 83–97 (2019). https://doi.org/10.1016/j.cities.2018.12.003
6. Moyano, A., Martínez, H.S., Coronado, J.M.: From network to services: a comparative accessibility analysis of the Spanish high-speed rail system. Transp. Policy **63**(November 2017), 51–60 (2018). https://doi.org/10.1016/j.tranpol.2017.11.007
7. Cui, C., Wu, X., Liu, L., Zhang, W.: The spatial-temporal dynamics of daily intercity mobility in the Yangtze River Delta: an analysis using big data. Habitat Int. **106**(September 2019), 102174 (2020). https://doi.org/10.1016/j.habitatint.2020.102174
8. Wang, Z., He, S.Y., Leung, Y.: Applying mobile phone data to travel behaviour research: a literature review. Travel Behav. Soc. **11**, 141–155 (2018). https://doi.org/10.1016/j.tbs.2017.02.005
9. NDRC. The Yangtze River Delta Transport Integration Development Plan (2020). http://www.gov.cn/zhengce/zhengceku/2020-04/29/content_5507368.htm. Accessed 15 Nov 2022
10. Shao, S., Tian, Z., Yang, L.: High speed rail and urban service industry agglomeration: evidence from China's Yangtze River Delta region. J. Transp. Geogr. **64**(August), 174–183 (2017). https://doi.org/10.1016/j.jtrangeo.2017.08.019
11. Chen, C.L., Loukaitou-Sideris, A., de Ureña, J.M., Vickerman, R.: Spatial short and long-term implications and planning challenges of high-speed rail: a literature review framework for the special issue. Eur. Plan. Stud. **27**(3), 415–433 (2019). https://doi.org/10.1080/09654313.2018.1562658
12. Wang, L., Zhang, S., Sun, W., Chen, C.L.: Exploring the physical and mental health of high-speed rail commuters: Suzhou-Shanghai inter-city commuting. J. Transp. Health **18**(May), 100902 (2020). https://doi.org/10.1016/j.jth.2020.100902

13. Li, T., Wang, J., Huang, J., Yang, W., Chen, Z.: Exploring the dynamic impacts of COVID-19 on intercity travel in China. J. Transp. Geogr. **95**, 103153 (2021). https://doi.org/10.1016/j.jtrangeo.2021.103153
14. Pan, H., Ye, S., Chen, M.: The influence of high-speed rail station site selection on travel efficiency: the case of the Shanghai Hongqiao Hub. In: High-Speed Rail and Sustainability, pp. 143–158. Routledge (2017)
15. Pitombo, C.S., de Souza, A.D., Lindner, A.: Comparing decision tree algorithms to estimate intercity trip distribution. Transp. Res. Part C Emerg. Technol. **77**, 16–32 (2017). https://doi.org/10.1016/j.trc.2017.01.009
16. Gaode Map: Route planning manual. https://lbs.amap.com/api/webservice/guide/api/newroute (2022). Accessed 30 July 2022
17. Zi, H.B.: Research on rail transit network planning in metropolitan area based on spatial coordination and mobility equality. Commun. Shipping (In Chinese) **2**(1), 10–15 (2021)
18. Janzen, M., Vanhoof, M., Smoreda, Z., Axhausen, K.W.: Closer to the total? Long-distance travel of French mobile phone users. Travel Behav. Soc. **11**(March 2017), 31–42 (2018). https://doi.org/10.1016/j.tbs.2017.12.001
19. Miller, E.J.: The trouble with intercity travel demand models. Transp. Res. Rec. **1895**, 94–101 (2004). https://doi.org/10.3141/1895-13
20. Yamaguchi, H., Nakayama, S.: Detection of base travel groups with different sensitivities to new high-speed rail services: non-negative tensor decomposition approach. Transp. Policy **97**(May), 37–46 (2020). https://doi.org/10.1016/j.tranpol.2020.07.012
21. Rojas, M.B., Sadeghvaziri, E., Jin, X.: Comprehensive review of travel behavior and mobility pattern studies that used mobile phone data. **2563**, 71–79 (2016).https://doi.org/10.3141/2563-11
22. Hengyu, G., Jianfa, S., Jun, C.: Understanding intercity mobility patterns in rapidly urbanizing China, 2015–2019: evidence from longitudinal poisson gravity modeling. Ann. Am. Assoc. Geogr. **0**(0), 1–24 (2022).https://doi.org/10.1080/24694452.2022.2097050
23. Martínez, H.S., Moyano, A., Coronado, J.M., Garmendia, M.: Catchment areas of high-speed rail stations: a model based on spatial analysis using ridership surveys. Eur. J. Transp. Infrastruct. Res. **16**(2), 364–384 (2016). https://doi.org/10.18757/ejtir.2016.16.2.3143
24. Breyer, N., Gundlegard, D., Rydergren, C.: Travel mode classification of intercity trips using cellular network data. Transp. Res. Proc. **52**, 211–218 (2021). https://doi.org/10.1016/j.trpro.2021.01.024
25. Huang, H., Cheng, Y., Weibel, R.: Transport mode detection based on mobile phone network data: a systematic review. Transp. Res. Part C Emerg. Technol. **101**(January), 297–312 (2019). https://doi.org/10.1016/j.trc.2019.02.008

The Benefits of Building High Speed Rail Stations at Airports

Paige Malott

Abstract In North America, the Cascadia megaregion has reached a tipping point where expanding its mobility choices to include high speed rail is critical to plan for long-term growth and address large-scale challenges such as climate change. The Cascadia megaregion is located in the Pacific Northwest, which includes British Columbia in Canada, and Washington and Oregon in the United States. The three major cities in this geographic area are Vancouver, British Columbia, Seattle, Washington, and Portland, Oregon. The Cascadia megaregion is forecast to grow by 3.6 million people by 2050; (Cascadia Innovation Corridor. Cascadia Vision 2050. p. 14. https://connectcascadia.com/wp-content/uploads/2020/09/Cascadia-Vision-2050_Published.pdf [1]) three times the size of the current population. Regional growth plans only account for 2.3 million of those people, leaving a gap of 1.3 million people unaccounted for where they will live and work, as well as adding further demand on the region's strained transportation network. Additionally, a study on regional growth in aviation (Puget Sound Regional Council. 2050 Forecasts for Aviation Demand. (2019). www.psrc.org/sites/default/files/rabs-2019jul16-aviationstudyforecasts_0.pdf [2]) found that SeaTac Airport in Seattle, Washington, will exhaust all expansions by 2027 and still be over capacity by three million passengers annually. Decision-makers are exploring building an additional airport to meet demand. Meanwhile, 60% of the region's carbon emissions (City of Seattle. Understanding Our Emissions—Environment. (2021). www.seattle.gov/environment/climate-change/climate-planning/performance-monitoring [3]) come from passenger transportation. Legislators have set a goal to reduce carbon emissions by 80% by 2050 in alignment with the Paris Climate Accord. On its current trajectory, Cascadia will miss this goal and only decrease emissions by 20–30% (Cascadia Innovation Corridor. Cascadia Vision 2050. p. 12. https://connectcascadia.com/wp-content/uploads/2020/09/Cascadia-Vision-2050_Published.pdf [4]) without a game-changing investment in fast, high-capacity, clean-energy transportation, such as high speed rail. This case study examines the economic, mobility, and climate benefits of building high speed rail at airports, and recommends how collaboration between aviation and high speed rail can improve outcomes for the Cascadia megaregion.

P. Malott (✉)
San José State University, San José, California, USA

F. Pagliara (ed.), *Socioeconomic Impacts of High-Speed Rail Systems*,
Springer Proceedings in Business and Economics,
https://doi.org/10.1007/978-3-031-26340-8_5

Keywords High speed rail · Intercity rail at airports · Growth planning

1 Overview of Case Study

The purpose of this study is to examine why high speed rail stations should be built at major airports, such as SeaTac Airport in Seattle, Washington, and how integration of fast trains at airports are good for growth planning, business, and climate goals.

In 2016, Governor Jay Inslee and the Washington State Legislature directed Washington State Department of Transportation (WSDOT) to study building high speed rail as a way to address the rapid population and economic growth in the Pacific Northwest by maximizing public transportation choices. WSDOT published three studies supporting the feasibility and business case for a 250 mph (402 kmh) high speed rail line along the Interstate 5 corridor. Train journey times would be competitive with air travel: one hour connections between Seattle and Vancouver, Canada, and one hour connections between Seattle and Portland, Oregon. The electric-powered system would carry up to 32,000 passengers per hour and support additional stops in intermediary locations between the three anchor cities [5]. The Interstate 5 corridor adjacent to SeaTac Airport could allow for multimodal collaboration with high speed rail as a solution to address aviation growth constraints and reduce carbon emissions, similar to rail/air partnerships in France and Germany. (see Fig. 1).

Fig. 1 A vision map for high speed rail in the Pacific Northwest. The North–South Corridor aligns with the existing Interstate 5 corridor and could accommodate a station location at Sea-Tac Airport, as indicated by the airport icon south of Seattle. Three other major airports could also be served by a high speed rail station: YVR in Vancouver, Canada, Payne Field in Everett, Washington, and PDX in Portland, Oregon. Vision map provided by Cascadia Rail

2 Good for Growth Planning

Regional growth trends are closely tied to demand for commercial aviation. In Greater Seattle, recent population, employment, and income growth is increasing aviation demand, a trend which has continued post-pandemic and is expected to surge over the next 30 years. In addition, SeaTac Airport is expanding as a hub for connections to Asia, further increasing demand for flights [2]. Puget Sound Regional Council forecast that passenger enplanements at SeaTac Airport will double by 2050, however near-term expansion solutions will only accommodate growth for the next five years. After exhausting all expansion options, the airport will be over capacity by three million annual passenger enplanements in 2027. By 2050, the airport will be over capacity by 27 million annual passenger enplanements [2] (see Fig. 2).

To remedy this, WSDOT is studying building an additional international airport to accommodate aviation demand. Results of a public engagement survey expressed that 79% of participants had significant concerns about noise and environmental impacts of building additional airports. 67% of respondents thought that more passenger service should be added to existing airports rather than building new airports, noting that they supported meeting travel demand while decreasing environmental and noise impacts [6].

Building a high speed rail station at SeaTac Airport is an opportunity to transition passengers from regional flights to high speed rail, managing passenger growth by freeing up more gates for long-haul, international flights and reducing need for building additional airports. With low carbon emissions, less noise than aircrafts, and less land use compared to building additional airports, high speed rail can provide a scalable, environmentally-friendly intercity travel solution that addresses the concerns of the community members.

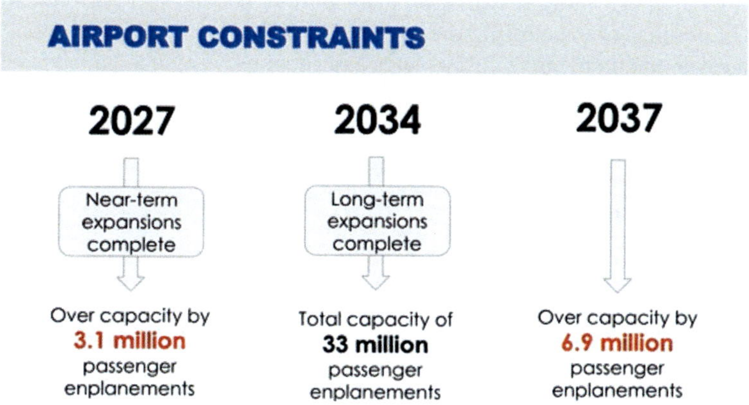

Fig. 2 Timeline of forecasted airport capacity constraints based on 2050 growth projection data from Puget Sound Regional Council

The WSDOT business case study states that high speed rail would capture a significant share of trips within the Interstate 5 corridor, and "reduce the future need for expanded terminals, allowing new airfield capacity to be more efficiently deployed in serving longer-distance markets." [7]. A subsequent public engagement survey by FM3 Research found that 67% of Washingtonians support connecting the major cities of the Cascadia region with high-speed rail [8].

3 Good for Business

3.1 Economic Investment

WSDOT notes that building high speed rail in Cascadia will create $355 billion in economic growth and cost $24–42 billion to build, serving three major cities, over a dozen intermediary cities, and up to four airports [9]. This is a fraction of the cost compared to building one additional airport in one city, within an entire region experiencing unprecedented growth. While data does not yet exist for land and facility costs of the prospective airport, an estimate can be referenced from recent international airport projects and ongoing capital projects at SeaTac Airport.

Near-term capital projects at SeaTac Airport, ranging from terminal improvements to opening a new 19-gate concourse will cost $8.2 billion [10] for construction between 2020 and 2027, according to the Sustainable Airport Master Plan. An investment of $10 billion is needed per runway and a minimum of $2 billion per terminal [9]. When evaluating costs of other major airport greenfield projects, the last international airport constructed in the United States was in Denver, Colorado, in 1995 at a cost of $4.8 billion. When adjusting for inflation, it would cost $9.4 billion in 2022 dollars [11]. More recently, in 2019, Beijing Daxing International Airport opened in Beijing, China, costing the equivalent of $11.4 billion [12]. Further analysis is needed on comparative cost savings and economic benefits, which is beyond the scope of this evaluation.

3.2 Operational Capacity

At SeaTac Airport, airlines lease gate spaces to operate flights. A gate that serves a regional aircraft can also accommodate several sizes of aircraft with larger passenger loads, up to 200 passengers. Regional flights tend to be money-losers for airlines as they seat an average of 70 passengers and have higher operating costs. By creating a high speed rail option, transitioning select flights to intercity trains opens up more gates for airlines to allocate long-distance flights, which are more profitable. High speed rail adds passenger capacity equivalent to 91 airport gates, two airport runways, and six highway lanes [13].

A code share agreement between train and plane operators would allow passengers to book tickets for both their flight and rail journey on one itinerary. In Germany, air carrier Lufthansa launched Lufthansa Express Rail, which connects 19 major cities by train to Frankfurt Airport. The move has transitioned 20% of airline passengers to fast trains. France also transitioned all domestic flights to passenger rail that can provide competitive journey times of 2.5 h or less. Having the option of high speed rail has shifted 40% of European airline passengers to choosing trains instead of flying [14]. In the United States, Amtrak's Northeast Corridor intercity train service has absorbed 75% of the air travel market share by providing a fast alternative to flying between New York City and Washington D.C [15].

3.3 *Job Creation*

High speed rail also brings labor benefits to aviation professionals as well as the transportation industry. The Cascadia high speed rail project is estimated to create 200,000 construction jobs, and an additional 800,000 jobs in other industries [9]. It can also address labor demands in the airline industry. With a seniority-based ranking system, airline pilots prefer long-haul flights as it allows for more credit towards their flying time, better pay, and a faster path to promotions. As the airline industry is experiencing a labor shortage, this can make Seattle a top choice for pilots seeking career advancement [16].

4 Good for the Environment

4.1 *Reducing Carbon Emissions*

The Puget Sound region has adopted climate goals of reducing emissions by 50% by 2030. In Seattle, 60% of the region's carbon emissions come from transportation, with aviation contributing a significant amount of pollution [3]. SeaTac Airport ranks ninth in the nation for highest carbon emissions per passenger. With over 30 flights per day between Portland and Seattle, and 30 flights per day between Seattle and Vancouver, creating a high speed rail line between these cities can absorb regional flights and offer a sustainable choice for intercity travel. Air travel emits 77 times more CO_2 per passenger than high speed rail on the same corridor, with regional flights being the most emissions-intensive [17]. For example, a one hour flight between Seattle and Vancouver is 2.6 times less efficient than a four-hour flight between Montreal to Miami. This is due to the most fuel being used in the "take off and climb" segments of a flight, with less fuel being used at cruising altitude. Shorter distance flights have more frequent "take off and climb" emissions and less cruising time, such as the flights shuttling passengers between Seattle, Portland, and Vancouver airports.

WSDOT estimates that Cascadia high speed rail can be powered by 100% renewable energy and reduce carbon emissions by six million metric tons [9].

4.2 Sustainable Solutions

The WSDOT aviation expansion study examined environmental benefits of creating "sustainable airport of the future" with their plans to build new airports to accommodate growth. Recommendations included electric planes, which are limited to carrying 11 passengers and are not yet commercially viable, as well as a 10% renewable fuel requirement on existing jets. The report acknowledged that existing fuel-powered aircrafts will likely be in service for the next 30 years [18]. These efforts are incompatible with climate goals. New airport construction will likely score low on environmental benchmarks for Federal grant funding, whereas high speed rail is uniquely positioned to receive dedicated grant funding from the Bipartisan Infrastructure Law and rank high environmentally.

Further, the three communities identified as potential locations in the WSDOT aviation expansion study have already voiced opposition to hosting an airport. High speed rail would offer similar capacity to an airport terminal with a smaller land use footprint. It would also have minimal neighborhood impacts, including less noise pollution and impacts to home values.

5 Recommendations

To address transportation growth demands for 2050, build high speed rail with station locations integrated at major airports. High speed rail has lower capital costs, is more sustainable, and has a higher return-on-investment than building a second airport. In addition to connecting the major cities of Seattle, Portland, and Vancouver with SeaTac Airport, high speed rail stations could also be built at the following locations along the I-5 corridor as an alternative to airport expansion:

- Payne Field, Everett, WA
- PDX Airport, Portland, OR
- YVR Airport, Vancouver, BC.

Connecting multiple airports via high speed rail will allow passengers more choices for their intercity travel. As the Puget Sound region expands, adding a station at SeaTac Airport as well as Payne Field, to the north of Seattle, creates more travel options for passengers to choose from as transportation demand increases.

As the region plans for growth in 2050, consider passenger outcomes. Airports and high speed rail can complement each other as intercity travel modes which make it convenient for people to travel long distances and connect with local transit. High speed rail can move up to 32,000 passengers per hour, generate $355 billion in economic growth, create over one million jobs, and operate on 100% clean energy to help reduce emissions and meet climate goals [9].

High speed rail and air travel go hand-in-hand. Consider how the rail project and airport expansion can collaborate and make a stronger case for WSDOT to pursue Federal grant funding as a joint project, rather than each being a stand-alone project. An airport can allocate more resources for scaling long-distance service, while high speed rail absorbs short haul flights. The sustainability benefits of high speed rail will help a joint project score higher on environmental benchmarks for Federal funding; those benchmarks cannot be achieved as a stand alone airport expansion.

As the saying goes, "The best time to start was yesterday. The next best time is now." By choosing to invest in high speed rail and station placements at airports, Cascadia can position itself favorably to prepare for growth for years to come.

References

1. Cascadia Innovation Corridor.: Cascadia Vision 2050, p. 14. https://connectcascadia.com/wp-content/uploads/2020/09/Cascadia-Vision-2050_Published.pdf. Accessed 21 Oct 2022
2. Puget Sound Regional Council.: 2050 Forecasts for Aviation Demand. www.psrc.org/sites/default/files/rabs-2019jul16-aviationstudyforecasts_0.pdf (2019). Accessed 21 Oct 2022
3. City of Seattle.: Understanding Our Emissions—Environment. www.seattle.gov/environment/climate-change/climate-planning/performance-monitoring (2021). Accessed 21 Oct 2022
4. Cascadia Innovation Corridor.: Cascadia Vision 2050, p. 12. https://connectcascadia.com/wp-content/uploads/2020/09/Cascadia-Vision-2050_Published.pdf. Accessed 21 Oct 2022
5. Washington State Department of Transportation.: Ultra High-Speed Ground Transportation Business Case Analysis, p. 67. Washington State Department of Transportation, Olympia, Washington (2019)
6. Puget Sound Regional Council.: Regional Aviation Baseline Study, p. 86. www.psrc.org/sites/default/files/psrc_aviation_baseline_study_-_final_report.pdf (2021). Accessed 21 Oct 2022
7. Washington State Department of Transportation.: Ultra High-Speed Ground Transportation Business Case Analysis, p. 45. Washington State Department of Transportation, Olympia, Washington (2019)
8. FM3 Research.: Results of a Survey of Oregon and Washington Voters on a High-Speed Rail Project, p. 2. FM3 Research, Seattle, Washington (2021)
9. Washington State Department of Transportation.: Ultra High-Speed Ground Transportation Business Case Analysis. Washington State Department of Transportation, Olympia, Washington (2019)
10. Conroy, B.: SeaTac is continuing to build for the future. Seattle Business Magazine. https://seattlebusinessmag.com/aerospace/sea-tac-airport-continuing-build-future-midst-coronavirus-crisis/ (2020). Accessed 21 Oct 2022
11. Mouawad, J.: Building boom takes hold at U.S. Airports. New York Times, New York. https://www.nytimes.com/2012/06/15/business/building-boom-takes-hold-at-us-airports.html (2012). Accessed 21 Oct 2022
12. Park, K.: Beijing's Daxing Airport helps China Rival U.S. in the Skies. Bloomberg. https://www.bloomberg.com/news/features/2019-09-25/beijing-s-daxing-airport-opens-helping-china-rival-us-in-skies (2019). Accessed 21 Oct 2022
13. California High Speed Rail Authority.: Capacity analysis 2022. California High Speed Rail Authority, Sacramento, California. https://hsr.ca.gov/wp-content/uploads/2022/01/Capacity-Analysis-Factsheet.pdf (2022). Accessed 21 Oct 2022

14. International Union of Railways (UIC).: Fast Track to Sustainable Mobility. International Union of Railways (UIC), Paris, France (2018)
15. Kamga, C.: Emerging travel trends, high-speed rail, and the public reinvention of U.S. transportation. Transp. Policy **37**, 111–120 (2015)
16. U.S. Bureau of Labor Statistics.: Airline and Commercial Pilots: Occupational Outlook Handbook, November 2021. U.S. Bureau of Labor Statistics, Washington D.C. www.bls.gov/ooh/transportation-and-material-moving/airline-and-commercial-pilots.htm (2021). Accessed 21 Oct 2022
17. Willsher, K.: France to ban some domestic flights where train available. The Guardian, Paris, France. www.theguardian.com/business/2021/apr/12/france-ban-some-domestic-flights-train-available-macron-climate-convention-mps (2021). Accessed 21 Oct 2022
18. Washington State Department of Transportation.: Airport investment study. Washington State Department of Transportation, Olympia, Washington. https://wsdot.wa.gov/sites/default/files/2021-10/aviation-ais-solutions.pdf (2021). Accessed 21 Oct 2022

The Impacts of High-Speed Rail Development on Territorial Cohesion: A Method with Two Case Studies in Italy

Pierluigi Coppola and Francesco De Fabiis

Abstract In the last decade, the concept of 'cohesion' has become increasingly important to study the distributional effects of infrastructure investment between different population groups, i.e., social cohesion, or between different zones inside a study area, i.e., territorial cohesion. Nevertheless, despite a growing interest, in the literature there is still a lack of consolidated methodologies and guidelines to assess the impacts on cohesion of transport infrastructure. In the attempt to bridge up this gap, this paper proposes a method based on the estimation of zonal accessibility variations as proxy for the effectiveness of the investment, and on statistical indices estimating accessibility dispersion as a proxy for the territorial cohesion. Two case studies are presented: the completion of the High-Speed Rail (HSR) line between Turin–Venice (Northern-Italy) and the construction of the new HSR line between Naples and Bari (Southern-Italy). The application shows that the investment in the more industrialized and densely populated areas of North Italy turns out to be more effective in terms of accessibility improvement than in less developed Southern ones. However, while the former increases inequalities in terms of accessibility and risks to amplify the gap between North and South of the Country, the latter (Naples–Bari) tends to reduce such disparities.

Keywords Transport equity assessment · Accessibility · Gini index

1 Introduction

In the last decades, the concept of cohesion has become increasingly important also because of the progressive enlargement of the European Union [1], due to the gradual inclusion of most of the European sovereign states. In this context, policies

P. Coppola (✉) · F. De Fabiis
Department of Mechanical Engineering, Politecnico Di Milano, Via G. La Masa 1, 20156 Milan, Italy
e-mail: pierluigi.coppola@polimi.it

F. De Fabiis
e-mail: francesco.defabiis@polimi.it

F. Pagliara (ed.), *Socioeconomic Impacts of High-Speed Rail Systems*,
Springer Proceedings in Business and Economics,
https://doi.org/10.1007/978-3-031-26340-8_6

and investments in the transport sector have been recognized as crucial to achieve the objectives of economic growth [2, 3], but also to promote cohesion within Countries and regions [4]. Decision-makers at different levels, from the international to the local one, are therefore increasingly demanding for the inclusion of cohesion effects in transportation project appraisal methodologies [5].

In the literature on transport planning, the term 'cohesion' usually refers to the distribution of the impacts of transportation policies or investments. In particular, many authors investigated it considering two different dimensions: the social and the territorial one. Social cohesion, also called "social equity" [6–8] or "vertical equity" [9–11], refers to the distribution of a certain variable among different population groups, while territorial cohesion, also called "spatial equity" [12–14] or "horizontal equity" [9, 11, 15], refers to the distribution of the same among different zones in a study area.

This study focuses on the latter and, as for other studies [16, 17], uses transportation accessibility as a variable to assess the impacts on territorial cohesion of new transport infrastructures. In the scientific literature, transportation accessibility has been identified as fundamental in view of a sustainable spatial development: in a society that is constantly on the move, facilitating the interaction among people living in different areas or among different economic activities spatially distributed over territories can be of crucial importance. Even though accessibility is only one of the necessary conditions for the economic progress [18], it is widely recognized that it represents a competitive advantage for regions, cities or neighborhoods, being a "desirable good" related to the welfare of an area [4]. The importance of transport accessibility is also proven by all the wider impacts that can be generated by its variation: redistribution of residences and economic activities [18, 19], changes in industries productivity [20] or changes in tourist destinations choices [19] are among the examples worth to be qualitative or quantitative considered in transport infrastructure planning. Therefore, given a study area, assessing whether a policy or a generic investment in the transport system generates or not a more equitable accessibility distribution among different zones is something that cannot be neglected in transportation planning.

To this aim, this paper uses an accessibility-based methodology to assess territorial cohesion impacts, i.e., comparative before and after analysis of accessibility values distribution, following the completion of two different High-Speed Rail (HSR) lines in Italy: these are peculiar since the first one is located in a very densely populated area with high accessibility in the reference scenario, whereas the second is placed in a sparsely populated one with low accessibility. Territorial cohesion impacts analysis is coupled with an efficiency impact analysis (i.e., average percentage variation of zonal transportation accessibility following the completion of a project) in order to assess potential trade-offs between the two impacts.

The subsequent sections of the manuscript are structured as follows: after this brief introduction, the next section provides a literature review on both HSR impacts and territorial cohesion, referring also to methodologies to assess them. Section 3 describes in detail the method that is therefore applied to the case studies described in Sect. 4. The fifth and final section contains a brief discussion and the concluding remarks of the research.

2 HSR Impacts and Territorial Cohesion: Literature Review

There's still a growing interest in the scientific community [20–22] about the High-speed Rail (HSR) due to a wide spectrum of unexplored impacts that can be analyzed, either internal or external.

Internal impacts are those related to level of service (transportation system supply) and to the HSR travelers (demand) directly affected by the interventions. In the evaluation of HSR investments, these kinds of impact are usually classified as first-order effects and derive from the fact that HSR services operates significantly faster (speeds exceeding 200–250 km/h, event tough there's not a unique international standard) than traditional train services, using particular rolling stocks, and usually dedicated tracks [23]. The level of service variation with respect to traditional railways can lead to a huge variation in spatial connectivity degree between different locations, together with changes in users travel behavior, users' mode choice and subsequent market share of other transport modes (for instance, air transport). The analysis of internal impacts is not limited to this, there are still other effects worth to be mentioned, such as the induced demand, i.e., the new mobility demand generated by the HSR network implementation. Cascetta and Coppola [24, 25], for instance, estimated 5.5 million induced trips in Italy from 2009 to 2013 (i.e., before and after HSR service introduction) consisting of both more frequent trips and new travelers moving mainly over distances in the range of 100–300 km. But this was not the only impact observed in Italy in the same period: with reference to mode choice, the same authors estimated a +20% of HSR market share at the expense of air and automobile modes. Similar evidence is reported also by other authors and with reference to other case studies, both in relation to induced demand phenomenon [26, 27], and to the analysis of the modal shift from other modes of transport towards HSR [28, 29].

External impacts are those impacts on members of population groups not directly interested by a policy or an intervention but that could positively (or negatively) be impacted by them, even if they will never use it. The literature on this topic is wide considering the large variety of impacts that can be considered (environmental, land use, economic, societal ones and so on so forth). From the environmental viewpoint, there's still a huge debate on HSR impacts on emissions savings and air quality. Considering the HSR between London and Manchester route in the UK, Miyoshi and Givoni [30] argue that HSR will have relatively limited potential for reduction in CO_2 emissions. On the contrary, other authors looked at this issue from a broader perspective and ignoring the overall emissions reduction magnitude, concluding that shifting traffic to rail will positively benefit the environment [31–33]. But the environmental impacts are not limited to local pollutants or global CO_2 emission reductions (i.e., those impacts easier to be taken into account from a quantitative viewpoint in assessment techniques). High-speed infrastructure development produces also wider environmental impacts such as those on land use, barrier effects or visual intrusion. In particular, the interaction between HSR investments and land use is something that many authors have focused on (see for instance [33, 34]). The Land Use/Transport Interaction traditional theory assumes that a not negligible increase in accessibility of a given area potentially generates a domino effect: increase in urbanization and real estate value, increase in property tax revenues and reallocation of economic activities are among the wider economic impacts that can be generated. However, there are complex mechanisms through which HSR investments interacts with land use and its economic substrate, which may lead to divergent conclusions depending on the context.

As previously mentioned, transport accessibility still remains a key variable to be evaluated in transport project appraisal. In particular, the impact on the spatial distribution of transport accessibility among different territories (i.e., territorial cohesion impact) is something that have been extensively studied with reference to several case studies. In a comparative perspective between different scenarios, most of the methodologies in the literature of transport planning evaluate these impacts by following two steps: first calculating one, or more, accessibility indicators and then analyzing the distribution of the latter through distribution indices. For instance, in their ex-post analysis, López et al. [4] analyze both a set of accessibility measures and a set of distribution indices (i.e., Coefficient of Variation, Gini coefficient, Thiel and Atkinson indices), to assess the investments carried out in the period 1992–2004 on the Spanish road and rail network. In particular, the study highlights how results interpretation can be strongly influenced by two factors: the chosen accessibility indicator and the level of development of the analyzed network with reference to the base year. As regard for the second factor, some warnings are provided: for instance, in an underdeveloped network, an initial polarization following firsts transportation investments can be expected, i.e., a negative impact on territorial cohesion. As pointed out by the authors, the latter is the case of the Spanish HSR in the analysis period 1992–2004. There are still other ex-post evidence, such as those related to the People's Republic of China HSR network development [35, 36]: the initial rise in accessibility inequalities within the study area will tend to decrease in a second phase with a more complete HSR development as planned in Western China. Similar results can be found in [37] with reference to the South Korean high-speed railway network: positive or negative cohesion impacts depend on the considered time horizon. By focusing on the European level, Kompil et al. [38] conduct an ex-post analysis of the impacts of road and rail transport infrastructure investments co-funded both by the Cohesion Fund and the European Regional Development Fund between 2000 and 2006: most of financed investments were located in peripheral European areas and the authors stress how the territorial cohesion impact of these investments, despite positive, is not sufficient to rebalance accessibility distribution among European Countries. Several worldwide studies [16, 17, 39–41] used a twofold approach, placing side by side a territorial cohesion and an efficiency analysis, intended as the average percentage accessibility variation, and highlighting potential tradeoffs between equity and efficiency considerations.

3 Methodological Approach

The methodological approach combines accessibility analysis with traditional transport (demand and supply) models (Fig. 1). Transportation network performances estimated in the project scenario allow for the comparison with a reference (non-project) scenario, in terms of overall accessibility variation in the study area (i.e., efficiency) and in terms of territorial cohesion, based on the distribution of zonal accessibility variations.

3.1 Accessibility Measures

Among all the accessibility indicators proposed in the literature (see [42, 46] for an overview of them), the potential accessibility indicator is among the most used in transport planning studies at national scale [5, 17]. The accessibility formulation used is the following:

$$PA_o = \sum\nolimits_d \mathrm{Add}_d^{\alpha} * e^{-\beta \cdot C_{od}} \quad (1)$$

where:

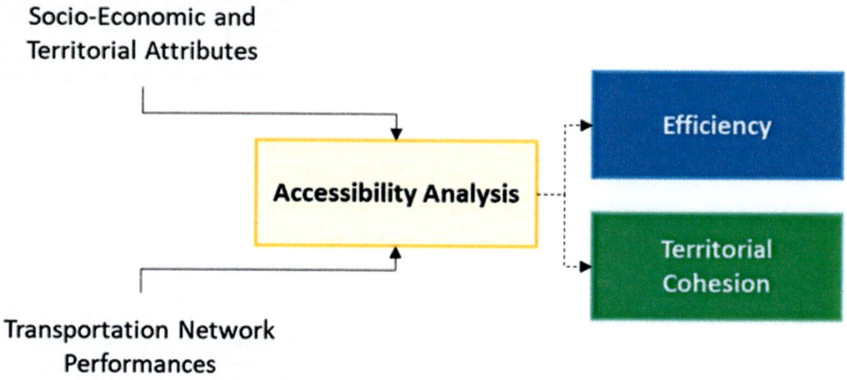

Fig. 1 Schematic representation of the proposed methodological approach

- PA_o is the accessibility for the origin zone o to destinations d;
- Add_d are the employees of the destination zone d;
- C_{od} is the generalized travel cost of the shortest path between the origin zone o and destination zone d;
- α and β are two estimated parameters, coming from a descriptive trip distribution model calibration which has as its denominator the same wording as accessibility formulation reported above.

The generalized travel cost C_{od} for the generic origin–destination pair (od) is computed as follows:

$$C_{od} = T_{b,od} + \beta_{cm} \cdot cm_{s,od} + \beta_{\varphi} \cdot \varphi_{od} \tag{2}$$

where:

- $T_{b,od}$ is the on-board time and depends on the commercial speed of the train service considered (HSR, intercity or regional rail services);
- $cm_{s,od}$ is the monetary cost expressed by the following formulation

$$cm_{s,od} = cm_{C,s,od} + cm_{Km} \cdot Km_{od} \tag{3}$$

  in which $cm_{C,s,od}$ is a constant that depends both on the type of service and on the od pair considered; Km_{od} is the distance, in kilometers, between the origin zone o and destination zone d and cm_{Km} is the unitary cost per kilometer (€/km);
- φ_{od} is the service frequency, considered as proxy of the waiting time for high-frequency services and of the scheduled early and late departure penalty for low-frequency ones [43];
- β_{cm} and β_{φ} are two calibrated parameters.

3.2 *Efficiency Analysis*

As in other studies in the literature [17], the efficiency analysis consists of the accessibility percentage variation calculation from the reference to the project scenario. The formulation is reported below:

$$\%Var\_PA_o = \frac{PA_{o,PS} - PA_{o,RS}}{PA_{o,RS}} \tag{4}$$

where:

- $\%Var\_PA_o$ is the accessibility percentage variation between project and reference scenario related to the origin zone o;
- $PA_{o,PS}$ is the accessibility indicator value of the origin zone o in the project scenario PS;

- $PA_{o,RS}$ is the accessibility indicator value of the origin zone o in the reference scenario RS.

The efficiency impact is computed averaging the accessibility percentage variation value among all the different zone o included in the study area: the higher is the average accessibility percentage variation, the higher is the efficiency of the intervention from an accessibility viewpoint.

3.3 Territorial Cohesion Analysis

Two are the statistical indices selected to measure the dispersion of accessibility values in each scenario. The first one is the Accessibility Dispersion Index (ADI), based on the coefficient of variation, and it is calculated using the following formula:

$$ADI^* = \frac{\sigma^*_{PA_o}}{\frac{\sum PA^*_o \cdot Pop_o}{\sum Pop_o}} \tag{5}$$

where:

- ADI^* is the Accessibility Dispersion index in scenario *;
- Pop_o is the population of the zone o;
- $\sigma^*_{PA_o}$ is the standard deviation of PA^*_o in scenario *, weighted by the population Pop_o.

The second one is the Gini index (GI), a measure commonly used as a gauge of economic inequality of income, or wealth, distribution among a population: as for other studies in the literature, this paper considers the accessibility indicator as the variable whose dispersion has to be measured. The Gini index mathematical formulation is reported below:

$$GI^* = 1 - \sum_o (cp_o - cp_{o-1}) \cdot (ca^*_o + ca^*_{o-1}) \tag{6}$$

where:

- GI^* is the GINI index of the scenario *;
- cp_o is the cumulative proportion of Pop_o, the population of the zone o, ranging from 0 to 1;
- ca_o is the cumulative proportion of PA_o, the accessibility for the origin zone o to destinations d, indexed in a non-decreasing order and ranging from 0 to 1.

Both of these measures were used for the same purpose in previous studies [4, 5, 38] and have the same reading key: the greater the value of these measures, the more polarized is the accessibility distributions; on the contrary, the smaller the value, the more equitable the distribution. To assess the impact on territorial cohesion, the percentage variation of each of these measures between different scenarios was calculated: if it is negative, there is a positive impact on territorial cohesion, while there is a negative impact on the contrary.

4 Application to Case Studies: Two New HSR Lines in Italy

The proposed methodology was therefore used to assess the impacts at national level due to the construction of two new HSR lines: the first one is between Brescia and Padova, to complete the Milan–Venice axis, the second one is between Naples and Bari.

In the reference scenario (Fig. 2), the Italian rail network is made of lines that extend for about 16,800 km. These are classified into main lines and complementary lines. For a length of approximately 6500 km, the main lines include international axes and connections between the main Italian cities. These are characterized by a high traffic density, unlike the complementary lines where there are minor flows. The main lines also include the approximately 1000 km of HSR line: those in service today are the Turin–Milan–Naples, with its extension towards Salerno, the Milan–Brescia and the Padova–Venice.

The entire railway network was modeled within a GIS software, building a graph of about 5000 nodes and about 10,000 arcs. Within the system, services have been implemented for about 9000 trains per day circulating throughout Italy. These services

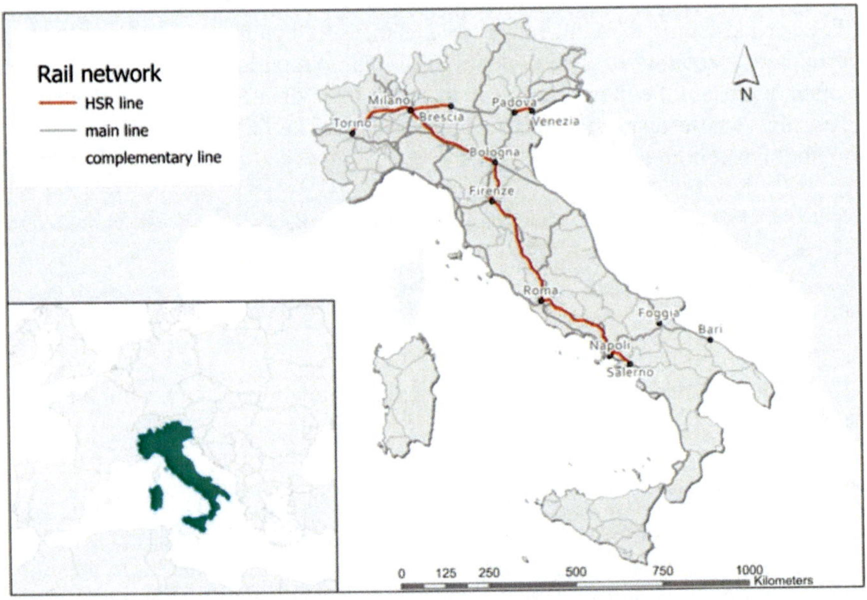

Fig. 2 Existing Italian railway network, with a focus on HSR infrastructure in the reference scenario

were divided by type (i.e., HSR, intercity or regional rail services) and characterized by timetable, fares and frequencies. The study area is the entire Italian national territory, divided up into approximately 270 zones. These are the sum of about 150 sub-provincial areas, 100 provincial capitals and few others large urban areas. Each zone, whether of origin o or destination d, was characterized in accordance with socio-economic data, set according to the base year of the analysis.

With reference to the do-nothing scenario, the results of the calculation of the potential accessibility indicator for each zone, as described above, are shown in Fig. 3. This suggests how the areas with high accessibility values are those close to the high-speed line, or those from which an HSR station can be easily reached. On the other hand, however, it is also worth noting the presence of "shadows areas" [44]: for instance, even though Florence and Rome are connected by the HSR line, there's a medium–low accessibility level area between them, due to the lack of any intermediate HSR station.

The map in Fig. 3 clearly shows a North–South accessibility pattern: the northern regions benefit from greater accessibility due to both a greater railway transportation network density and to their structural geographical location, i.e., geometrically more central to the distribution of the European population (and employees as well, used as a mass element in the accessibility calculation) than peripheral southern Italian areas.

4.1 The Brescia–Padova HSR Line

The Brescia–Padova HSR line (see Fig. 4), completing the Milan–Venice axis, is a project of strategic importance both at national and European level and is located in northern Italy in a very densely populated area with high transport accessibility. This corridor, in fact, in addition to being affected by important passenger and freight traffic, is an integral part of the Mediterranean Corridor, which represents the east–west link of the TEN-T network south of the Alps. In addition to the removal of interference with other services already present and the increase in the regularity of services, the reduction of travel times is among the main benefits expected from the completion of the project: for instance, from the Milano Centrale to the Venezia Santa Lucia station it is estimated a recovery of about 20 min compared to the current 2 h and 25 min, according to the Commercial Plan of the Italian railway manager (i.e., Rete Ferroviaria Italiana—RFI S.p.A.).

Only with reference to those Italian major cities which benefit most from the construction of this line (i.e., accessibility value relative percentage variation greater than 0.5%), Table 1 shows the accessibility values and its absolute and relative variations compared to the reference scenario. The selected cities benefit from an average accessibility (159.3 millions) higher than the national average (104.9 millions) in the reference scenario and the realization of the Brescia–Padova HSR line project will lead to a different efficiency impact for these selected cities (+7.6% average percentage variation) and for the whole of Italy (+2.8% average percentage variation).

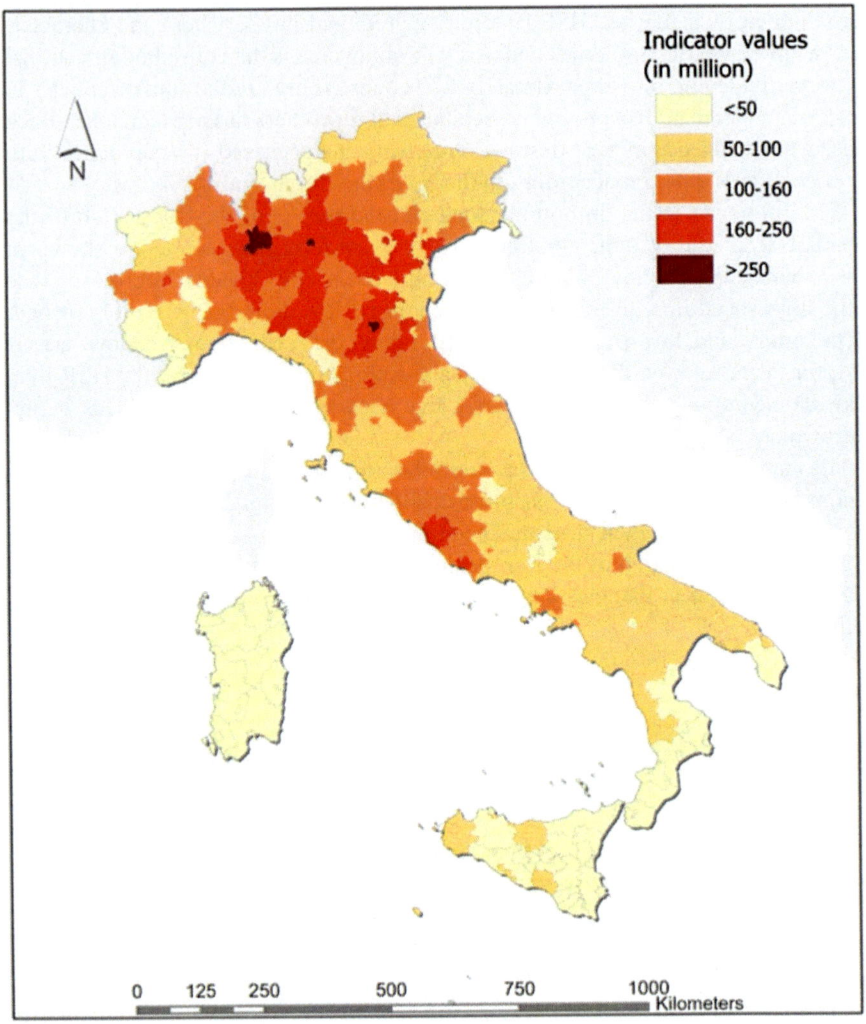

Fig. 3 Railway accessibility map: reference scenario

Relative changes in the accessibility value for each zone of the study area (i.e., the Italian territory) are shown in Fig. 5: the Brescia–Padova HSR line project will significantly impact northern Italian areas, north–east ones in particular. For instance, cities such as Vicenza, Padova and Venice will benefit relative percentage changes of more than 30% from the reference to the project scenario. Regarding the two distribution indices used as a proxy of territorial cohesion impact, results are shown below. The ADI index goes from 0.690 in the reference scenario to 0.709 in the project scenario simulating the completion of the Brescia–Padova HSR line: this increase corresponds to a positive percentage change of +2.8%. On the other hand,

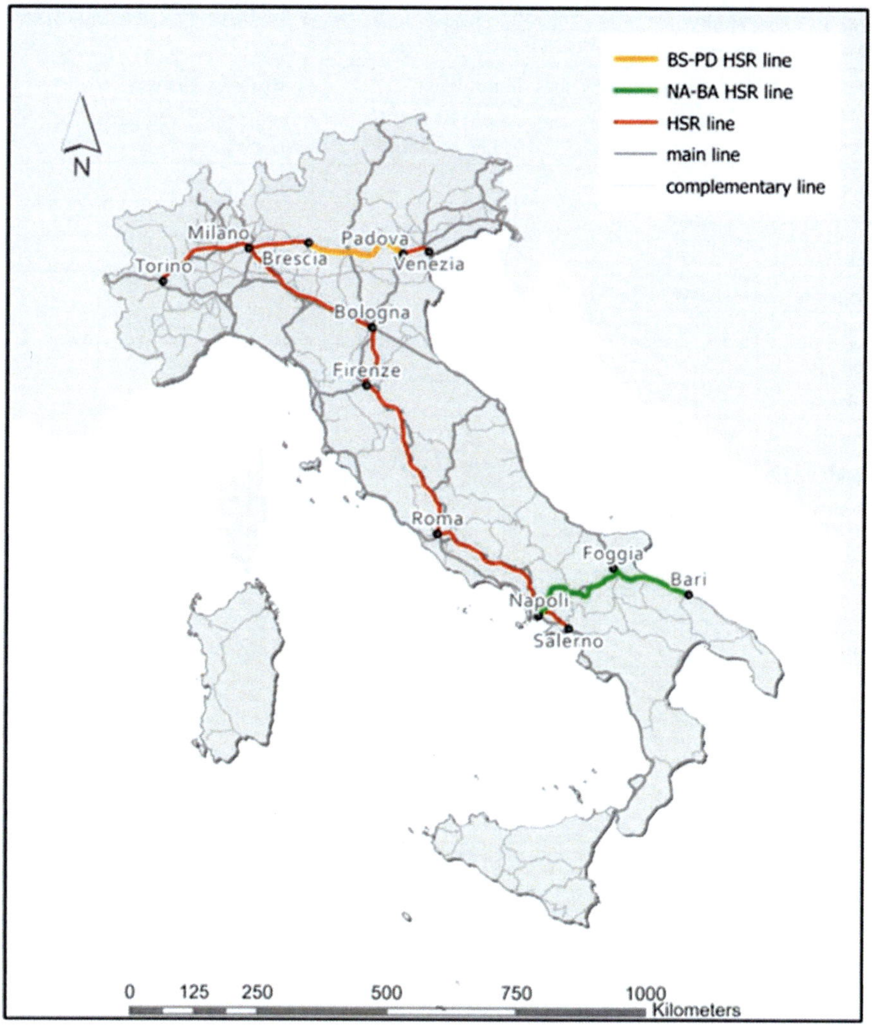

Fig. 4 HSR existing network, together with the two HSR projects to be simulated

the GINI index switches from 0.245 in the reference scenario to 0.253 in the project one, showing an increase of +3.5%. The positive percentage change of both indices indicates a negative impact of the new HSR line on territorial cohesion at National level, i.e., an increase in the polarization of the distribution of accessibility values.

Table 1 Accessibility indicator values and changes between the reference and "w/BS-PD HSR line" project scenario

City	Accessibility values (in millions)		Accessibility changes	
	Reference scenario	w/BS-PD HSR line scenario	Absolute (in thousand)	Relative (%)
Alessandria	125.5	126.9	1462.1	1.2
Asti	105.7	106.6	863.3	0.8
Belluno	45.6	48.9	3251.9	7.1
Bergamo	210.3	219.4	9079.5	4.3
Biella	64.6	67.0	2412.5	3.7
Bologna	333.1	338.4	5328.4	1.6
Bolzano	79.2	84.8	5631.1	7.1
Brescia	263.3	318.4	55015.6	20.9
Como	167.9	173.3	5375.1	3.2
Cremona	147.9	158.4	10540.1	7.1
Cuneo	42.9	43.5	596.1	1.4
Ferrara	195.1	206.9	11796.1	6.0
Firenze	212.0	213.8	1778.7	0.8
Forlì	147.3	148.6	1277.0	0.9
Genova	107.2	109.8	2592.4	2.4
Gorizia	38.3	41.3	3049.4	8.0
Imperia	22.3	22.5	246.4	1.1
Lecco	171.5	176.9	5378.0	3.1
Lodi	233.7	240.2	6457.4	2.8
Mantova	140.8	151.1	10370.7	7.4
Milano	381.3	394.7	13361.3	3.5
Modena	218.7	220.3	1638.5	0.7
Novara	180.5	190.7	10153.1	5.6
Padova	233.5	307.5	73977.7	31.7
Parma	178.9	182.6	3681.2	2.1
Pavia	225.3	232.6	7240.4	3.2
Piacenza	214.6	218.0	3390.9	1.6
Pordenone	92.8	103.8	10977.5	11.8
Ravenna	101.1	103.5	2432.7	2.4
Reggio Emilia	269.3	275.7	6406.1	2.4
Rovigo	197.9	217.0	19086.8	9.6
Savona	58.2	59.0	807.5	1.4
Torino	182.1	186.7	4618.4	2.5
Trento	109.4	121.0	11518.2	10.5

(continued)

Table 1 (continued)

City	Accessibility values (in millions)		Accessibility changes	
	Reference scenario	w/BS-PD HSR line scenario	Absolute (in thousand)	Relative (%)
Treviso	144.2	174.3	30164.9	20.9
Trieste	36.3	43.4	7117.1	19.6
Udine	71.0	80.0	9052.1	12.8
Varese	184.5	190.1	5629.9	3.1
Venezia	178.9	234.3	55344.1	30.9
Verbania	115.5	119.3	3818.1	3.3
Vercelli	149.8	161.8	11974.6	8.0
Verona	245.7	289.6	43848.8	17.8
Vicenza	226.8	300.0	73142.7	32.2
Average for selected cities	159.3	172.1	12834.5	7.6
Nation-based average	104.9	109.0	4082.0	2.8

4.2 The Naples–Bari HSR Line

The Naples–Bari HSR line (see Fig. 4) is included in the Scandinavian–Mediterranean Corridor of the TEN-T network and it is located in southern Italy, in a low-populated area with low accessibility. The realization of the project will guarantee the interconnection and the interoperability within the European rail corridors and will allow to integrate the railway infrastructure of Puglia and of the inner areas of Campania, with the lines of connection to the North of the Country and with Europe. Among the main expected benefits, there is the reduction of travel times: for instance, the Rome–Bari rail route will recover about 35 min of the current 3 h and 40 min. There will be also an increase in line service frequency, going from the current 4 trains/h, in both directions, to 10 trains/h when the project will be completed. In Table 2 are listed all the major Italian cities which experience an increase of more than 0.5% following the implementation of this HSR project. The selected cities have an average accessibility value (101.5 millions) slightly lower than the national average (104.9 millions) in the reference scenario and the completion of the Naples–Bari HSR line project will lead to a +4.7% average percentage accessibility variation for the selected cities and a +0.8% for Italy as a whole.

As shown in Fig. 6, impacts in terms of relative accessibility changes in accessibility are mainly located in central-southern Italy, in particular in the Campania (the city of Benevento experience a +26.3% accessibility variation) and Puglia (the city of Foggia experience a +13%) regions. The calculation of the ADI and GINI indices with reference to entire the study area produced the following results: starting from the value 0.690 in the reference scenario, the ADI decrease by 0.6%; the same goes

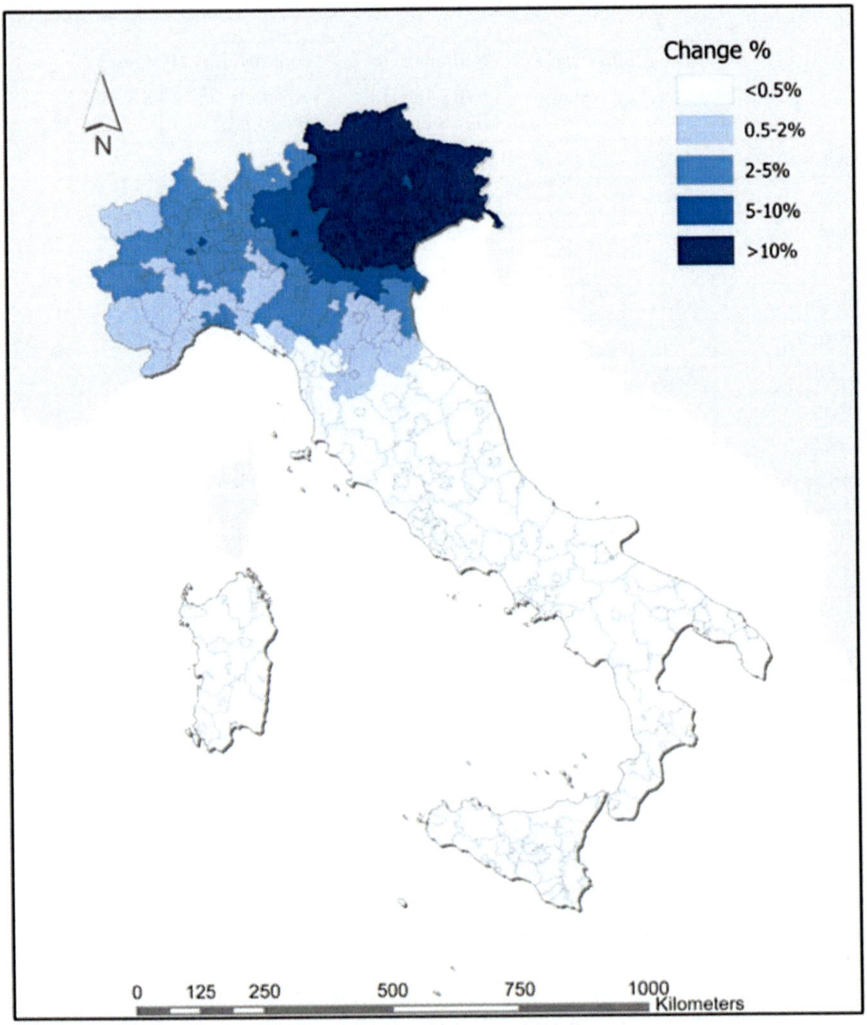

Fig. 5 Accessibility relative changes (%) between the reference and "w/BS-PD HSR line" project scenario

for the GINI index, that decrease by 1.2% going from 0.245 to 0.242. In this case the percentage variation of the distribution indices is negative: the completion of the Naples–Bari HSR line will therefore make a positive contribution on territorial cohesion.

Table 2 Accessibility indicator values and changes between the reference and "w/NA-BA HSR line" project scenarios

City	Accessibility values (in millions)		Accessibility changes	
	Reference scenario	w/NA-BA HSR line scenario	Absolute (in thousand)	Relative (%)
Avellino	96.5	104.5	8019.4	8.3
Bari	90.8	96.9	6138.0	6.8
Benevento	85.2	107.6	22413.7	26.3
Brindisi	72.6	75.5	2956.7	4.1
Campobasso	90.9	92.0	1103.5	1.2
Caserta	136.8	147.4	10668.9	7.8
Chieti	88.5	90.2	1746.7	2.0
Foggia	112.3	127.5	15245.9	13.6
Frosinone	109.4	110.2	740.3	0.7
Isernia	44.6	46.2	1557.6	3.5
Latina	164.8	166.1	1302.0	0.8
Lecce	65.8	68.9	3120.7	4.7
Napoli	149.7	152.1	2354.9	1.6
Pescara	73.2	73.9	670.1	0.9
Potenza	40.5	40.8	259.8	0.6
Roma	220.3	222.1	1823.3	0.8
Salerno	110.7	111.5	781.9	0.7
Taranto	75.0	75.8	725.5	1.0
Average for selected cities	101.5	106.1	4534.9	4.7
Nation-based average	104.9	105.6	680.9	0.8

5 Discussion and Conclusions

This paper analyzes the impacts on both efficiency (i.e., average accessibility percentage variation) and territorial cohesion (i.e., spatial distribution of accessibility values) following the construction of two new HSR lines in Italy: the first between Brescia and Padova and the second between Naples and Bari.

Results reported in Table 3 suggest both projects have a positive impact at National level in terms of average accessibility variation (i.e., investment efficiency): +2.8% average accessibility percentage variation for the Brescia–Padova HSR line, +0.8% for the Naples–Bari one. The different efficiency magnitude between the two projects can be explained with the so called "network effect" [45]: this effects describes the phenomenon that benefits of a link improvement goes up faster with the density of a transport network. This is because the number of origin–destination relations

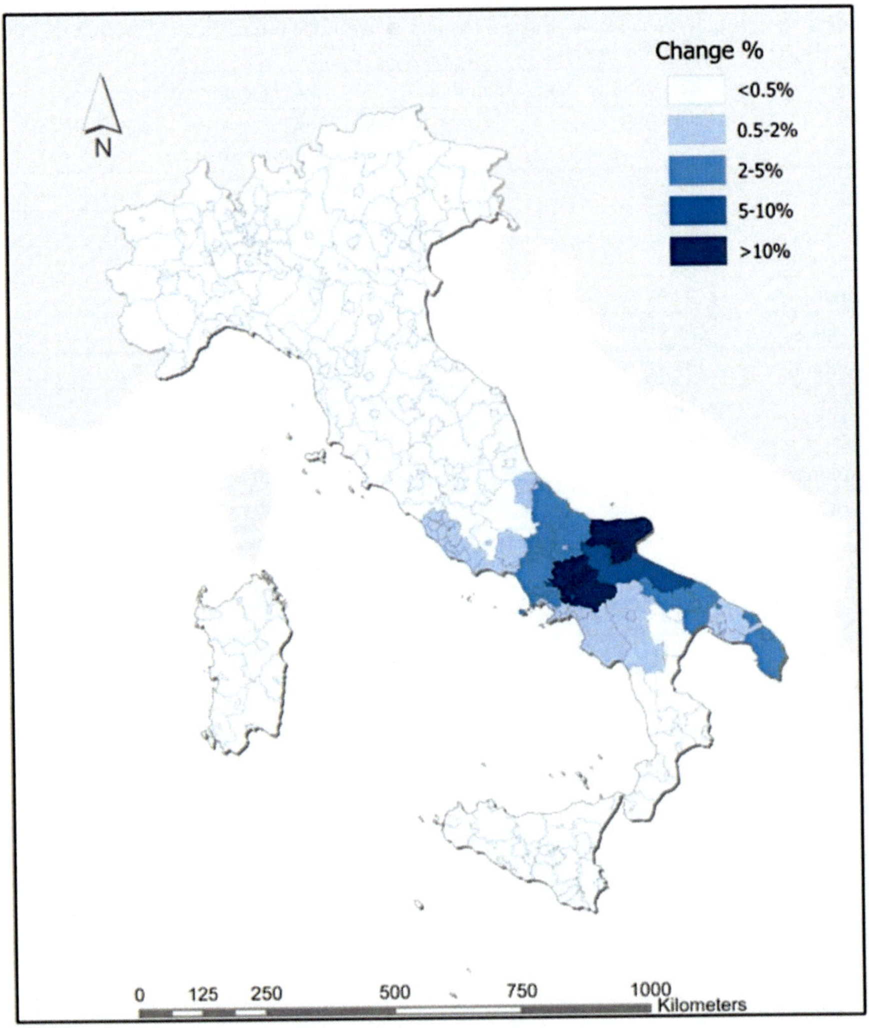

Fig. 6 Accessibility relative changes (%) between the do-nothing and "w/NA-BA HSR line" scenarios

increases exponentially with the increase of connecting links and network density, creating an implicit advantage for areas with highly dense transportation networks (such as northern Italy, where the Brescia–Padova HSR line is located).

On the other hand, different conclusions can be drawn by analyzing both projects from a territorial cohesion viewpoint: the Brescia–Padova HSR line has a negative impact on territorial cohesion since it contributes to an accessibility polarization in the North of Italy (where accessibility levels are higher than in the South), while the Naples–Bari has a positive outcome, in contributing to balance the railway accessibility gap of Southern regions with respect to the Northern ones.

Table 3 Results summary with reference to both case studies analyzed: the Brescia–Padova and the Naples–Bari HSR lines

Impact category	Indicator	w/BS-PD HSR line % change (%)	Impact magnitude	w/NA-BA HSR line % change (%)	Impact magnitude
Efficiency	Accessibility % variation	+2.8	++	+0.8	+
Territorial Cohesion	ADI	+2.8	−−	−0.6	+
	GINI	+3.5		−1.2	

As a matter of fact, the results presented in this paper show that projects prioritization can highly depends on the criteria on which it is based: efficiency-based or equity-based. In particular, an efficiency-based assessment might promote project increasing the overall accessibility but that risk to amplify territorial gaps, as this is the case for the North–South gaps in Italy in terms of rail infrastructural endowment. Even though transportation accessibility is only one of the necessary conditions for territorial cohesion, these results and their subsequent interpretation claim for a holistic approach to project appraisal, whose absence risks to exacerbate territorial inequalities.

References

1. European Commission: My region, My Europe, our future: the seventh report on economic, social and territorial cohesion (2017)
2. Wang, C., Lim, M.K., Zhang, X., et al.: Railway and road infrastructure in the Belt and Road Initiative countries: Estimating the impact of transport in-frastructure on economic growth. Transp. Res. Part Policy Pract. **134**, 288–307 (2020). https://doi.org/10.1016/j.tra.2020.02.009
3. Farhadi, M.: Transport infrastructure and long-run economic growth in OECD countries. Transp. Res. Part Policy Pract. **74**, 73–90 (2015). https://doi.org/10.1016/j.tra.2015.02.006
4. López, E., Gutiérrez, J., Gómez, G.: Measuring regional cohesion effects of large-scale transport infrastructure investments: an accessibility approach. Eur. Plan. Stud. **16**, 277–301 (2008). https://doi.org/10.1080/09654310701814629
5. Ortega, E., López, E., Monzón, A.: Territorial cohesion impacts of high-speed rail at different planning levels. J. Transp. Geogr. **24**, 130–141 (2012). https://doi.org/10.1016/j.jtrangeo.2011.10.008
6. Ren, X., Chen, Z., Wang, F., et al.: Impact of high-speed rail on social equity in China: evidence from a mode choice survey. Transp. Res. Part Policy Pract **138**, 422–441 (2020). https://doi.org/10.1016/j.tra.2020.05.018
7. Behbahani, H., Nazari, S., Jafari Kang, M., Litman, T.: A conceptual framework to formulate transportation network design problem considering social equity criteria. Transp. Res. Part Policy Pract. **125**, 171–183 (2019). https://doi.org/10.1016/j.tra.2018.04.005
8. Manaugh, K., Badami, M.G., El-Geneidy, A.M.: Integrating social equity into urban transportation planning: a critical evaluation of equity objectives and measures in transportation plans in North America. Transp. Policy **37**, 167–176 (2015). https://doi.org/10.1016/j.tranpol.2014.09.013

9. Camporeale, R., Caggiani, L., Ottomanelli, M.: Modeling horizontal and vertical equity in the public transport design problem: a case study. Transp. Res. Part Policy Pract. **125**, 184–206 (2019). https://doi.org/10.1016/j.tra.2018.04.006
10. Peters, J.R., Kramar, J.K.: Just who should pay for what? Vertical equity, transit subsidy and road pricing: the case of New York City. J. Public Transp. **15**, 117–136 (2012). https://doi.org/10.5038/2375-0901.15.2.6
11. Delbosc, A., Currie, G.: Using Lorenz curves to assess public transport equity. J. Transp. Geogr. **19**, 1252–1259 (2011). https://doi.org/10.1016/j.jtrangeo.2011.02.008
12. Hosein Mortazavi, S.A., Akbarzadeh, M.: A framework for measuring the spatial equity in the distribution of public transportation benefits. J. Public Transp. **20**, 44–62 (2017). https://doi.org/10.5038/2375-0901.20.1.3
13. Jang, S., An, Y., Yi, C., Lee, S.: Assessing the spatial equity of Seoul's public transportation using the Gini coefficient based on its accessibility. Int. J. Urban Sci. **21**, 91–107 (2017). https://doi.org/10.1080/12265934.2016.1235487
14. Bröcker, J., Korzhenevych, A., Schürmann, C.: Assessing spatial equity and efficiency impacts of transport infrastructure projects. Transp. Res. Part B Methodol. **44**, 795–811 (2010). https://doi.org/10.1016/j.trb.2009.12.008
15. Bandegani, M., Akbarzadeh, M.: Evaluation of horizontal equity under a distance-based transit fare structure. J. Public Transp. **19** (2016). https://doi.org/10.5038/2375-0901.19.3.10
16. Li, T., Zhang, S., Cao, X., Witlox, F.: Does a circular high-speed rail network promote efficiency and spatial equity in transport accessibility? Evidence from Hainan Island, China. Transp. Plan. Technol. **41**, 779–795 (2018). https://doi.org/10.1080/03081060.2018.1504186
17. Monzón, A., Ortega, E., López, E.: Efficiency and spatial equity impacts of high-speed rail extensions in urban areas. Cities **30**, 18–30 (2013). https://doi.org/10.1016/j.cities.2011.11.002
18. Martínez Sánchez-Mateos, H.S., Givoni, M.: The accessibility impact of a new High-Speed Rail line in the UK—a preliminary analysis of winners and losers. J. Transp. Geogr. **25**, 105–114 (2012). https://doi.org/10.1016/j.jtrangeo.2011.09.004
19. Pagliara, F., La Pietra, A., Gomez, J., Manuel Vassallo, J.: High Speed Rail and the tourism market: evidence from the Madrid case study. Transp. Policy **37**, 187–194 (2015). https://doi.org/10.1016/j.tranpol.2014.10.015
20. Yan, L., Tu, M., Chagas, A.L.S., Tai, L.: The impact of high-speed railway on labor spatial misallocation—based on spatial difference-in-differences analysis. Transp. Res. Part Policy Pract. **164**, 82–97 (2022). https://doi.org/10.1016/j.tra.2022.07.017
21. Cascetta, E., Cartenì, A., Henke, I., Pagliara, F.: Economic growth, transport accessibility and regional equity impacts of high-speed railways in Italy: ten years ex post evaluation and future perspectives. Transp. Res. Part Policy Pract. **139**, 412–428 (2020). https://doi.org/10.1016/j.tra.2020.07.008
22. (Ato) Xu W, Huang Y (2019) The correlation between HSR construction and economic development—empirical study of Chinese cities. Transp. Res. Part Policy Pract. **126**, 24–36.https://doi.org/10.1016/j.tra.2019.05.017
23. Cartenì, A., Pariota, L., Henke, I.: Hedonic value of high-speed rail ser-vices: quantitative analysis of the students' domestic tourist attractiveness of the main Italian cities. Transp. Res. Part Policy Pract. **100**, 348–365 (2017). https://doi.org/10.1016/j.tra.2017.04.018
24. Cascetta, E., Coppola, P., Velardi, V.: High-Speed rail demand: before-and-after evidence from the Italian market. DisP Plan. Rev. **49**, 51–59 (2013). https://doi.org/10.1080/02513625.2013.826539
25. Cascetta, E., Coppola, P.: New high-speed rail lines and market competition: short-term effects on services and demand in Italy. Transp. Res. Rec **2475**, 8–15 (2015). https://doi.org/10.3141/2475-02
26. Givoni, M., Dobruszkes, F.: A review of ex-post evidence for mode substitution and induced demand following the introduction of high-speed rail. Transp. Rev. **33**, 720–742 (2013). https://doi.org/10.1080/01441647.2013.853707
27. D'Alfonso, T., Jiang, C., Bracaglia, V.: Would competition between air transport and high-speed rail benefit environment and social welfare? Transp. Res. Part B Methodol. **74**, 118–137 (2015). https://doi.org/10.1016/j.trb.2015.01.007

28. Bergantino, A.S., Madio, L.: Intermodal competition and substitution. HSR versus air transport: understanding the socio-economic determinants of modal choice. Res. Transp. Econ. **79**, 100823 (2020). https://doi.org/10.1016/j.retrec.2020.100823
29. Shi, J., Hussain, M., Kong, X.P.: Factors affecting travel mode choice between high-speed railway and air transportation among university students for tourism—evidence from China. J. China Tour. Res. **18**, 106–120 (2022). https://doi.org/10.1080/19388160.2020.1802802
30. Miyoshi, C., Givoni, M.: The environmental case for the high-speed train in the UK: Examining the London-Manchester route. Int. J. Sustain. Transp. **8**, 107–126 (2013). https://doi.org/10.1080/15568318.2011.645124
31. Chester, M., Horvath, A.: High-speed rail with emerging automobiles and aircraft can reduce environmental impacts in California's future. Environ. Res. Lett. **7**, 034012 (2012). https://doi.org/10.1088/1748-9326/7/3/034012
32. Bueno, G., Hoyos, D., Capellán-Pérez, I.: Evaluating the environmental performance of the high speed rail project in the Basque Country, Spain. Res. Transp. Econ. **62**, 44–56 (2017). https://doi.org/10.1016/j.retrec.2017.02.004
33. Robertson, S.: The potential mitigation of CO_2 emissions via modal substitution of high-speed rail for short-haul air travel from a life cycle perspective—an Australian case study. Transp. Res. Part Transp. Environ. **46**, 365–380 (2016). https://doi.org/10.1016/j.trd.2016.04.015
34. Guirao, B., Lara-Galera, A., Campa, J.L.: High Speed Rail commuting impacts on labour migration: the case of the concentration of metropolis in the Madrid functional area. Land Use Policy **66**, 131–140 (2017). https://doi.org/10.1016/j.landusepol.2017.04.035
35. Shaw, S.-L., Fang, Z., Lu, S., Tao, R.: Impacts of high speed rail on railroad network accessibility in China. J. Transp. Geogr. **40**, 112–122 (2014). https://doi.org/10.1016/j.jtrangeo.2014.03.010
36. Yang, J., Guo, A., Li, X., Huang, T.: Study of the impact of a high-speed railway opening on China's accessibility pattern and spatial equality. Sustainability **10**, 2943 (2018). https://doi.org/10.3390/su10082943
37. Kim, H., Sultana, S.: The impacts of high-speed rail extensions on accessibility and spatial equity changes in South Korea from 2004 to 2018. J. Transp. Geogr. **45**, 48–61 (2015). https://doi.org/10.1016/j.jtrangeo.2015.04.007
38. Kompil, M., Demirel, H., Christidis, P.: Accessibility and territorial cohe-sion: ex post analysis of Cohesion Fund infrastructure projects. In: Chapters. Edward Elgar Publishing, pp. 210–242 (2016)
39. Stepniak, M., Rosik, P.: Accessibility improvement, territorial cohesion and spillovers: a multi-dimensional evaluation of two motorway sections in Poland. J. Transp. Geogr. **31**, 154–163 (2013). https://doi.org/10.1016/j.jtrangeo.2013.06.017
40. Stępniak, M., Rosik, P.: From improvements in accessibility to the impact on territorial cohesion: the spatial approach. J. Transp. Land Use **9**, 1–13 (2016). https://doi.org/10.5198/jtlu.2015.570
41. Monzon, A., Lopez, E., Ortega, E.: Has HSR improved territorial cohesion in Spain? An accessibility analysis of the first 25 years: 1990–2015. Eur. Plan. Stud. **27**, 513–532 (2019). https://doi.org/10.1080/09654313.2018.1562656
42. Reggiani, A.: Accessibility, Trade and Locational Behaviour. Routledge, London (2019)
43. Cascetta, E., Coppola, P.: Assessment of schedule-based and frequency-based assignment models for strategic and operational planning of high-speed rail services. Transp. Res. Part Policy Pract **84**, 93–108 (2016). https://doi.org/10.1016/j.tra.2015.09.010
44. Vickerman, R., Spiekermann, K., Wegener, M.: Accessibility and economic development in Europe. Reg. Stud. **33**, 1–15 (1999). https://doi.org/10.1080/00343409950118878
45. Martens, K., Di Ciommo, F.: Travel time savings, accessibility gains and equity effects in cost–benefit analysis. Transp. Rev **37**, 152–169 (2017). https://doi.org/10.1080/01441647.2016.1276642
46. Geurs, K.T., van Wee, B.: Accessibility evaluation of land-use and transport strategies: review and research directions. J. Transp. Geogr **12**, 127–140 (2004). https://doi.org/10.1016/j.jtrangeo.2003.10.005

UIC Youth/Student HSR Innovation Challenge Pilot Program Proposal

Beverly A. Scott, Yaqeline Castro, and Isaiah Chauhan

Abstract At Introducing Youth to American Infrastructure Inc. ("Iyai"), we believe that *inclusive* "people development" is key to collective success. Globally motivating and providing opportunities for *all* young people to become tomorrow's infrastructure leaders, innovators, entrepreneurs, and skilled workforce—with emphasis on increasing the participation of young women and other historically under-represented groups in these careers. In the transport sector, women, people of color, indigenous groups and others are significantly under-represented in employment and business utilization. While there has been progress in recent years, significant inequities continue to persist—underscoring the importance of building greater youth awareness of transportation/mobility careers and the critical role they play in building and shaping communities, exposure to industry professionals/mentoring and active community engagement. On that note, we believe that valuing "youth voice" and "learned" experience in decision-making, active mentoring, and exposure to diverse professional development networks—as well as the opportunity to "learn by doing"—active youth engagement beyond the more traditional classroom and employer-based career-readiness programs is an important motivator. The emphasis of this research project is the expansion of Next Generation Talent Development and Diversification as an "actionable" initiative in support of The UIC Alliance of Universities' overarching goals to improve education in High-Speed Rail (HSR); to exchange and disseminate know-how between Universities and UIC members—to include active youth awareness and engagement regarding the effects, both on economy and on society, of investments in HSR systems. This presentation highlights two efforts that are developing in the California area: the Iyai North American

B. A. Scott (✉)
San Francisco Bay Area, CA 94706, USA
e-mail: beverly@beverlyscottassociates.com

Y. Castro
Fresno, CA 93722, USA
e-mail: yaqeline.castro@hsr.ca.gov

I. Chauhan
Niles, IL 60714, USA
e-mail: isaiahchauhan17@gmail.com

F. Pagliara (ed.), *Socioeconomic Impacts of High-Speed Rail Systems*,
Springer Proceedings in Business and Economics,
https://doi.org/10.1007/978-3-031-26340-8_7

Innovation Challenge, and California High Speed Rail Authority's (Authority) *I Will Ride* program; both are designed to go "beyond" a traditional teaching experience. First, exposing youth to knowledgeable, experienced, and diverse professionals at all levels providing an opportunity to build and expand their own networks; learn about and gain access to important industry-specific and community career resources). And as importantly actively engage and encourage youth to give "voice" and "visibility" to their ideas, solutions, and personal "learned" experience(s) to help tackle locally identified sustainable mobility challenges in the facilities, systems, and network of services they use every day.

Keywords Infrastructure · Diversity · Inclusion · Equity · Workforce development · Youth · Human resources

1 Introducing Youth to American Infrastructure

Iyai Founder, Dr. Beverly A. Scott opened this presentation with a brief Overview of the Organization's mission, goals, and programs—advancing Page | 2greater transport sector career diversity and inclusion at all levels. She advanced the recommendation to include a "pilot" Next Generation Talent Development and Diversification" youth programming element in the 2023 HSR Program as an "actionable" initiative. Noting the alignment with The UIC Alliance of Universities overarching goal to include active youth awareness and engagement regarding the effects, both on economy and on society, of investments in HSR systems.

Dr. Scott concluded her remarks by introducing two (2) youth presenters who briefly highlighted two (2) "equity-centered" transport sector youth programs—both providing meaningful and complementary opportunities beyond the traditional classroom experience—for transport sector career awareness, professional and personal networking, and active civic engagement.

- Isaiah Chauhan, Iyai Fellow and Alum; University of Illinois @ Chicago graduate
- Yaqeline Castro, California High Speed Rail (CAHSR) Outreach and Engagement Specialist; Mineta Transportation Institute, San Jose State University (Master's Program).

1.1 *Encouraging Youth Voice and Valuing Learned Experience*

Introducing Youth to American Infrastructure (Iyai) [1], emphasizes exposing youth to the community-building aspects of transport/infrastructure careers, including opportunities to interact with industry professionals and community stakeholders at all levels; and meaningful youth engagement to address "real-world" challenges and opportunities in their local communities as a great way to "get their feet wet."

Iyai Alumnus, Isaiah Chauhan spoke about one of the Iyai Youth programming themes introduced in 2022, *Re-Imagine Normal.* This belief squarely advances the criticality of embracing transformative societal change. Youth as well as many others do not want to "go back to normal to a pre-Covid society, the same society that suffered from a digital divide, a history of discriminatory practices and inequities." Instead, Iyai Youth opt for an "equity-centered" New Normal—focused on community-building, inclusion, shared prosperity and sustainable mobility for *all* people and communities.

1.2 "Sustainable Mobility Careers Awareness and Innovation Challenge"

Isaiah went in-depth into Iyai's upcoming programming, the 2023 North American Sustainable Mobility Careers Awareness and Innovation Challenge [2]. The Challenge is a countrywide event designed to build greater youth awareness of mobility, transportation careers, and the critical role they play in building and shaping communities. The second element of the Challenge involves youth voice and engagement in tackling important locally identified "sustainable mobility" challenges. The final element is a Future Workplace Local Dialogue/Listening Session including both youth and industry professionals on ways to improve Next Generation career awareness (image and marketing), recruitment, selection, and retention of a diverse pool of workforce entrants.

This Iyai pilot program will commence in 2023, anchored by Local Organizers in up to 20 locations within the U.S. and Canada.

2 California. High-Speed Rail Authority

2.1 "I Will Ride" Student Outreach

California High Speed-Rail Authority's (Authority) Outreach and Student Engagement Specialist, Yaqeline Castro, spoke on the Authority's *I Will Ride* [3] program. *I Will Ride* is a transportation student leadership development outreach program designed to inform, engage, inspire and connect students to the nation's first high-speed rail system in the United States. The program provides internship, fellowship, and career opportunities in addition to a host of student resources and opportunities to discuss and learn about high-speed rail in a classroom or community setting.

Upon completion of the California high-speed rail system, it will connect all the major population centers across the state. The first initial operating segment is planned for the Central Valley of California. The presentation covered the importance of California high-speed rail and its capacity in reducing pollution and carbon emissions, vehicular and air traffic reduction, creating green quality jobs, and shared economic prosperity—including jobs and expanded business opportunities, and overall sustainability benefits.

The *I Will Ride* program includes three major elements; networking, education, and digital media that further expands outreach and participation opportunities. From showcasing benefits of signing up for the Authority's free membership, to directly connecting students to the nation's first high-speed rail system currenly under construction, Castro displayed the success of *I Will Ride*, its growth, and how it has positively impacted its student participants.

Moving forward, they aim to continue to partner with similarly aligned organizations that prioritize meaningful youth involvement, hopefully support a UIC youth participation initiative in future UIC Socioeconomic Impacts of HSR convenings, and partner with others to enhance and expand student programs and pathways for careers in high-speed rail including global exchange/collaboration, mentoring, and other professional development programming opportunities for youth.

References

1. Iyai Homepage: https://www.iyai.org/. Accessed 21 Nov 2022
2. Iyai Innovation Challenge Homepage: https://www.iyai.org/innovation-challenge/. Accessed 21 Nov 2022
3. CAHSR Homepage: https://hsr.ca.gov/. Accessed 21 Nov 2022
4. CAHSR I Will Ride: https://hsr.ca.gov/i-will-ride/. Accessed 21 Nov 2022

HSR and Land Use Impacts 1

A Review of Key Socio-economic Factors Affecting High-Speed Rail Station Location Selection

Malavika Jayakumar and Avijit Maji

Abstract High-Speed Railway (HSR) location selection is a multidimensional process involving choosing the best station locations, corridors and alignment connecting the stations. Only by strategically placing the HSR stations, can the advantages of shorter intercity travel times, comfort, and safety associated with HSR be fully realized. Therefore, choosing a suitable HSR station location is essential while planning an HSR facility. The significant aspects affecting station location selection are ridership, connectivity to other modes of transportation, the costs associated with the stakeholders, socio-economic impacts, environmental feasibility, and inter-station spacing. The limitation of the existing literature is that only a few combinations of these aspects were considered in the optimization formulation to simplify the problem and make it computationally less expensive. Few studies have taken into account these aspects for urban rails, which may not be directly applicable to HSR. Based on the existing literature in this field, this paper discusses the important factors to consider when choosing an HSR station location.

Keywords High-speed rails · Station location · Multi-objective optimization · Socio-economic impacts

1 Introduction

1.1 General

Investments in transportation infrastructure, particularly railways, are closely linked to a country's economic growth. High-speed rails (HSRs) have gained popularity recently because of their favourable characteristics like speed, high frequency, low cost compared to other long-distance travel modes, improved comfort, and safety.

M. Jayakumar (✉) · A. Maji
Department of Civil Engineering, Indian Institute of Technology Bombay, Mumbai, Maharashtra 400076, India
e-mail: malujaykay@gmail.com

F. Pagliara (ed.), *Socioeconomic Impacts of High-Speed Rail Systems*,
Springer Proceedings in Business and Economics,
https://doi.org/10.1007/978-3-031-26340-8_8

HSR is a greener mode than air transport and cars in its operation perspective [1]. The equivalent energy consumption of HSR is 14 times lower than aircraft and 19 times lower than cars, and its CO_2 emission is 8.7 times lower than aircraft and 12 times lower than cars for a 600 kilometers trip [2]. HSRs are also safer considering the low passenger fatalities per billion passenger kilometres [3]. One of the primary features promoting the introduction of HSRs in many developing countries is their ability to reduce travel time between megacities located 200 to 750 km apart (up to a maximum of 1000 km in some cases), providing commuters with access to distant opportunities resulting in economic development [4, 5].

Considering the high number of passengers who use highways and airlines for intercity travel, HSR provides numerous benefits, such as relief from traffic congestion on roads and airports, reduced consumption of fossil fuels, lower emissions, improved accessibility, safety, comfort, and environmental benefits. These benefits contribute to a modal shift from other modes to HSR [6]. Even though the development of HSR involves a massive amount of land and capital, it has helped enhance job opportunities, brought together cities and allowed people to access megacities without living in the city [5]. HSR can only have a positive impact if the stations are in strategic locations that maximize the benefits to the economy and people. This makes station positioning at the macro and micro levels critical.

The construction of a new HSR station needs to consider several complexities, such as initial feasibility factors, reducing environmental damage, intermodal integration through access to other transport facilities, reducing right-of-way costs, and strategic positioning in terms of nearness to the city centre. Some existing studies have not considered or over-simplified these aspects while selecting station locations [7–10]. Hence, the solutions obtained through these methodologies may not be optimal, as a balance between all the aspects is necessary to get the optimal HSR station location [6]. Therefore, there is a need to understand these complex factors in detail before formulating the optimization problem. This chapter examines the key factors affecting HSR station location choice, focusing on the micro-level analysis based on the existing literature.

2 Literature Review

2.1 HSR Systems

Based on the relationship with conventional rails, HSR systems are of two types: (1) dedicated systems with an exclusive right-of-way for HSR and (2) interoperable systems where the HSR trains share the track with conventional trains. The two ways to develop an HSR system are either by upgrading conventional railways or building new systems. Existing stations may not always ensure the requisite ridership potential and inter-station spacing for the successful operation of HSR, even though it reduces land acquisition expenditures. Constructing a new line also comes with challenges like finding demand points, determining the station location and finding the best corridor and alignment connecting them. Therefore, careful planning is necessary when choosing the best site for the facility.

2.2 Station Location

The potential use of high-speed rail depends on the station's location within a city and its positioning concerning the network [11]. HSR stations can be classified primarily into three types based on their position concerning the city, as shown in Fig. 1.

Central stations are located at the edge of the urban centre, are easily accessible and can efficiently serve pedestrian traffic. They attract new users and activities due to the well-established feeder modes providing easy station access. Stations located at the edge of the city are strategically positioned since the increased land availability attracts development. Such stations are easily accessible by foot and by private or public modes of transportation. Externals stations are located on the outskirts of the cities and can only be accessed by private cars without well-established public transportation infrastructure [12]. Therefore, improving the existing public transit system or establishing new ones is necessary to bridge this transportation gap.

HSR station location selection is a challenging real-world facility location problem (FLP). In FLPs, the strategic location of a facility is chosen from a set of potential locations based on specified objectives and (or) criteria to reduce transportation costs [13]. FLPs are multi-objective optimization problems typically solved using heuristic approaches due to their NP-hard nature [14]. The solutions obtained are a trade-off between the desired objectives. Optimization techniques for solving FLPs can be grouped into exact and approximation methods. Exact methods provide optimal solutions to the problem at hand, but the time required to solve the problem increases as the problem size increases. Approximation methods or heuristic methods are problem-dependent. They start with a (set of) solution(s) and try to identify a near-optimal solution for the problem [15]. These methods are typically used when

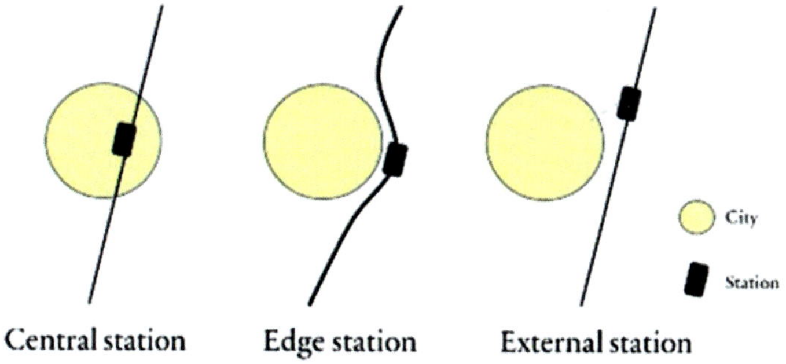

Fig. 1 Station position concerning the city [12]

approximate solutions are acceptable and exact methods are computationally expensive. Artificial intelligence (AI)-based techniques centred on the behaviours of living organisms, such as particle swarm optimization (PSO), genetic algorithm (GA) and ant colony optimization (ACO), are also used in solving FLP. The technique used depends on the type of problem to be solved.

Commonly used methods to solve FLP include the branch-and-bound method, mixed-integer programming, Lagrangian relaxation heuristic, tabu search and several nature-inspired approximation methods [15]. Previously, branch-and-bound method has been used to locate long-term care facilities [16]. GA, greedy algorithm combined with geographic information system (GIS), has been used in determining healthcare facilities' locations [17–19]. The location of bus stations and bus rapid transit stations have been decided using PSO, GA and ACO [20, 21]. Rail station location (and alignment) have been determined using mixed-integer programming [9, 10, 22], tabu search [8], PSO [23], ACO [24], GA [24–26], etc. The AI-based methods require less time to find the optimal solution and are widely used in optimization.

Station location selection is a complex facility location problem due to different stakeholder objectives that may be conflicting. Due to increased computational times, exact methods, as mentioned previously, may not be the best method for this problem if there are many feasible locations or if the study area is vast. Exact methods can be conveniently used for a problem with few feasible locations where it is easier to compare alternatives and select one alternative over another. Hence, heuristic methods may better handle large and complex station location selection problems where a near-optimal solution can be obtained within a reasonable time frame.

2.3 *Station Location Optimization*

The literature mentions two approaches in station location selection (Fig. 2) based on when the locations are optimized:

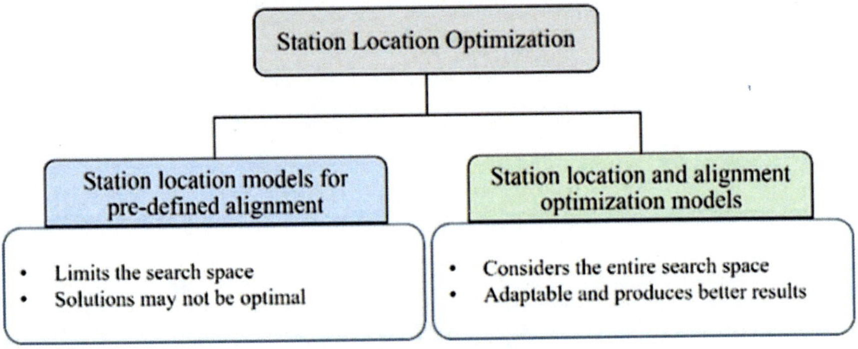

Fig. 2 Station location optimization approaches

- Station locations models for pre-defined alignment
- Station location and alignment optimization models.

In the first technique, station locations are optimized along an already-established alignment. This approach condenses the search space into a narrow strip of land along the alignment where the station can be placed. Studies have tackled this problem based on different objectives, such as maximizing ridership and minimizing cost [8–10, 27]. The drawback of this method is that the alignment was already chosen, which limits the method's effectiveness in finding the optimal station location. Also, by omitting the initial feasibility phase study for locating the stations, the problem was simplified to a great extent. Therefore, these models are limited by the practicality of the suggested approach for actual implementation since station location optimization is only possible when the railway routes are identified.

The second technique determines the station locations and the alignment that connects them. This strategy is more adaptable and produces better outcomes as it does not limit the study area. In the sequential method, the optimal number and station location are found, and the railway alignment connecting the stations is found in the next stage [22, 28, 29]. In concurrent or simultaneous optimization, the station locations are found while searching for the railway alignment [23, 25, 26, 30, 31]. A summary of the literature on station location optimization is presented in Fig. 3.

For a greenfield station facility, the problem can be solved as a combined station location and alignment optimization model, as it can easily incorporate various conflicting objectives while considering the entire study area rather than just a narrow strip. This method may be more intensive in terms of the usage of computational

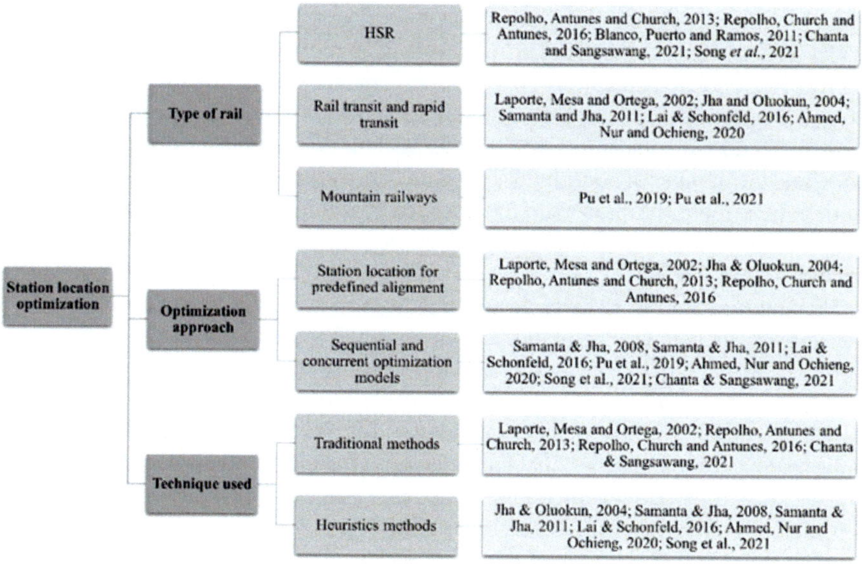

Fig. 3 Summary of literature on station location optimization

power. However, it is more comprehensive and exhaustive as it considers all the feasible locations in the study area rather than just a part.

2.4 Factors Influencing Station Location Selection

According to the literature, the key factors that should be addressed while selecting a station location are ridership potential, connectivity, costs, environmental feasibility, spacing constraints and socio-economic impact [25, 26, 32], as explained below.

Ridership. Ridership on public transportation denotes the number of people who use the transit service. The HSR system's success depends on the number of people it attracts; if the ridership is high, it will deliver higher benefits [3, 4]. HSR services are more prevalent on high-demand lines linking megacities. Ridership depends on the location of the stations. Ideally, terminal stations should be located close to the city centre since HSR is not a door-to-door service. If the stations are located on urban fringes, passengers have to travel long distances from the city, reducing ridership. Cities that are dense in population or employment are thus more suited for station location than those that are spread out, as the extra access time to the stations may negate the time savings that HSR provides [6, 33]. HSR stations can be located near intermediate-sized cities, potential employment districts, and activity centres like universities, hospitals and recreational areas, which are key trip generators [34]. Intermediate stations can be in suburban areas to give room for future development. Hence, the strategic positioning of stations is necessary for improving ridership.

Demand for HSR has three main components: (1) diverted demand resulting from commuters switching to HSR from other modes such as cars, airlines, or rail services, (2) induced demand occurring due to changes in commuters' trip frequency and lifestyle choice, and (3) economy-based demand growth due to changes in the economic system, either nationally or internationally [35]. Studies have also analyzed ridership by maximizing surrogate variables such as population density, job opportunities, and the number of households based on data availability [25, 32]. According to the literature, there are three main methods to incorporate ridership in the optimization problem formulation: (1) population/employment coverage models, (2) gravity model (3) logit model.

The station location is taken as a point in the study area in coverage models. The number of people or employment opportunities (denoting demand) within a threshold distance from the station location is considered to incorporate ridership. Points with high-demand coverage are regarded as potential station locations. Traditional and partial coverage models can be used for this purpose, as shown in Fig. 4. In the traditional coverage model, demand within the threshold distance will only be covered. In the partial coverage model, demand will be covered with a probability as its distance increases from the station point, as shown in Fig. 4. The disadvantage of this method is that it is quite a simplified method.

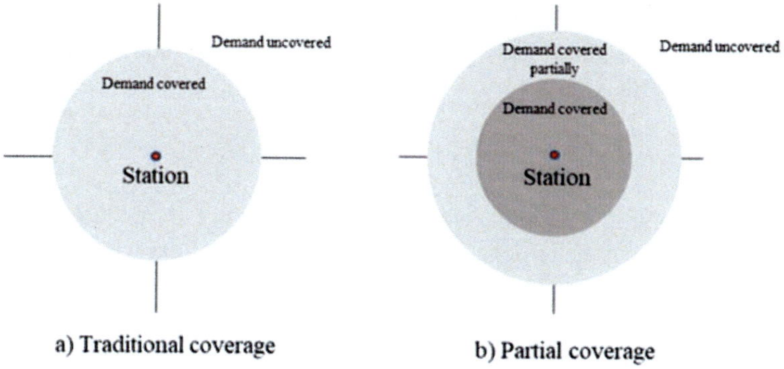

Fig. 4 Coverage types

In the gravity model, the passenger flow between two points i and j is expressed as a function of the attractiveness (population, employment etc.) of the points (A_i and A_j) and the travel impedance (T_{ij}) between them, as in Eq. 1.

$$Q_{ij} = \alpha \frac{A_i.A_j}{F(T_{ij})^{\beta}} \tag{1}$$

Logit models are a widely accepted method to forecast transit ridership. The commuter's preference for a mode depends on the utility of the mode (U_m), which depends on mode-specific parameters (X) as given in Eq. 2. Probability of choosing the mode (P_m) is given by Eq. 3.

$$U_m = \gamma_1 X_{1m} + \gamma_2 X_{2m} + \cdots + \gamma_n X_{nm} + \varepsilon \tag{2}$$

$$P_m = \frac{e^{(U_m)}}{\sum_m e^{(U_m)}} \tag{3}$$

These are some of the methods found in the literature to incorporate ridership into the station location optimization problem.

Connectivity. Transport connectivity shows the effectiveness of the network in transporting commuters from one location to another. It measures the degree of connection from one node to another in a network [36]. Connectivity stimulates economic activity and regional development by allowing seamless movement within and between cities through various travel modes [37]. Communities without HSR links can only benefit from its services if suitable access options are available to connect the HSR stations, thus increasing ridership [38]. Connectivity is measured based on the number of bus stops/train stations/airports present within a threshold distance from the station location. The threshold is represented based on Euclidian distance or network distance [25, 26, 32] using network data. If the study area is divided into cells (C_i) then each cell will be assigned a binary value (0 or 1) if the number of transport facilities (TF) within a threshold distance (D_{sp}) is more than the specified number of facilities (TF_{sp}) as given in Eq. 4.

$$C_i = 1, if\{(TF|D_{sp}) > TF_{sp}\}, otherwise\, C_i = 0 \tag{4}$$

Another technique is assigning values between 0 and 1 to locations within a study area, considering the closeness of transport facilities to the selected locations based on distance decay functions. Hence, based on the literature, connectivity to highways and other transport infrastructure is critical in ensuring adequate access to HSR facilities.

Costs. The cost components associated with various stakeholders that should be considered are: (1) location cost, (2) construction cost, (3) facility cost, (4) environmental cost and (5) life-cycle cost.

The location cost, also known as the land cost, is a major component of the station-related cost. Extensively developed areas or cities may have higher land prices, and the process of land acquisition by the government may become difficult [6, 30, 32]. The unit cost of station construction varies with its position, whether at surface level, underground or elevated section [30]. The station facility cost is a fixed cost for all fixed facilities, such as switches, signal systems, track elements and other devices, which are assumed constant [23, 26, 30].

Environmental cost includes the impact of noise and vibration. The operation of trains in an area may lead to excessive noise and vibration in the surroundings of the station area. Life-cycle cost includes all the costs, such as operation cost, maintenance cost and user cost per year, which are incurred throughout the facility's life, which may be about 60–100 years [39]. Operation and maintenance cost (operator's cost) refers to all costs associated with the operation and maintenance of facilities for the entire design life of the infrastructure. User cost depends on the access time and unit cost of access to the station, which should ideally be a minimum. The life cycle cost is brought to the present value using Eq. 5, where P is the present value, C is the lifecycle cost, i interest rate per annum and n is the number of years or the life of structure which may be taken to be 60–100 years for HSR.

$$P = C * \frac{(1+i)^n - 1}{i(1+i)^n} \tag{5}$$

It was observed that several studies considered costs while formulating the problem, but all components were not considered. For example, user and operator costs were not considered in certain studies [9, 10, 22, 30]. Since each cost component is attributed to stakeholders (owners, operators, and users) with different interests, all cost components must be considered in the problem formulation.

Environmental feasibility. The area's ecosystem and natural environment should not be adversely affected by the construction of the HSR station. Forests, wetlands, lakes, and rivers are examples of environmentally sensitive regions which should be avoided. Historical, religious, and culturally significant sites, such as monuments, places of worship, burial grounds, etc., should not be considered for HSR station development [1, 32]. Therefore, infeasible sites should be excluded, leaving just the feasible locations in the study area for investigation so that the integration of HSR stations into the city does not affect the community's sentiments and the quality of the environment. By overlaying a station cell layer over an environmentally sensitive region layer to create a feasible layer utilizing digital maps and GIS tools, it is possible to ensure the environmental feasibility of a given area.

Spacing constraint. Adequate spacing between nearby stations is required to serve passenger demand while successfully ensuring traffic safety. Inter-station spacing of at least 20 km is necessary to accommodate the high-speed train's acceleration and deceleration. Hence, the spacing is crucial for efficient HSR operation [10, 23]. This constraint is more severe in the case of urban rails, where nearby stations are closer together than in the case of intercity trains.

Socio-economic impacts. The introduction of HSR in a region is linked with changes in the region's economy, society, and environment. Studies on rail infrastructure acknowledged that HSR facilities impact accessibility [33, 38, 40], travel time, economic development (e.g., gross domestic product (GDP), gross value added, number of firms, foreign direct investment, and housing values) [38, 41–45], environment [46], and other aspects. The socio-economic impacts of HSR include its effect on the economy, labour market, tourism, housing, and the environment. Studies have not previously considered the socio-economic effects of locating HSR station locations. Based on the literature so far, the major socio-economic impacts of HSR can be divided into direct impacts (improvements in accessibility) and wider economic impacts (WEI), as shown in Fig. 5. [47].

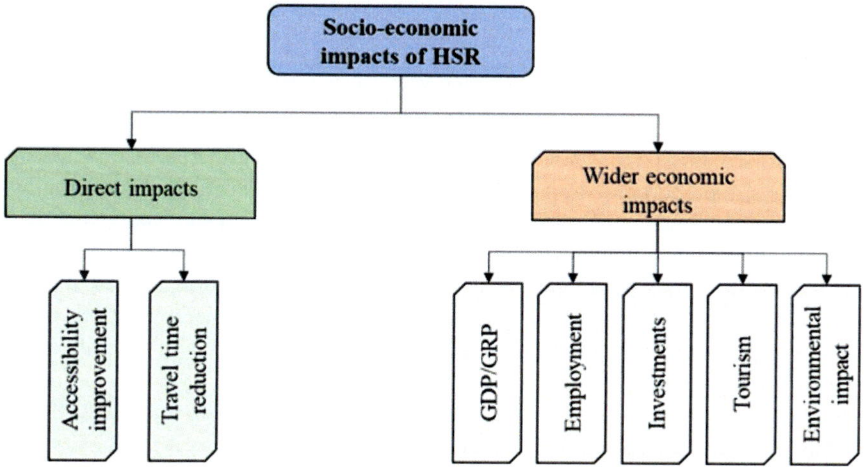

Fig. 5 Socio-economic impacts of HSR

The direct impact of HSR denotes the improvement in accessibility. Several studies have quantified the changes in accessibility after the development of HSR in a region [36, 38, 40, 48–50]. While comparing the accessibility changes of Chinese cities for the years 2006 and 2014 based on the 266 prefecture-level cities, it was observed that HSR contributed to 25–45% of the accessibility improvements [38]. The extent of accessibility enhancement varies based on the city's location concerning the HSR and the transportation modes in the selected locality. It was observed that accessibility impacts were more pronounced in the core cities than in the peripheral areas [40].

The WEI of HSR mainly reflects the increase in national GDP, gross regional product (GRP) of an area, and increased job opportunities and investments. Further, the improvement in GDP/GRP is also brought about as the result of reduced transportation costs and better accessibility. This attracts firms and workers from other areas to less congested areas near cities, thus making them grow. This is known as the spillover effect of HSR. On a similar note, HSR may also attract firms and workers away from other regions, leading to less development in these regions. This is known as the siphon effect [51]. Thus, the start of HSR in an area facilitates cross-regional labour mobility and expands the labour pool in the region. It also attracts more investments in fixed assets and business firms and creates more job opportunities due to improved accessibility [52]. But these effects are not consistent throughout all the HSR cities. The extent of development varies with the area analyzed. Core cities may have more benefits than intermediate-level and small cities. Some studies have reported a positive effect on economic productivity; however, others pointed out the adverse effects. Most studies report that HSRs positively affect the GDP per capita in core cities of Japan, China, Italy, and Germany, due to reduced rail time, employment growth, and foreign direct investments [38, 53–55].

In contrast, a few studies have reported a negative influence on economic development. In peripheral cities, due to the migration of people and the flow of financial investments to core cities, there is reduced use of locally available amenities and local transport, negatively affecting economic development. For example, the peripheral cities of the Yangtze River delta experienced the negative influence of HSR [56]. The existing literature primarily evaluated the socio-economic effects of HSR in ex-ante and ex-post scenarios. It has not been considered in station location selection. Maximizing the socio-economic impact can be beneficial in the micro-level positioning of the station, which would help ensure the facility's long-term success.

These are primary factors to consider in station location selection. The major drawback in the existing literature is the lack of studies considering all these factors in station location selection [8, 27]. Some of these factors and combinations were considered in previous literature, but not all (Table 1), which greatly simplified the problem.

Therefore, incorporating all these components can give better station locations. Existing studies do not explicitly evaluate the initial feasibility of stations which may be due to difficulty in the formulation and computational complexities [9, 10, 30]. Specific studies that considered initial feasibility, ridership and cost were in the context of the urban rail system [25, 26]. These studies may not be directly

Table 1 Critical factors in locating stations from literature (arranged chronologically)

Publications	Rail type	Factors					
		Ridership/Coverage	Connectivity	Environmental feasibility	Costs	Spacing	Socio-economic impacts
Vuchic [57]	RT	✓					
Laporte et al. [8]	RAPT	✓				✓	
Jha and Oluokun [27]	RT				✓		
Schöbel [58]	RT	✓					
Samanta and Jha [28]	RT				✓		
Blanco et al. [59]	HSR	✓			✓		
Laporte et al. [60]	RAPT	✓			✓		
Samanta and Jha [29]	RT	✓			✓		
Repolho et al. [10]	HSR				✓	✓	
Costa et al. [61]	HSR				✓		
Lai and Schonfeld [26]	RT	✓	✓	✓	✓	✓	
Repolho et al. [9]	HSR				✓	✓	
Roy and Maji [32]	HSR	✓	✓	✓	✓		

(continued)

Table 1 (continued)

Publications	Rail type	Factors					
		Ridership/Coverage	Connectivity	Environmental feasibility	Costs	Spacing	Socio-economic impacts
Pu et al. [30]	MR			✓	✓	✓	
Ahmed et al. [25]	RT	✓	✓	✓	✓	✓	
Chanta and Sangsawang [22]	HSR	✓			✓		
Song et al. [23]	HSR	✓		✓	✓	✓	
Pu et al. [62]	MR				✓	✓	

RAPT—Rapid transit; RT—Rail transit; HSR—High-speed rail; MR—Mountain railway applicable to HSR due to changes in the network density and associated variations in interstation spacing and geometric elements. With the advent of several AI-based advanced methods in optimization, we may be better equipped to incorporate all these factors into the problem formulation, thereby giving better solutions.

3 Conclusions

High-Speed Railway location optimization is a multifaceted process involving many complexities. Finding a good station location for HSR is critical to the facility's success. Significant factors affecting station location choice are the ridership it attracts, connectivity, the cost associated with the stakeholders, the environmental feasibility of the location, spacing constraints, and socio-economic impact. As observed from the literature review, many previous studies have focused on incorporating various combinations of these factors using different optimization techniques. For example, ridership and spacing constraints using the graph method [8], user cost and spacing constraints using mixed-integer optimization [9, 10], ridership and cost components using mixed-integer optimization, GA, and GIS [22, 29]. Some studies have failed to incorporate critical components such as owners' and operators' costs [8–10, 22], user cost [30], local feasibility study [8–10, 27] and connectivity to other transport modes [22, 23, 30] in formulating the problem.

As far as a greenfield facility is concerned, based on the literature review, it can be concluded that the locations can be established using the station location-alignment models, wherein station locations and alignment are found sequentially or simultaneously. Ideally, stations should be close to the city centre with ample access to other transportation facilities to ensure adequate ridership. The interests of owners, operators, users, and the community must be satisfied by considering all cost factors and environmental viability while locating the stations. The spacing constraint between stations is critical to ensure adequate safety for the trains. These major components were often considered in previous studies, but only in parts, not together, to simplify the problem. The solutions thus obtained might not be optimal.

There is little doubt that the HSR will provide socio-economic benefits while improving accessibility in the cities it serves. However, policymakers and planners need to carefully address the intricacies involved in developing the HSR infrastructure to enhance the overall benefits of the HSR. Therefore, future studies should incorporate all the necessary aspects of station location optimization in problem formulation to obtain the best possible solution.

References

1. Dobruszkes, F., Givoni, M.:Competition, integration, substitution: myths and realities concerning the relationship between high-speed rail and air transport in Europe. Transp. Sustain. **4**, 175–197 (2013)
2. Leboeuf, M.: High-speed rail: fast track to sustainable mobility. Int. Union Railways (2018)
3. Campos, J., de Rus, G.: Some stylized facts about high-speed rail: a review of HSR experiences around the world. Transp. Policy **16**(1), 19–28 (2009). https://doi.org/10.1016/j.tranpol.2009.02.008
4. Sun, X., Zhang, Y., Wandelt, S.: Air transport versus high-speed rail: an overview and research agenda. J. Adv. Transp. **2017** (2017). https://doi.org/10.1155/2017/8426926
5. Zheng, S., Kahn, M.E.: China's bullet trains facilitate market integration and mitigate the cost of megacity growth. Proc. Natl. Acad. Sci. USA **110**(14), 1248–1253 (2013). https://doi.org/10.1073/pnas.1209247110
6. Givoni, M.: Development and impact of the modern high-speed train: a review. Transp. Rev. **26**(5), 593–611 (2006). https://doi.org/10.1080/01441640600589319
7. Laporte, G.: Maximizing trip coverage in the location of a single. Ann. Oper. Res. 49–63 (2005)
8. Laporte, J., Mesa, A., Ortega, F.A.: Locating stations on rapid transit lines. Comput. Oper. Res. **29**(6), 741–759 (2002). https://doi.org/10.1016/S0305-0548(00)00013-7
9. Repolho, H.M., Church, R.L., Antunes, A.P.: Optimizing station location and fleet composition for a high-speed rail line. Transp. Res. Part E Logist. Transp. Rev. **93**, 437–452 (2016). https://doi.org/10.1016/j.tre.2016.06.006
10. Repolho, H.M., Antunes, A.P., Church, R.L.: Optimal location of railway stations: the Lisbon-Porto high-speed rail line. Transp. Sci. **47**(3), 330–343 (2013). https://doi.org/10.1287/trsc.1120.0425
11. Martínez, H.S., Moyano, A., Coronado, J.M., Garmendia, M.: Catchment areas of high-speed rail stations: a model based on spatial analysis using ridership surveys. Eur. J. Transp. Infrastruct. Res. **16**(2), 364–384 (2016). https://doi.org/10.18757/ejtir.2016.16.2.3143
12. Menéndez, J.M., et al.: New high-speed rail lines and small cities: locating the station. Sustain. City II **54**, 811–820 (2002). https://doi.org/10.2495/URS020771
13. Owen, S.H., Daskin, M.S.: Strategic facility location: a review. Eur. J. Oper. Res. **111**(3), 423–447 (1998). https://doi.org/10.1016/S0377-2217(98)00186-6
14. Xu, Y., Xu, D., Du, D., Zhang, D.: Approximation algorithm for squared metric facility location problem with nonuniform capacities. Discret. Appl. Math. **264**, 208–217 (2019). https://doi.org/10.1016/J.DAM.2019.03.013
15. Adeleke, O.J., Olukanni, D.O.: Facility location problems: models, techniques, and applications in waste management. Recycling **5**(2), 1–20 (2020). https://doi.org/10.3390/recycling5020010
16. Kim, D.G., Kim, Y.D.: A branch and bound algorithm for determining locations of long-term care facilities. Eur. J. Oper. Res. **206**(1), 168–177 (2010). https://doi.org/10.1016/J.EJOR.2010.02.001
17. Shariff, S.S.R., Moin, N.H., Omar, M.: Location allocation modeling for healthcare facility planning in Malaysia. Comput. Ind. Eng. **62**(4), 1000–1010 (2012). https://doi.org/10.1016/j.cie.2011.12.026
18. Zhang, W., Cao, K., Liu, S., Huang, B.: A multi-objective optimization approach for health-care facility location-allocation problems in highly developed cities such as Hong Kong. Comput. Environ. Urban Syst. **59**, 220–230 (2016). https://doi.org/10.1016/j.compenvurbsys.2016.07.001
19. Gu, T., Li, L., Li, D.: A two-stage spatial allocation model for elderly healthcare facilities in large-scale affordable housing communities: a case study in Nanjing City. Int. J. Equity Health **17**(1), 1–13 (2018). https://doi.org/10.1186/s12939-018-0898-6
20. Umair, M., ur Rehman, S., Sohail, A., Khattak, A.: Computation of optimal spacing and density of bus rapid transit stations using evolutionary algorithms. Arab. J. Sci. Eng. **46**(5), 5179–5202 (2021). https://doi.org/10.1007/S13369-020-05199-7/FIGURES/16

21. Li, Y.: A novel quantum ant colony algorithm for the choice for the site of the urban public transport station problem. In: Proceedings—2012 3rd International Conference on Digital Manufacturing and Automation ICDMA, pp. 116–120 (2012). https://doi.org/10.1109/ICDMA.2012.28
22. Chanta, S., Sangsawang, O.: Optimal railway station locations for high-speed trains based on partial coverage and passenger cost savings. Int. J. Rail Transp. **9**(1), 39–60 (2021). https://doi.org/10.1080/23248378.2020.1719372
23. Song, T., Pu, H., Schonfeld, P., Zhang, H., Li, W., Hu, J.: Simultaneous optimization of 3D alignments and station locations for dedicated high-speed railways. Comput. Civ. Infrastruct. Eng. 1–22 (2021). https://doi.org/10.1111/mice.12739
24. Samanta, S., Jha, M.K.: Applicability of genetic and ant algorithms in highway alignment and rail transit station location optimization. Int. J. Oper. Res. Inf. Syst. **3**(1), 13–36 (2012). https://econpapers.repec.org/RePEc:igg:joris0:v:3:y:2012:i:1:p:13-36. Accessed 31 Oct 2021
25. Ahmed, C., Nur, K., Ochieng, W.: GIS and genetic algorithm based integrated optimization for rail transit system planning. J. Rail Transp. Plan. Manage. **16**(xxxx) (2020). https://doi.org/10.1016/j.jrtpm.2020.100222
26. Lai, X., Schonfeld, P.: Concurrent optimization of rail transit alignments and station locations. Urban Rail Transit **2**(1), 1–15 (2016). https://doi.org/10.1007/s40864-016-0033-1
27. Jha, M.K., Oluokun, C.: Optimizing station locations along transit rail lines with geographic information systems and artificial intelligence. Adv. Transp. **15**, 11–20 (2004)
28. Samanta, S., Jha, M.K.: Identifying feasible locations or rail transit stations: two-stage analytical model. Transp. Res. Rec. **2063**, 81–88 (2008). https://doi.org/10.3141/2063-10
29. Samanta, S., Jha, M.K.: Modeling a rail transit alignment considering different objectives. Transp. Res. Part A Policy Pract. **45**(1), 31–45 (2011). https://doi.org/10.1016/J.TRA.2010.09.001
30. Pu, H., Zhang, H., Li, W., Xiong, J., Hu, J., Wang, J.: Concurrent optimization of mountain railway alignment and station locations using a distance transform algorithm. Comput. Ind. Eng. **127**(January 2018), 1297–1314 (2019). https://doi.org/10.1016/j.cie.2018.01.004
31. Roy, S.: Optimization of Multi-Objective Station Locations and Alignment Problem for High Speed Rail By. Indian Institute of Technology Bombay (2021)
32. Roy, S., Maji, A.: High-speed rail station location optimization using customized utility functions. IEEE Intell. Transp. Syst. Mag. 2–11 (2022). https://doi.org/10.1109/MITS.2022.3207411
33. Cao, J., Liu, X.C., Wang, Y., Li, Q.: Accessibility impacts of China's high-speed rail network. J. Transp. Geogr. **28**, 12–21 (2013). https://doi.org/10.1016/j.jtrangeo.2012.10.008
34. FRA: Stations location and access analysis technical memorandum (2015). https://docslib.org/doc/1265014/appendix-b-7-stations-location-and-access-analysis-tm
35. Cascetta, E., Coppola, P.: High Speed Rail (HSR) induced demand models. Proc. Social Behav. Sci. **111**, 147–156 (2014). https://doi.org/10.1016/j.sbspro.2014.01.047
36. (Ato) Xu, W., Zhou, J., Yang, L., Li, L.: The implications of high-speed rail for Chinese cities: connectivity and accessibility. Transp. Res. Part A Policy Pract. **116**(April), 308–326 (2018). https://doi.org/10.1016/j.tra.2018.06.023
37. Liu, S., Wan, Y., Zhang, A.: Does China's high-speed rail development lead to regional disparities? A network perspective. Transp. Res. Part A Policy Pract. **138**(October 2019), 299–321 (2020). https://doi.org/10.1016/j.tra.2020.06.010
38. Liu, L., Zhang, M.: High-speed rail impacts on travel times, accessibility, and economic productivity: a benchmarking analysis in city-cluster regions of China. J. Transp. Geogr. **73**(September), 25–40 (2018). https://doi.org/10.1016/j.jtrangeo.2018.09.013
39. Chang, B., Kendall, A.: Life cycle greenhouse gas assessment of infrastructure construction for California's high-speed rail system. Transp. Res. Part D Transp. Environ. **16**(6), 429–434 (2011). https://doi.org/10.1016/J.TRD.2011.04.004
40. Kim, H., Sultana, S.: The impacts of high-speed rail extensions on accessibility and spatial equity changes in South Korea from 2004 to 2018. J. Transp. Geogr. **45**(2015), 48–61 (2015). https://doi.org/10.1016/j.jtrangeo.2015.04.007

41. Murakami, J., Cervero, R.: California high-speed rail and economic development: station-area market profiles and public policy responses. Symposium. University of California, Berkeley, pp. 1–33 (2010). http://www.its.berkeley.edu/volvocenter/Cervero/California_High_Speed_Rail_and_Economic_Development_Paper_-murakami_cervero2.pdf
42. (Ato) Xu, W., Huang, Y.: The correlation between HSR construction and economic development—empirical study of Chinese cities. Transp. Res. Part A Policy Pract. **126**(September 2018), 24–36 (2019). https://doi.org/10.1016/j.tra.2019.05.017
43. Deng, T.: Impacts of transport infrastructure on productivity and economic growth: recent advances and research challenges. Transp. Rev. **33**(6), 686–699 (2013). https://doi.org/10.1080/01441647.2013.851745
44. Diao, M.: Does growth follow the rail? The potential impact of high-speed rail on the economic geography of China. Transp. Res. Part A Policy Pract. **113**(August 2017), 279–290 (2018). https://doi.org/10.1016/j.tra.2018.04.024
45. Chen, Z., Haynes, K.E.: Impact of high speed rail on housing values: an observation from the Beijing-Shanghai line. J. Transp. Geogr. **43**, 91–100 (2015). https://doi.org/10.1016/j.jtrangeo.2015.01.012
46. Givoni, M. : Environmental benefits from mode substitution: comparison of the environmental impact from aircraft and high-speed train operations. **1**(4), 209–230 (2007). https://doi.org/10.1080/15568310601060044
47. Cheng, J., Chen, Z.: Socioeconomic impact assessments of high-Speed rail: a meta-Analysis. Transp. Rev. **42**(4), 467–502 (2022). https://doi.org/10.1080/01441647.2021.1979689
48. Liu, Z., Zhu, A.X., Zhang, W., Ren, M.: An improved potential-based approach to measuring the daily accessibility of HSR. Transp. Res. Part A Policy Pract. **150**, 271–284 (2021). https://doi.org/10.1016/J.TRA.2021.06.009
49. Zhong, C., Bel, G., Warner, M.: High-Speed rail accessibility: what can California learn from Spain? Working Paper. Cornell University Department of City and Regional Planning (2012)
50. Zhu, Y., Diao, M., Fu, G.: The evolution of accessibility surface of China in the high-speed-rail era. Environ. Plan. A **48**(11), 2108–2111 (2016). https://doi.org/10.1177/0308518X15619935
51. Zhou, Z., Zhang, A.: High-speed rail and industrial developments: evidence from house prices and city-level GDP in China. Transp. Res. Part A Policy Pract. **149**, 98–113 (2021). https://doi.org/10.1016/J.TRA.2021.05.001
52. Shi, W., Lin, K.C., McLaughlin, H., Qi, G., Jin, M.: Spatial distribution of job opportunities in China: evidence from the opening of the high-speed rail. Transp. Res. Part A Policy Pract. **133**(January), 138–147 (2020). https://doi.org/10.1016/j.tra.2020.01.006
53. Cascetta, E., Cartenì, A., Henke, I., Pagliara, F.: Economic growth, transport accessibility and regional equity impacts of high-speed railways in Italy: ten years ex post evaluation and future perspectives. Transp. Res. Part A Policy Pract. **139**(June), 412–428 (2020). https://doi.org/10.1016/j.tra.2020.07.008
54. Wetwitoo, J., Kato, H.: High-speed rail and regional economic productivity through agglomeration and network externality: a case study of inter-regional transportation in Japan. Case Stud. Transp. Policy **5**(4), 549–559 (2017). https://doi.org/10.1016/j.cstp.2017.10.008
55. Yu, D., Zhang, Y., Wu, X., Li, D., Li, G.: The varying effects of accessing high-speed rail system on China's county development: a geographically weighted panel regression analysis. Land Use Policy **100**(July), 2021 (2020). https://doi.org/10.1016/j.landusepol.2020.104935
56. Gao, Y., Song, S., Sun, J., Zang, L.: Does high-speed rail connection really promote local economy? Evidence from China's Yangtze River Delta. Rev. Dev. Econ. **24**(1), 316–338 (2020). https://doi.org/10.1111/rode.12642
57. Vuchic, V.R.: Rapid transit interstation spacings for maximum number of passengers. **3**(3), 214–232 (1969). https://doi.org/10.1287/TRSC.3.3.214
58. Schöbel, A.: Locating stops along bus or railway lines-a bicriteria problem. Ann. Oper. Res. **136**(1), 211–227 (2005). https://doi.org/10.1007/s10479-005-2046-0
59. Blanco, V., Puerto, J., Ramos, A.B.: Expanding the Spanish high-speed railway network. Omega **39**(2), 138–150 (2011). https://doi.org/10.1016/J.OMEGA.2010.05.003

60. Laporte, G., Mesa, J.A., Ortega, F.A., Perea, F.: Planning rapid transit networks. Socioecon. Plann. Sci. **45**(3), 95–104 (2011). https://doi.org/10.1016/j.seps.2011.02.001
61. Costa, A.L.D., Cunha, M., Coelho, P.A.L.F., Einstein, H.H.: Decision support systems for real-world high-speed rail planning. J. Transp. Eng. **142**(5), 04016015 (2016). https://doi.org/10.1061/(ASCE)TE.1943-5436.0000837
62. Pu, H., et al.: Concurrent optimization of mountain railway alignment and station locations with a three-dimensional distance transform algorithm incorporating a perceptual search strategy. IEEE Access **9**, 34736–34754 (2021). https://doi.org/10.1109/ACCESS.2021.3061587

Exploring Aspects of Sustainable Rail Infrastructure Development Process and Alternative TOD Financing

Dheeraj Joshi, Shikha Saini, and Vivek Joshi

Abstract Transport infrastructure is a catalyst of economic growth and development as it creates better access to markets and ideas. The history of major developed geographies such as Japan, Western Europe and the United States shows that the period of construction of transport infrastructure like railroads harmonize with the period of rapid economic growth. Particularly in Japan where this economic growth further led to development of high-speed rails. Literatures have shown that high-speed rails have significant impact in the serving regions in terms of social, economic and environmental aspects, the three pillars of sustainability. India having more than a billion populations has a large and diverse, but energy inefficient transport network and thereby has a large potential to invest in energy efficient transport infrastructure. Sustainable transport systems are key drivers of sustainable development as identified in the Sustainable Development Goals of the United Nations. Ministry of Railways in India has identified various freight, high-speed corridors besides multitude of metro projects in the National Rail Plan 2021 as an effort towards modernization. This study focuses on the alternatives to rail infrastructure development processes to lead towards sustainable development with focus on spatial planning. The study utilizes temporal statistical and spatial remote sensing data towards comparing global rail infrastructure development with lessons for rail infrastructure development in India which will be useful for development of not only socio-economically viable but sustainable transport systems and further, a discussion is done on the alternative TOD financing approaches for development of rail infrastructure in India.

Keywords Sustainable development · Transit oriented development · Night-lights · NDVI · Railways

D. Joshi (✉) · S. Saini
Ministry of Railways, Delhi 110001, India
e-mail: dheeraj.joshi86@gov.in

V. Joshi
National Rail & Transportation Institute, Vadodara 390004, India

F. Pagliara (ed.), *Socioeconomic Impacts of High-Speed Rail Systems*,
Springer Proceedings in Business and Economics,
https://doi.org/10.1007/978-3-031-26340-8_9

1 Introduction

Rail infrastructure in India is the main transport mode for passenger and freight transportation and can easily be called the backbone of the economy in India. There exists huge potential to invest in energy efficient transport infrastructure in India and banking on energy-efficient technologies in National Rail Plan, Ministry of Railways has recommended new rail corridors for development after Mumbai-Ahmedabad High Speed Corridor. But such projects requiring high public expenditure should be developed in a sustainable manner besides propelling the economy of India.

Quality infrastructure is not only confined to infrastructure development, but it is an all-inclusive concept including factors as sustainability, safety and resilience, local job creation, and human resource development. The United Nations, Sustainable Development Goals also list "sustainable infrastructure projects" which includes efficient transportation services as one of its several objectives [1].

The need for infrastructure development with a focus on consideration of the nexus between the economic, social and environmental dimensions of sustainable development was reaffirmed at the 2012 UN Conference on Sustainable Development [2]. The argument for considering integration and nexus between the economic, social and environmental dimensions of sustainable development was reaffirmed at the 2012 UN Conference on Sustainable Development [3]. The word *nexus* as defined in the Oxford Dictionary stands, in general, for "a connection or series of connections linking two or more things" [4]. The core argument is that human development cannot be stable if seen as independent from its integration in the overall earth system. For instance, when governments only seek to enhance economic growth in order to reduce poverty levels, then drastic environmental impacts like climate change cannot be avoided [5, 6]. Therefore, sustainability implies that any development path that leads to an overall reduction of the stocks of natural capital (or, specially, to a decline below the minimum threshold) fails to be sustainable even if other forms of capital increase [7].

In this study, global rail infrastructures, like high-speed rail corridors, are compared on socio-economic-environmental development process on spatio-temporal scale to learn and move towards sustainable development of upcoming rail infrastructure and further, a discussion is done on the alternative TOD financing approaches for development of rail infrastructure in India.

2 Literature Review

2.1 HSR as an Emerging Alternative in Transport Services

Please Transport infrastructure is a catalyst of economic growth and development as it creates better access to markets and ideas. The history of major developed geographies such as Japan, Western Europe and the United States shows that the period of construction of transport infrastructure like railroads harmonize with the period of rapid economic growth [8]. As observed from Japan, the rapid economic growth created further travel demand and led to the birth of High-Speed Rail when Shinkansen or "bullet train" started operations with the Tokaido line in 1964 [9]. This transport mode has rapidly expanded and by 2025, there will be 43, 322 km of high-speed lines in 23 different countries [10].

High Speed Rail is a complex system involving interplay of various elements of infrastructure, rolling stock, energy and operations and cross-sectoral issues of finance, commerce, socio-economic, human factors and managerial components [11]. HSR as a technical concept defined vide directive 96/48/EC [12], is high speed infrastructure and rolling stock enabling minimum speed of 250 km/h on dedicated high-speed lines and 200 km/h on upgraded conventional lines including different models of HSR systems and technological specificities [10, 13, 14].

But the High-Speed Rail system is more than a technical concept and the anticipated impacts from HSR are omnibus [15]:

> High speed rail is a tool for political integration: linking territories, encouraging the modernisation of other transport modes, and improving accessibility to broader geographic areas. High speed increases the mobility of people and, as a metro network organises the city, high speed rail organises the territory. High speed performance invites people to move by a cleaner means of transport and improves quality of life.

2.2 HSR and Economy

Economic benefits of HSR can be significant in two ways, the first benefit accruing from time saving by travelers due to improved accessibility and if it leads to more people getting access to more productive jobs then it can have an economic productive impact and secondly through addition and changes in economic activity leading to enhanced economic productivity and market expansion but it is not always predictable and varies from project to project [16]. In the context of Japan [9] found that regions with HSR stations have higher economic productivity. In the context of China, [17] found the investment in HSR had positive effect on national economic growth. Regional economic productivity is also affected by agglomeration and diffusion effect [18]. Marginal effect of HSR on small and medium cities is much larger than that on large cities and HSR can stimulate economic growth in less agglomerated area located between major cities as a result of diffusion effect [19]. Assessment of impact of HSR on economy has been studied extensively due to the investment of large-scale public funds [17]. Economic indicator like GDP has some source of discrepancies as some informal industries are left out in such National and prefectural level statistics [18, 20]. The spatial pattern of economic development of the region is important for understanding the economic impacts and [21] suggested that satellite night datas are useful proxy for economic activity at spatio-temporal scales. First such method of mapping city lights using digital data from OLS was shown by Christopher et al. [22]. Such link between economic activity and night light data is established in the studies of [23] and [24]. Impact assessment of HSR on urban economy has been done within China using night light data of Chinese cities showing positive effect on urban expansion [19]. Impact of HSR on urban economy has been investigated using night light data of Chinese cities by Chunyang et al. [18] with high-speed rail construction exerted a greater influence on the economic development level of big cities and had no significant impact on the economic development level of small and medium-sized cities along the high-speed rails.

2.3 HSR and Environment

As there are no natural prices for environment, landscape and life, the economic analysis requires creation of artificial ones [25]. How much economic analysis is inappropriate for such elements is understood from the study of [26] wherein cost benefit analysis of smoking is done for both smoker and society. Therein he suggested smokers are assets to society by dying early and the society benefits by the costs saved on their pensions and social security. Even he discounted the studies on environmental tobacco smoke by stating that time lags are not recognized in these studies as people are not killed instantaneously. Due to such cost benefit analysis in favour of smoking, he concluded that cigarettes are fit case for subsidies rather than taxation. The study of [27] highlighted the shortcomings of cost–benefit analysis in environmental impact assessment of HSR infrastructure to landscape, natural habitat and emissions of harmful pollutants. HSR is being projected as an environmentally friendly alternative to other modes of transport as HSR fares better than road and air transportation systems in the CO_2 emissions (per 100 passenger-kilometers), energy efficiency (passenger-kilometers carried per unit of energy) and in the land use aspect it fares better than road transportation due to reduced impact on account of high transport capacity [15]. Akerman [28] in his Europabanan HSR emission study found out emission reduction in HSR and Freight Measure scenario in 2025/2030 against reference scenario (No HSR) in 2025/2030. Impact of HSR on economy and environment has been studied locally within China by computable general equilibrium approach by Zhenhua et al. [17] and the study suggested HSR as an environmentally friendly alternative to other modes of transport but the environmental impact of such an infrastructure needs assessment on account of large public investment and benefit involved.

2.4 HSR and Society

Population as identified vide Agenda 21 of the United Nations is a crucial social indicator affecting long term sustainability and has implications for elements related to education, infrastructure and employment [29]. Role of transportation systems like Railways plays important role in developing cities and population growth of people in cities though the transition pattern differs from city to city as was pointed by Seina Uchida et al. [30] in the spatial data study of socio-economic impact of Trans-Siberian Railways. HSR requires high volume of demand to compensate for high construction costs in urban areas, which is sensitive to population density. Cities with high population along the corridor improves benefit to people through lower the cost of travel as more users are sharing the fixed cost of capacity [31].

2.5 HSR and Transit Oriented Development

Transit oriented development as a concept is understood from the seminal study of Hansen wherein, he defined accessibility as the potential of opportunities for interaction with spatially distributed activities in a given location [32]. Transit oriented development is a land-use and transport planning to optimise transport services by concentrating urban development around transit stations [33]. Location of HSR station is strategic for the success of the system to take advantage of the reduced travel times by passengers including access and egress. HSR stations with intramodality have the potential for attracting retail shops, hotels, leisure activities, business activities grouping together, coincidental presence of energy producers and consumers and also serving as remote co-working office [11]. Thereby making area surrounding HSR transit stations suitable for transit-oriented development. Rastogi and Rao [34] have found the modal walking distance as 1250 m for commuters in Mumbai. Pornraht and Hisashi [35] suggested pedestrian radius as 500-1000 m walking distance to rail station and used radius of 1000 m of rail stations for their study. Arasan et al. [36] found the walking distance value to be 1.7 km (1.3 km for females) and 5.2 km for bicycling for deciding location of transport related facilities in urban areas in Indian cities context. Katie and Dorina [37] used buffer radius as 800 m based on TOD guide prepared by Queensland Government. Therefore, TOD buffer zone around HSR station is important for understanding the impact of HSR stations on HSR city.

2.6 A Literature Review Done for the Selection of Indicators to Be Used in This Study

Sustainable development means designing the right mix of economic, social and environmental designs for today and for tomorrow. As measurement of sustainable development help in evaluating the effects of specific actions and policies not only for current generation but also for future generations. To compare sustainability across different societies, development of such indicators is necessary which account for holistic sustainability evaluation in a coherent and consistent manner [38].

Over half the world's population is living in urban areas but the economic performance across regions and nations is impaired by the absence of data on urban Gross

Domestic Product (GDP). To overcome this scarcity of global urban information, night-time light data is used as a proxy for economic activity in urban areas [24] as economic indicator like GDP has some source of discrepancies in urban GDP calculations [18, 20]. Henderson et al. [21] suggested that satellite night datas are useful proxy for economic activity at spatio-temporal scales. Impact of HSR on urban economy has been investigated using night light data of Chinese cities by Chunyang et al. [18]. Therefore, night-light data is used as an economic indicator for this study.

The urbanization impact on vegetation cover is captured by NDVI trends to observe urban greening as pointed by Igor et al. [39] and will be of immense help in finding trend of disturbances in city cores and subsequently help to establish alternative strategies for minimizing environmental impact. NDVI as an environmental indicator is convenient for spatial land use change analysis. Environmental assessment of urban development suggests measurement of land-use changes with respect to NDVI as an environmental indicator to assess the greening trends for a city as a measure of environmental friendliness [39].

HSR requires high volume of demand to compensate for high construction costs in urban areas, which is sensitive to population and population growth rate as identified vide Agenda 21 of the United Nations is a crucial social indicator affecting long term sustainability and has implications for elements related to education, infrastructure and employment [29]. Therefore, the population growth rate will serve as an appropriate social indicator for this study.

3 Methodology and Data Sources

UIC has classified the HSR history as gradual progression from the "Rocket" locomotive by George Stephenson reaching 50 km/h in nineteenth century to Tokaido Shinkansen operating at 210 km/h in twentieth century. With twenty-first century commercial HSR speeds have seen a jump reaching speeds above 320 km/h [11], though HSR is still coherently defined with the definition given by 96/48/EC European directive [12] and a new dimension has been added with the huge development of HSR lines especially in Asia with HSR network length across the globe seeing quantum jump from the year 2000 onwards to almost reaching 50,000 km by 2020 [11, 40]. Therefore, the period from the year 2000 to the year 2020 can be seen as great leap forward by the HSR systems across the globe. Thereby, this study examines some global HSR systems developed after the year 2000 for analysis.

Data Sources. The data is derived for HSR corridors namely Madrid–Barcelona, Aomori–Hokuto, Osong–Mokpo, Beijing–Shanghai and for Mumbai-Ahmedabad route for understanding nexus of indicators of social, economic and environment trends and summarized in Table 1.

Night-time light data. The night light images are obtained from the time series data archive for the years 2000 onwards of nighttime lights available on Google Earth Engine Data Catalog derived from dataset provider American National Oceanic and Atmospheric Administration (NOAA) for DMSP OLS and VIIRS Nighttime

Table 1 Description of data used

Data	Data description	Year
DMSP-OLS NTL	Annual product—stable lights composite	2001–2013
NPP-VIIRS DNB	Monthly product—"vcmcfg"	2013–2019
LANDSAT-7	Landsat 7 ETM + sensor, SR product	2001–2019
Population	Census data	2001–2020
	CIESIN data	2000–2020

Day/Night Band Composites Version 1 [41]. Time series data archive from the year 1992 to 2014 of nighttime stable lights available on Google Earth Engine Data Catalog derived from dataset provider American National Oceanic and Atmospheric Administration (NOAA) for DMSP OLS: Nighttime Lights Time Series Version 4, Defense Meteorological Program Operational Linescan System has been utilised. Its spatial resolution is 30 arc seconds. Further the data archive from the year 2012 onwards of nighttime lights for VIIRS Nighttime Day/Night Band Composites Version 1 is also hosted on Google Earth Engine Data Catalog derived from the American National Oceanic and Atmospheric Administration (NOAA) has been utilised in this study. The spatial resolution is 15 arc seconds [41].

Population Data. Population counts and their spatial distribution for the selected cities were obtained from the Socioeconomic Data and Applications Center hosted by CIESIN at Columbia University [42]. The dataset used is UN WPP-Adjusted Population Count, v4.11 (2000, 2005, 2010, 2015, 2020), with relative spatial distribution being consistent with those of national censuses. Population data is assigned in GeoTiff file format with spatial resolution of 30 arc-seconds (~1 km at the equator).

Landsat-7 data. The Landsat-7 Enhanced Thematic Mapper Plus (ETM+) derived from United States Geological Survey Earth Resources Observation and Science Centre (EROS) is also available on Google Earth Engine. Landsat-7/ETM+ imagery is used to produce NDVI composites vide Eq. (1). The spatial resolution is 30 m [41].

$$NDVI = \frac{NIR - RED}{NIR + RED} \tag{1}$$

Population growth data. The Population growth data is derived from the official statistics (as available on the official census websites of each country) for the respective HSR corridor cities upto the latest census. The rate of population growth (herein used as PGR), r, between two time points, t_1 and t_2, is calculated as an exponential rate of growth [29] in Eq. (2):

$$\mathrm{r} = 100 \ln (\mathrm{P_2}/\mathrm{P_1})/(\mathrm{t_2} - \mathrm{t_1}) \tag{2}$$

Administrative boundaries. Database of Global Administrative Areas hosted on DIVA-GIS [43].

Coordinates of the HSR stations. The coordinates of the stations for linking above data is based on WGS_1984 Coordinate System hosted on Google Maps [44]. The coordinates of Mumbai-Ahmedabad HSR stations are converted from UTM coordinates to WGS84 coordinates.

Development of the indicators. The socio-economic-environmental indicators are then developed to understand the nexus of social, economic and environmental trends.

Economic indicator. The Night light images are used as proxy for economic activity for calculating the DN (Digital Number) values of the HSR cities administrative boundaries and TOD buffer zones around the HSR stations.

Environmental indicator. The NDVI values are used for spatial land use change analysis to assess the greening trends for the HSR region administrative boundaries and TOD buffer zones.

Social indicator. The population growth rate of the HSR cities is used to assess the social growth trends in the HSR regions.

Decomposition of the night-light (economic) and NDVI (green) composites into buffers for TOD impact assessment. Based on literature reviews done, TOD buffer zones are utilised for 500 m, 2.5 km, 5 km, 10 km and 25 km around HSR station besides administrative boundary of HSR regions for decomposing night-light and NDVI data and understanding the impact of HSR stations on HSR region.

The methodology adopted in this study for data research and analysis is shown in Fig. 1.

The data is derived for HSR corridors being developed after the year 2000 for developing socio-economic-environmental indicators in HSR cities at two levels:

(a) City level—Here in this step, the indicators are developed by taking HSR corridor end cities as two nodes and between nodes cities.
(b) TOD level—The indicators are developed for TOD buffer zones of 500 m, 2.5 km, 5 km, 10 km and 25 km around HSR station. Again, in this step, the indicators are developed for TOD buffer zones around HSR stations of end nodes cities and between nodes cities.

4 Results and Discussions

The socio-economic-environmental indicators developed by the consolidated global HSR framework are analysed by finding the threshold through medians for population growth rate data, temporal comparison for night-light data, NDVI threshold for green space, for the zones: HSR city and TOD buffer zones around the HSR stations.

Environmental trends. Quantitative identification and the development of urban green space is a critical step in planning sustainable urban development [45, 46]. Green spaces play critical role in preventing surface runoff [47] which pose risks in urban areas and affect well-being of the communities nearby too. The urban

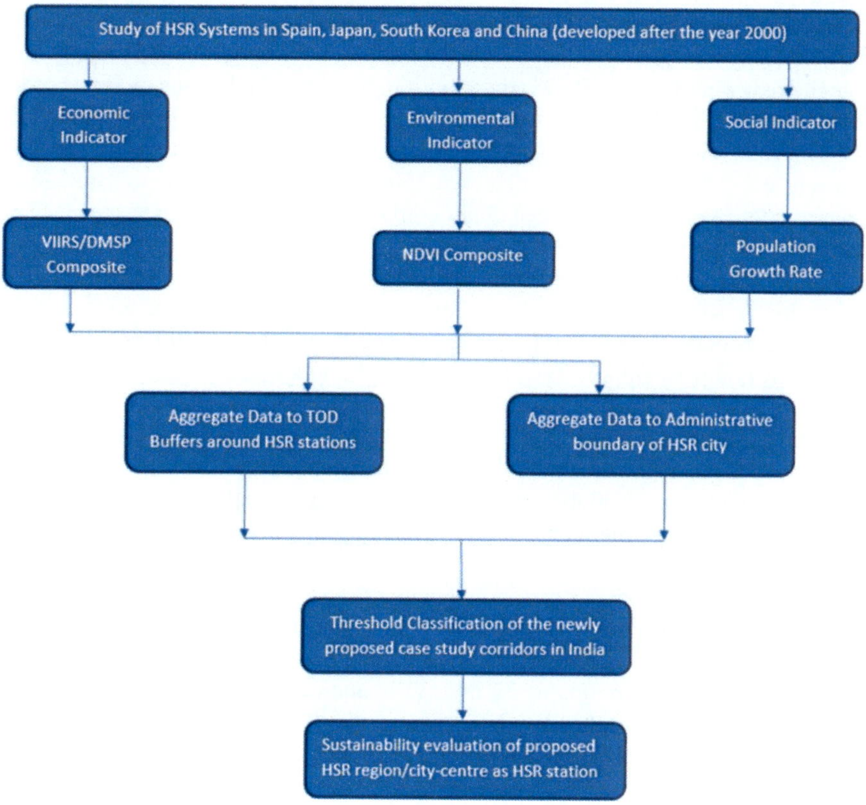

Fig. 1 The process flow showing the methodology in the study

green-spaces is thereby a necessity and the minimum requirement of green-spaces as laid out in various guidelines should be provided in the first step whether the area is affected by risks like flooding due to surface runoffs and preventing associated risks like urban heat island effect and air pollutions to the communities nearby. The requirement of minimum green-space availability of 0.5–1 hectare (ha) is supported by the [48] towards design of urban green spaces. The guidelines of Ministry of Urban Development in India [49] recommended a whole to part approach in spatial planning for delineation and preparation spatial development plan for each city such that 10–12 m^2 per person is desirable with minimum area requirement of 0.5 ha.

The NDVI threshold (used for assessment of green-spaces) is adopted under two criteria, first one for Mumbai-Ahmedabad HSR (MAHSR) corridor cities for *analysing availability of "high green-space*" with minimum 1-hectare area (as the MAHSR corridor is a green field project and is expected to have more green space) and *identification of at least "medium green-space*" to analyse the performance on conservation and improvement of such green-spaces around the global HSR corridor cities.

The NDVI values are assumed representative of vegetation fraction and the values of NDVI varies between −1 and +1. NDVI values in the range 0.5–0.75 are taken for high green quality and NDVI values above 0.75 as very high green quality as per [50]. The threshold of 0.5 was also assumed for green-space analysis in the study of [51]. The value of NDVI for threshold of minimum vegetation is assumed in various studies as 0.2 [52]. Further, Abutaleb et al. [50] used the classification of 0.25–0.5 as representative of moderate green quality representative of green open spaces like public parks. Therefore, in this section of study, NDVI value of 0.5–0.75 is assumed for high green-space, NDVI values above 0.75 as very high green-space and 0.25 is used as the threshold for classification of moderate green-space. The performance of the urban environmental indicator of green-space around global HSR framework are analysed by comparing the NDVI values with threshold of 0.25 for green space. For similarity purpose, the HSR cities of MAHSR corridor are also analysed on revised NDVI threshold criteria and compared with global HSR corridors. The NDVI values are obtained from the LANDSAT-7 satellite time-series data for the years 2000 onwards on ETM+ as available on Google Earth Engine Dataset Catalog derived from dataset provider American National Oceanic and Atmospheric Administration [41].

Though the resolution is 30 m and we may miss very small green areas, but we are assuming this will not significantly impact the results. Then twelve connected pixels counts are used to get minimum contiguous area of 1 ha as the 1 pixel of LANDSAT image has an area of 30 m * 30 m and 12 connected pixels represent contiguous area of minimum 1 ha. Therefore, in this study, green-space availability for MAHSR corridor is checked in terms of green-space per capita (minimum of 1 ha of green-space).

The analysis of MAHSR corridor shows the results for green-space availability in the Fig. 2.

The results for green-space availability (NDVI threshold > 0.5) for MAHSR corridor shows that only 6 HSR station cities out of 12 have the green-space availability above threshold in all the zones of consideration in this section of the study. The TOD zones of 1 km, 2.5 km and 5 km are critical as only 6 HSR cities, 9 HSR cities and 9 HSR cities have minimum green-space availability respectively.

The results for identification of green-space scenario (NDVI threshold of 0.25) as in Fig. 3, shows that green-space conservation in Madrid-Barcelona, Honam and Hokkaido HSR corridors is quite promising. For comparative purpose with global HSR corridors with NDVI threshold of 0.25, the green-space identification in upcoming Mumbai-Ahmedabad HSR (MAHSR) corridor was carried out in similar manner and 6 HSR cities were found to have green-space availability above the threshold of NDVI 0.25 value. Besides, the number of HSR cities have green-space (NDVI threshold 0.25 basis) falling temporally in TOD zone of 2.5 and 5 km and such HSR cities require green-space conservation strategies. Also, out of 12 proposed HSR cities in MAHSR corridor, there are cities having green-space below the threshold of NDVI of 0.25 in TOD zones of consideration especially 4 cities in TOD zone of 0.5 km, 5 cities in TOD zone of 2.5 km and 6 cities in TOD zone of 5 km and such HSR cities require green-space improvement strategies.

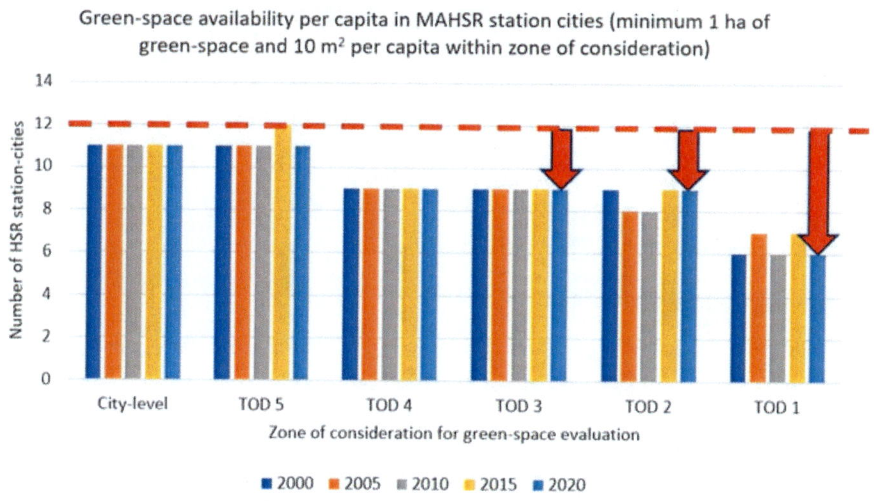

Fig. 2 Green-space availability status for HSR cities in MAHSR corridor

This holds relevance that such planned rail infrastructure needs comprehensive coordinated planning by urban and transport planners for improving and conserving the green-space around the rail infrastructure and beyond the rail infrastructure boundaries towards mitigating some of the urban risks like surface runoffs affecting rail infrastructure besides associated risks like urban heat island effect affecting the well-being of the communities around.

Economic trends. The results for economic growth trends vide Fig. 3 shows that in many Beijing-Shanghai corridor cities though the economic development is rising as compared to previous years in all the zones of consideration around the HSR cities but the environmental performance in most of the cities is falling below the NDVI threshold especially for critical zones of TOD levels of 0.5, 2.5 and 5 km around the cities analysed. The temporal trend of night-light data in Madrid–Barcelona, Honam and Hokkaido HSR also shows positive trend over previous years for most of the zones of consideration around the HSR cities and is reflective of economic development in such areas. Thereby, it shows that urban green-space conservation strategies can be consonant with the economic development of the area.

Social trends. In social growth trends for global corridors, it is observed that for big cities like Beijing and Shanghai, the population growth rate (PGR) was high when HSR opened but now PGR is showing decreasing trend but for other cities it is almost constant. For MAHSR corridor, the official statistics are utilised for the census values up to year 2011 for which the PGR values are computed and analysed. In the social indicator analysis, 6 out of 9 cities have PGR above threshold value and Mumbai city though having highest population is having lowest PGR. But the adjoining city of Thane has very high PGR showing population adjoining to big city as a result of "dispersion effect" as the night-light data is showing positive trends and NDVI values are below threshold value.

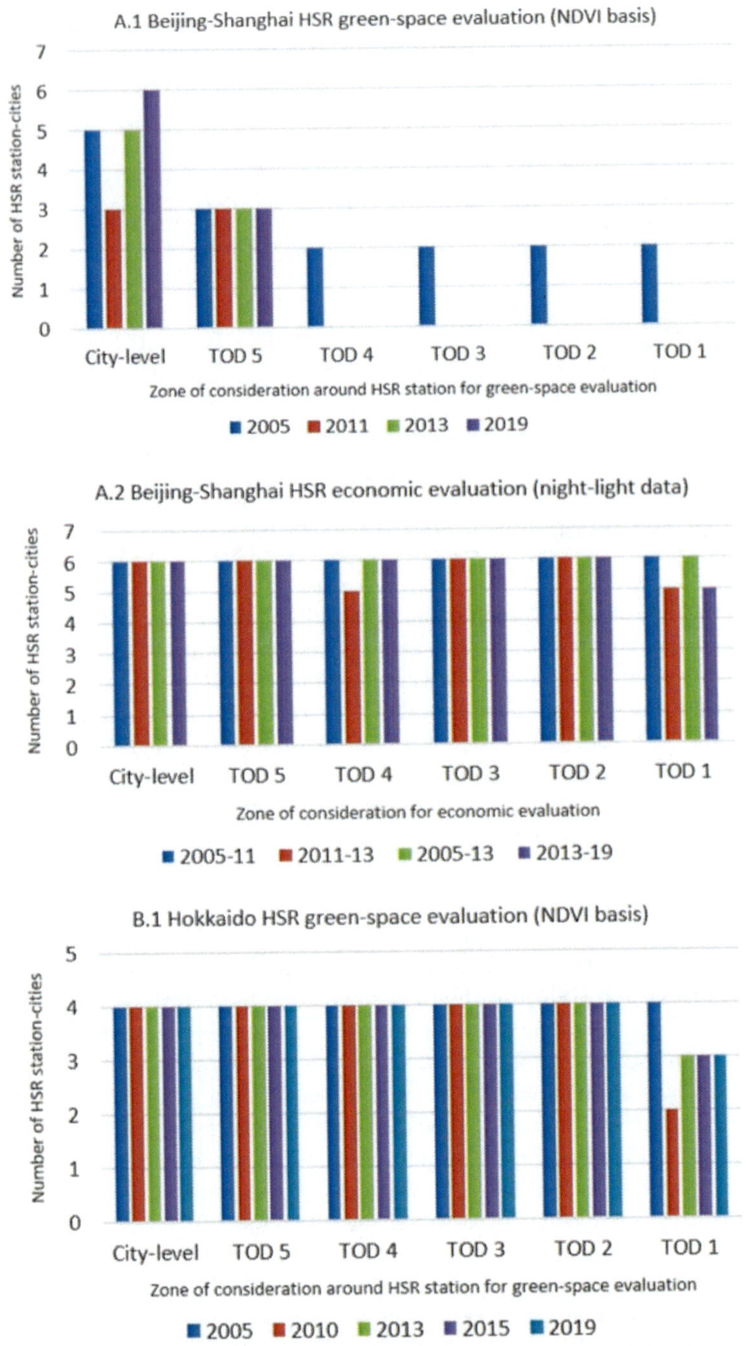

Fig. 3 **a** Green-space and economic evaluation (Beijing-Shanghai HSR corridor). **b** Green-space and economic evaluation (Hokkaido HSR corridor). **c** Green-space and economic evaluation (Honam HSR corridor). **d** Green-space and economic evaluation (Madrid-Barcelona HSR corridor). **e** Green-space and economic evaluation (Mumbai-Ahmedabad HSR corridor)

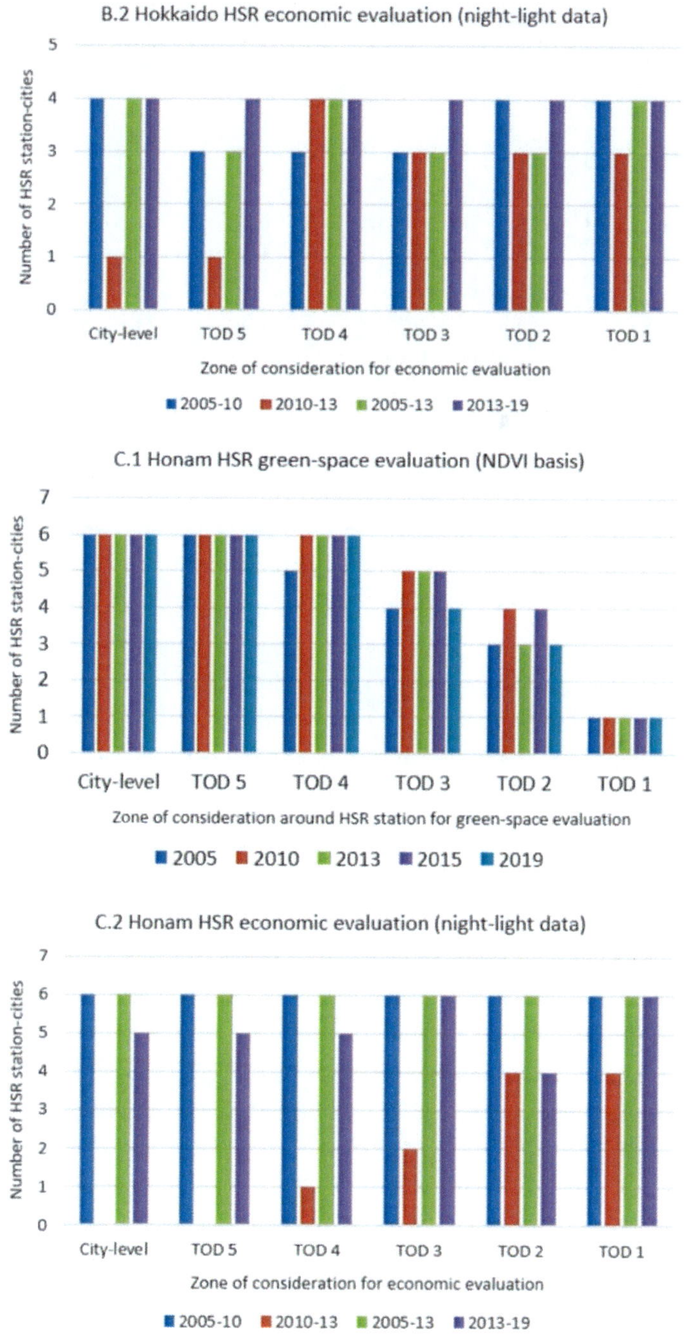

Fig. 3 (continued)

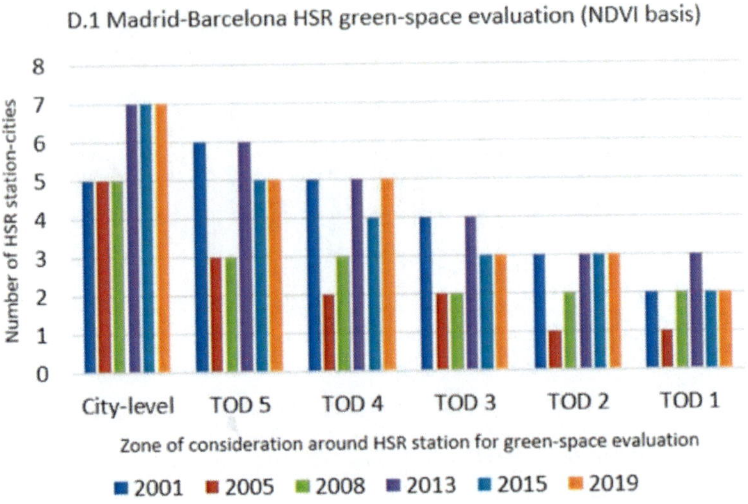

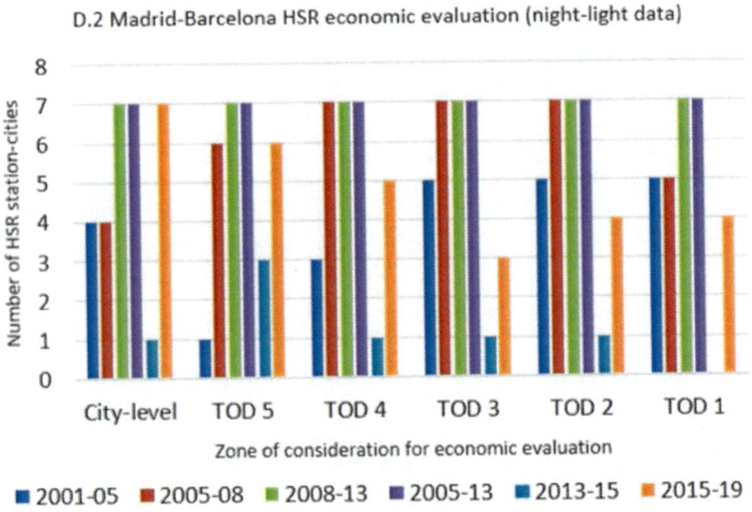

Region: Administrative boundary of the city
TOD 1: 500m buffer around station coordinates
TOD 2: 2.5Km buffer; TOD 3: 5Km buffer
TOD 4: 10Km buffer; TOD 5: 25Km buffer

Fig. 3 (continued)

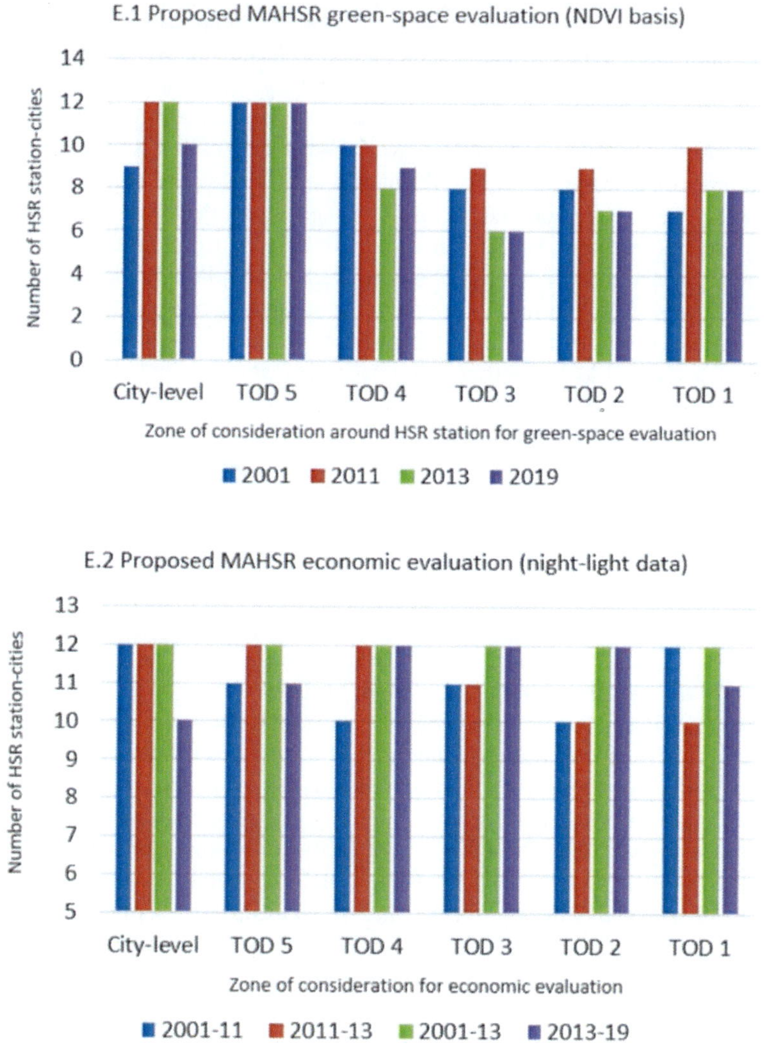

Fig. 3 (continued)

Alternative TOD financing. Post-analysing the rail infrastructure development process towards a sustainable transport system, the study explores the alternative financing model towards Value Capture Financing, which is based on the principle that private land and buildings benefit from public investments in infrastructure and policy decisions of the Government are very much relevant for TOD transit systems. The investment in transit system as well as increase in FAR and provision for mixed use development would result in increase in value of land within the influence zone. Land Value Capture can be used as a mechanism to finance the required upgradation of infrastructure and amenities within the influence zone and expansion of the public transport system. In TOD influence zones, land value capture can be done through enhanced or additional land value tax (like in Tamil Nadu and Maharashtra) or one time betterment levy (as in MMRDA Act 1974 and Hyderabad Municipal Corporation Act 1955 provides for such provisions), development charges or impact fee (in Gujarat, Maharashtra, Tamil Nadu, Madhya Pradesh and Andhra Pradesh), transfer of development rights (Maharashtra, Karnataka and Gujarat), or other such mechanisms which have been adopted in various states across the country and abroad. The economic potential of TOD zones can be understood from the study by NIUA [53] showing construction potential of around 0.28–1.06 billion of USD per station in metro-cities based on permissible FAR with estimated job generation of 6998 to 10,497 per station in the TOD zone of influence in India. Other global examples include the cases in Tokyo, wherein such FAR rights were sold to buildings in TOD zones around stations to finance redevelopment of Tokyo Railway Station including the building of Tokyo Railway Station.

5 Conclusion

To enable sustainable development process for rail infrastructure development in line with SDG goals it is critical to learn from the negatives of negative spill overs of economy as seen from the adverse environmental impact in Beijing-Shanghai HSR corridor. Thus, through this novel integrated approach, cities and transit zones around HSR or even metro station locations can be identified where socio-economic-environmental aspects especially environmental indicator is critical especially open and green spaces around HSR stations (WHO guidelines stipulate 9 m^2 per capita [48] and Ministry of Urban Development [49] calls for making conscious efforts for adhering to the WHO norms). The approach of environmental assessment is checked for some of the global HSR corridors to understand the availability of greens-spaces in different zones of consideration around HSR cities in a simplified manner by just checking the NDVI values of a region above a threshold of 0.25. The results show that even though in Beijing-Shanghai corridor cities the economic development is rising as compared to previous years in all the zones of consideration around the HSR cities but the environmental performance in most of the cities is falling below the NDVI threshold of 0.25, especially for critical zones of TOD levels of 0.5, 2.5, and 5 km around HSR stations. But the results of other global corridors like Hokkaido HSR, Honam HSR and Madrid-Barcelona HSR show promising trend of green-space conservation in consonance with economic development in HSR cities. The results of MAHSR cities shows green-space availability only 6 HSR cities in all the zones of consideration. The green-space conservation and improvement strategies are called for in the critical TOD zones of 1, 2.5, and 5 km around the HSR stations. As the strategies for conservation and improvement of green-spaces extend beyond the boundaries of the railway infrastructure around the HSR stations, this requires coordinated planning of urban and transport planners to work towards improving the natural green space to learn from the green-space conservation strategies from other global urban scenarios. The strategy can provide twin benefits through improving resiliency against floods and simultaneously improving the well-being of the communities nearby through such natural ecosystems.

Further, this study explores alternative financing models like development charges or betterment levy which can be levied on users and is apt for TOD zones. It is important to understand that implementation of TOD involve preparation of master plans such that suitable interventions can be planned through coordinated planning by urban and transport planners for integrated sustainable transport systems development and designing right mix of alternative financing approaches.

Acknowledgements The authors are grateful to the Ministry of Railways, Government of India for supporting us in this endeavor.

References

1. Purendra J.: Japan's foreign aid and 'Quality' infrastructure projects: the case of the bullet train in India". JICA-RI. No. 184 (2019)
2. UN Report on Sustainable Development: https://www.un.org/ga/search/view_doc.asp?symbol=A/CONF.216/16&Lang=E (2012). Accessed 29 June 2020
3. RIO+20, United Nations Conference on Sustainable Development.: The future we want, p. 1 (2012)
4. Oxford Dictionaries.: http://www.oxforddictionaries.com/definition/english/nexus. Accessed July 2020
5. Ingrid, et al.: Cross-sectoral strategies in global sustainability governance: towards a nexus approach. Int. Environ. Agreements **16**, 449–464 (2016). https://doi.org/10.1007/s10784-016-9321-1
6. Young et al.: Earth system challenges and a multi-layered approach for the sustainable development goals. In: Post2015/UNU-IAS Policy Brief No. 1, Tokyo: UNU-IAS (2014)
7. Gallopin.: A systems approach to sustainability and sustainable development. ECLAC/ Government of the Netherlands Project NET/00/063 Sustainability Assessment in Latin America and the Caribbean, United Nations Publication (2003)
8. Banerjee, A.V., Duflo, E., Qian, N.: On the road: access to transportation infrastructure and economic growth in China. Social Science Electronic Publishing 11(1), 1–53 (2012)
9. Jetpan, W., Hironori, K.: High-speed rail and regional economic productivity through agglomeration and network externality: a case study of inter-regional transportation in Japan. Case Stud. Transp. Policy (2017). https://doi.org/10.1016/j.cstp.2017.10.008
10. Maddi, G., Cecilia, R., José, M.U.: High speed rail: implication for cities. Cities (2012). https://doi.org/10.1016/j.cities.2012.06.005
11. UIC.: High speed rail. UIC High Speed Department. https://uic.org/IMG/pdf/uic_high_speed_2018_ph08_web.pdf (2018). Accessed 31 Jan 2020
12. Council of the European Union.: Directive 96/48/EC of 23 July 1996 on the interoperability of the trans-European high-speed rail system. OJ L 235, 17.9.1996, p. 6–24 (1996)
13. Campos, J., De Rus, G.: Some stylized facts about high-speed rail: a review of HSR experiences around the world. Transp. Policy **16**, 19–28 (2009)
14. Moshe, G.: Development and impact of the modern high-speed train: a review. Transp. Rev. **26**(5), 593–611 (2006). https://doi.org/10.1080/01441640600589319

15. UIC.: High speed rail, UIC High Speed Department. https://uic.org/IMG/pdf/uic_high_speed_2018_ph08_web.pdf (2012). Accessed 31 Jan 2020
16. Rus, D. (ed.): Economic analysis of high-speed rail in Europe, Revised Version. Fundacion BBVA, Bilbao (2012)
17. Zhenhua, C., Junbo, X., Adam, Z.R., Kingsley, E.H.: The impact of high-speed rail investment on economic and environmental change in China: a dynamic CGE analysis. Transp. Res. Part A (2016)
18. Chunyang, W., Weidong, M., Xinshuo, H.: The impact of high-speed rails on urban economy: an investigation using night lighting data of Chinese cities. Res. Transp. Econ. (2020). https://doi.org/10.1016/j.retrec.2020.100819
19. Fenjie, L., Longfei, Z., Zhida, S.: High-speed rail and urban expansion: an empirical study using a time series of nighttime light satellite data in China. J. Transp. Geogr. (2018). https://doi.org/10.1016/j.jtrangeo.2018.08.011
20. Misra, P., Takeuchi, W.: Analysis of air quality and nighttime light for Indian urban regions. IOP Conf. Ser. Earth Environ. Sci. **37**, 012077 (2016)
21. Henderson, J.V., Adam, S., David, N.W.: Measuring economic growth from outer space. Am. Econ. Rev. (2012)
22. Christopher, D.E., et al.: Mapping city lights with nighttime data from the DMSP operational linescan system. Photogram. Remote Sens. **63**(6), 727–734 (1997)
23. Katayama, N., Takeuchi, W.: Comparison between nighttime light and socioeconomic indicators on an international scale using VIIRS day-night band. In: 35th Asian Conference on Remote Sensing 2014, ACRS 2014: Sensing for Reintegration of Societies (2014)
24. Charlotta, M., José, L., Kevin, S., Zara, M.: Night-Time light data: a good proxy measure for economic activity?. PLOS One (2015). https://doi.org/10.1371/journal.pone.0139779
25. Ackerman, F., Heinzerling, L.: Pricing the Priceless. Cost–Benefit Analysis of Environmental Protection. Georgetown Environmental Law and Policy Institute, Georgetown University Law Center. https://www.globalpolicy.org/socecon/encronmt/2002/CostBen2002.htm. (2002)
26. Viscusi, K.W.: Cigarette Taxation and The Social Consequence of Smoking. National Bureau of Economic Research (1994)
27. van Wee, B., et al.: Environmental impacts of high-speed rail links in cost–benefit analyses: a case study of the Dutch Zuider Zee line. Transp. Res. Part D (2003). https://doi.org/10.1016/S1361-9209(03)00017-8
28. Akerman, J.: "The role of high-speed rail in mitigating climate change – The Swedish case Europabanan from a life cycle perspective". Transportation Research Part D (2011). https://doi.org/10.1016/j.trd.2010.12.004
29. United Nations, Department of Economic and Social Affairs (DESA). https://www.un.org/esa/sustdev/natlinfo/indicators/methodology_sheets/demographics/population_growth_rate.pdf. Accessed 02 Feb 2020
30. Uchida, S. et al.: Socio-economic impact of Trans-Siberian railway after the collapse of Soviet Union by integrated spatial data analysis. IOP Conf. Ser. Earth Environ. Sci. **37**, 012057 (2016). https://doi.org/10.1088/1755-1315/37/1/012057
31. Gines et al.: Is investment in high-speed rail socially profitable?. J. Transp. Econ. Policy **41**(1), 3–23 (2007)
32. Hansen, W.: How accessibility shapes land use. J. Am. Inst. Plan. (1959). https://doi.org/10.1080/01944365908978307
33. Anna, I., Gonçalo, H., Cecília, S., Antonio, P.A.: Transit-oriented development: a review of research achievements and challenges. Transp. Res. Part A (2019). https://doi.org/10.1016/j.tra.2019.10.018
34. Rastogi and Rao: Travel characteristics of commuters accessing transit: case study. J. Transp. Eng. (2003). https://doi.org/10.1061/(ASCE)0733-947X(2003)129:6(684)
35. Pornraht, P., Hisashi, K.: Factors affecting residents living near transit stations to use non-motorized access mode: case study about daily travel in Bangkok, Thailand. J. East. Asia Soc. Transp. Stud. **12** (2017). https://doi.org/10.11175/easts.12.854

36. Arasan, V.T., Rengaraju, V.R., Rao, K.V.K.: Characteristics of trips by foot and bicycle modes in Indian city. J. Transp. Eng. (1994). https://doi.org/10.1061/(ASCE)0733-947X(1994)120:2(283)
37. Katie, Y., Dorina, P.: A decade of transit oriented development policies in Brisbane, Australia: development and land-use impacts. Urban Policy Res. ISSN 0811-1146 (Print) 1476–7244 (2017). https://doi.org/10.1080/08111146.2017.1294537
38. Tracey, et al.: Sustainable development—linking economy, society and environment. OECD Insights (2008). https://doi.org/10.1787/9789264055742-en
39. Igor et al.: Trends in normalized difference vegetation index (NDVI) associated with urban development in northern West Siberia. Atmos. Chem. Phys. (2016). https://doi.org/10.5194/acp-16-9563-2016
40. EEIS, Environment and Energy Study Institute.: Fact sheet high speed rail worldwide. https://www.eesi.org/files/FactSheet_High_Speed_Rail_Worldwide.pdf (2018). Accessed 31 Jan 2020
41. Google Earth Engine Dataset Catalog. https://developers.google.com/earth-engine/datasets/catalog. Accessed 07 July 2021
42. Center for International Earth Science Information Network—CIESIN—Columbia University.: Gridded population of the world, Version 4 (GPWv4): population count adjusted to Match 2015 revision of UN WPP country totals, revision 11. NASA Socioeconomic Data and Applications Center (SEDAC), Palisades, NY (2018). Accessed 18 Feb 2021
43. DIVA-GIS website: https://www.diva-gis.org/gdata. Accessed 16 Feb 2020
44. Coordinates from Google Maps: https://www.google.com/maps. Accessed 25 July 2022
45. Beiranvand, A., Eslambonyad, A., Soosani, J., Beiranvand, A.: Evaluation of Khorramabad' s physical changes and its green space using remote sensing data 3(3), 431–438 (2013)
46. Shekhar, S., Shirish, K.: Assessing the quality of urban environment through urban neighbourhood green index. In: ISPRS TC VIII International Symposium on Operational Remote Sensing Applications: Opportunities, progress and challenges, Hyderabad, India (2014)
47. Ramaiah, M., Avtar, R.: Urban green spaces and their need in cities of rapidly urbanizing India: a review. Urban Sci. 3(3), 94 (2019)
48. WHO: Urban green spaces: a brief for action. Regional office for Europe, 24 (2017)
49. Urban Greening Guidelines, Ministry of Urban Development, Government of India (2014). https://mohua.gov.in/upload/uploadfiles/files/G%20G%202014(2).pdf
50. Abutaleb, K., Freddy, M.M., Nkongolo, N., Newet, S.W.: Estimating urban greenness index using remote sensing data: a case study of an affluent vs poor suburbs in the city of Johannesburg. Egypt. J. Remote Sens. Space Sci. (2020)
51. Giuliani, G., Petri, E., Interwies, E., Vysna, V., Guigoz, Y., Ray, N., Dickie, I.: Modelling accessibility to urban green areas using open earth observations data: a novel approach to support the urban SDG in four European cities. Remote Sens. 13(3) (2021)
52. Wong, M.M.F., Fung, J.C.H., Yeung, P.P.S.: High-resolution calculation of the urban vegetation fraction in the Pearl River Delta from the Sentinel-2 NDVI for urban climate model parameterization. Geosci. Lett. 6(1) (2019)
53. NIUA: Your guide to transit oriented development (2016). https://niua.org/tod/files/guidetod.pdf

Site Selection of High-Speed Railway Stations on Travel Efficiency: An Example of Shanghai Hongqiao Hub

Haixiao Pan, Xinyi Wang, Ya Gao, Song Ye, and Minglei Chen

Abstract High-speed rail (HSR) construction is rapidly advancing in China to improve rail competitiveness in the passenger market and facilitate rapid contact between different areas of China. In order to ensure the progress of HSR construction in a relatively short time-frame, and to reduce the property right-of-way cost related with removal of housing or industry, most HSR stations have been placed in suburban areas, far away from city centers. However, the city center remains the main origin and ultimate destination of most HSR passengers. The practice of establishing large-scale HSR stations in the suburbs is not conducive to travel efficiency even up to 500 km, especially for passengers from city centers where there is great potential for people to travel by HSR for business. It is suggested that instead of constructing one costly, huge, high-speed station in the suburban areas of mega cities, building multiple stations close to city centers will greatly reduce the connecting time to high-speed train stations, and hence greatly improve the travel efficiency of high-speed rail.

Keywords High-speed train · Shanghai · Travel efficiency · Station site selection

H. Pan · X. Wang · Y. Gao (✉)
Department of Urban Planning, Tongji University, Shanghai 200092, China
e-mail: gaoya@tongji.edu.cn

H. Pan
e-mail: hxpank@online.sh.cn

S. Ye
China Academy of Urban Planning and Design, Beijing 100044, China

M. Chen
Shanghai Weizhu Construction Design Co., Ltd, Shanghai 200010, China

F. Pagliara (ed.), *Socioeconomic Impacts of High-Speed Rail Systems*,
Springer Proceedings in Business and Economics,
https://doi.org/10.1007/978-3-031-26340-8_10

1 Introduction

China is a large country, and with its recent rapid economic development, there is a large demand for land transport between different regions. Rail transportation plays a very important role in this development. Since 1990s, the proportion of rail transit in the inter-city passenger market has declined because of the development of highway and aviation. But there is growing recognition that excessive reliance on highway transport would cause a number of environmental, energy, and safety problems. Weather factors like fog and road congestion can reduce the efficiency and reliability of highway transport as well [1]. Therefore, how to provide fast and reliable inter-regional rail service has become an important issue for government.

This chapter reviews the development of different modes of intercity transport in China since 1978. It is shown that the use of rail, measured in rail passenger-kilometers, has grown far less than the other modes, especially the highway mode. To provide a better rail option, China began experimenting with high-speed rail in the late 1990s. Running speed of HSR has increased from its initial 200 km/h to its current maximum 350 km/h.

The Chinese government has planned four "vertical" (north–south) and four "horizontal" (east–west) passenger corridors for the country and upon their completion, the national HSR system will be over 16,000 km [2]. Meanwhile, coastal regions like the Yangtze River Delta and the Pearl River Delta are developing regional high-speed rail networks.

Many researchers believe that HSR can bring economic, environmental, and social benefits to the regions and cities it serves [3, 4]. Researchers have also asserted that areas situated outside the HSR network but efficiently linked up to it could benefit from the diffuse effects of major urban agglomerations [5]. US researchers have argued that in addition, HSR can enable big cities to connect further into the hinterland where housing and commercial space is more affordable [6]. However, capturing these benefits requires attention to station location and design [7], because the improvement of regional accessibility due to the arrival of HSR stations plays an essential role in the urbanization process and in the accessibility that results.

Researchers have observed that actions to improve HSR station accessibility, whether via private car or by public transit, not only affects the access to stations, but is a key factor in efforts to change the market shares of the competing modes in the intercity corridor [8]. Station location and accessibility plans also can make a big difference in passengers' choice of modes for station access.

Chen and Peter found that China's HSR services have had substantial and demonstrable effects in aiding the economic transition of cities that are now within a 2-h travel from the country's major urban regions, helping to generate renewed economic growth [9]. The important point here is the time factor. In this regard both in-vehicle and access times must be considered and if the access time is too long or is unreliable, it will affect negatively on door-to-door travel efficiency.

However, in order to accelerate the construction of China's high-speed rail network and reduce the difficulty of land acquisition for HSR construction, many new station sites in China are located in suburbs distant from large urban centers. This means that access to the stations can require a long trip from the city center where many activities are located. This can affect not only the mode used to get to the station but the decision on whether to use HSR or a different mode.

In the study presented here, the authors investigated and analyzed HSR passenger travel behavior, using the Shanghai Hongqiao Station as a case example. The research explores the origin points of HSR passengers, their mode of travel access to the Hongqiao HSR Station, door-to-door transit time for HSR travel, and changes in access time before and after the opening of the Hongqiao HSR Station. The results show that HSR stations located distant from the city center require long access times for the majority of passengers, even when significant investments in new rail and highway links to the HSR station have been made. These long access trips reduce overall transport efficiency and partly offset the fast speeds of the HSR.

2 High-speed Rail Network Planning and Construction in China

In China the construction of high-speed rail started at the beginning of the twenty-first century. The Qinhuangdao–Shenyang line, with a top speed of 200 km/h, started service in October 2003 and has been the cornerstone of China's HSR age. In order to speed up rail development, the state council executive meeting discussed and passed the "Mid-term and Long-term Railway Network Plan" in January 2004. The State Council again adjusted the HSR planning program in 2008, expanding the national railway program to produce over 120,000 km of national railway by 2020, with over 16,000 km of HSR. At the same time, they decided that inter-city HSR should cover the main cities in economically developed and densely populated areas, such as the Yangtze River Delta, the Guangdong-Hong Kong-Macao Greater Bay Area (hereafter the Greater Bar Area) and the Shandong Peninsula (see Fig. 1). In July 2016, the existing "Mid-term and Long-term Railway Network Plan (2016–2030)" was published by National Development and Reform Commission (NDRC), Ministry of Transport, China State Railway Group Co., LTD jointly [10]. The plan proposed a high-speed railway network of eight vertical and eight horizontal lines, which is used as the main skeleton, regional connecting lines as connections, and intercity railways as supplements. In September 2019, the Central Committee of the Communist Party of China and the State Council issued the Out-line of China to enhance global competitiveness in transport, proposing the construction of an urban agglomeration transportation network with smooth borders, and the integrated development of "four railway networks" of arterial railway, intercity railway, suburban railway, and urban rail transit lines [11]. By December 2021, China's HSR development had reached 41.583 km (see Fig. 2).

The Yangtze River Delta is one of the most developed areas in China with 20% of the national GDP output, an area of 109,600 km^2, and a high urbanization rate. In 2005

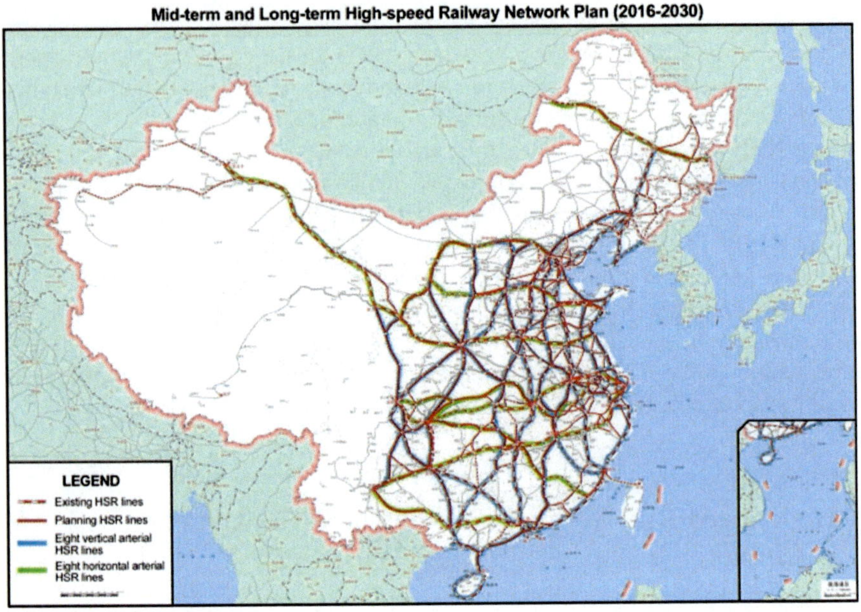

Fig. 1 China HSR network plan

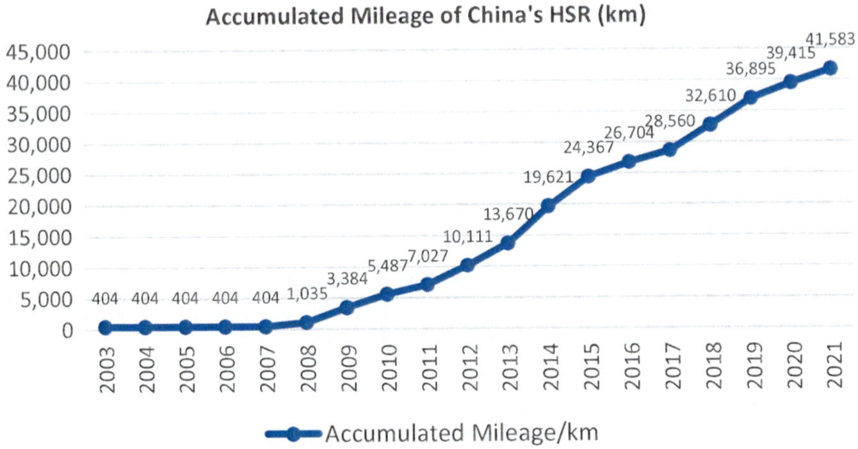

Fig. 2 Expansion of HSR in China

and 2010, the National Development and Reform Commission issued and revised the "Yangtze River Delta Intercity Rail Transit Planning," announcing the establishment of an intercity high-speed rail network with Shanghai, Nanjing, Hangzhou, and Hefei as the four major cities [12]. The current network planning, "Multi-Level Rail Transit Planning in Yangtze River Delta," was published in June 2021 and will take five cities

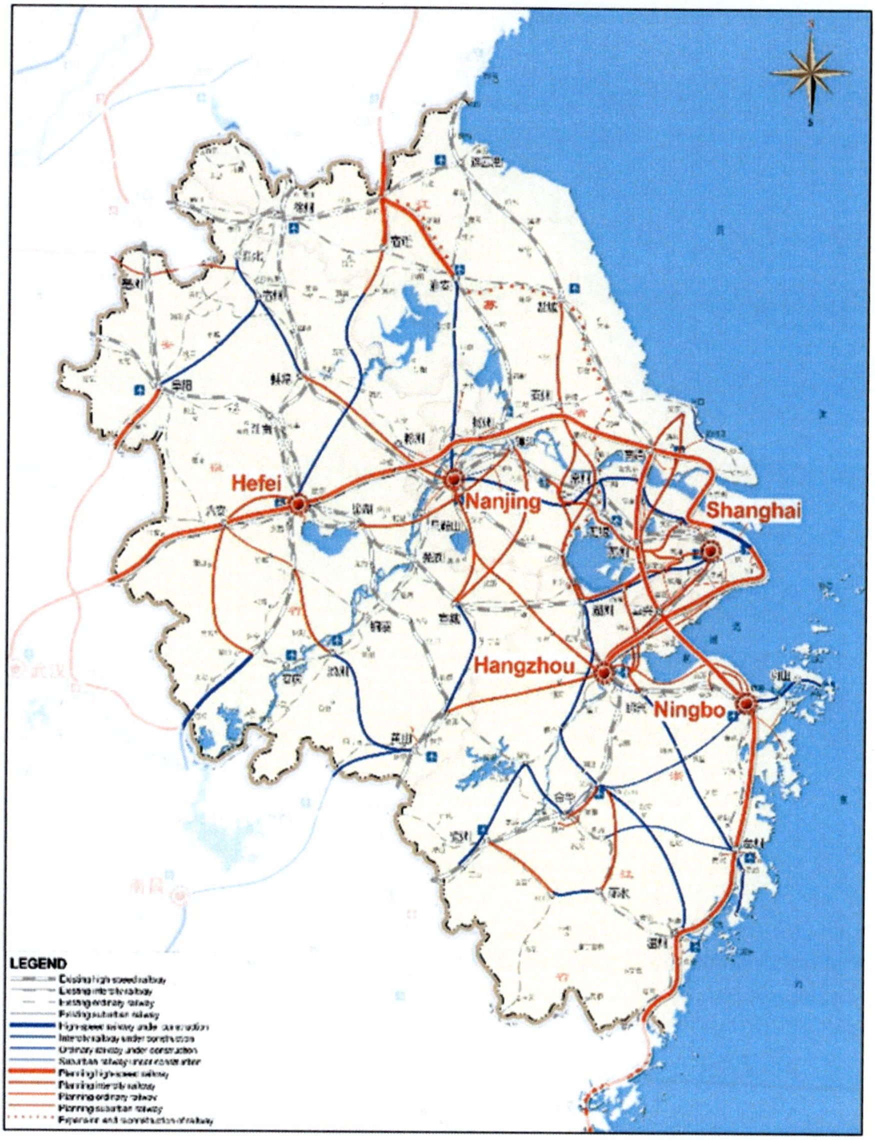

Fig. 3 HSR network plan for the Yangtze River Delta. (*Source* [13])

as centers: Shanghai, Nanjing, Hangzhou, Hefei, and Ningbo (see Fig. 3) [13]. The system design is for a 0.5–1-h "commute traffic circle" between five major cities and their adjacent towns, and is for a 1–1.5-h "intercity traffic circle" between hub cities and their adjacent cities. Such a network could allow the five cities and their smaller counterparts to function as a single integrated urban megaregion.

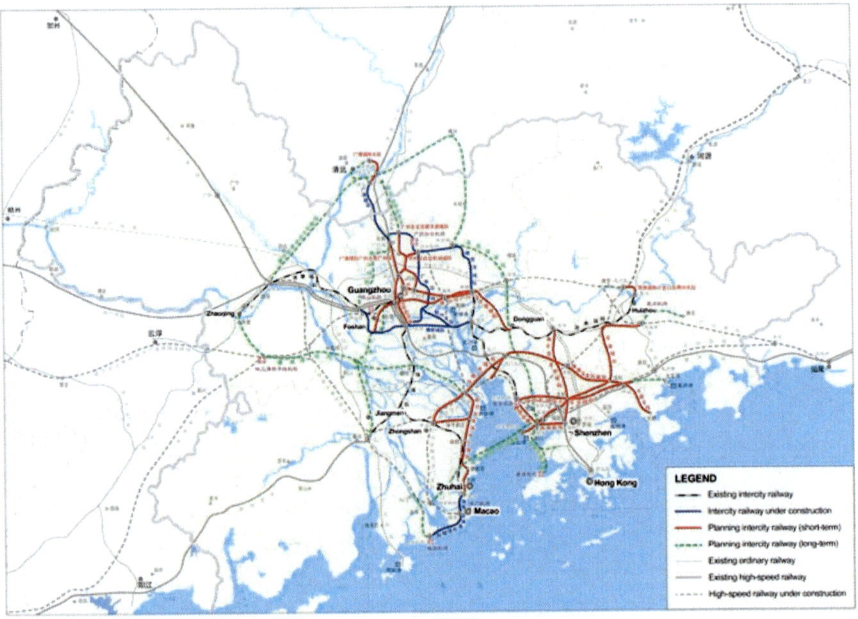

Fig. 4 HSR network planning of the Greater Bay Area. (*Source* [15])

In 2009, the National Development and Reform Commission also passed the "Inter-city Rail Transportation Network Plan for the Pearl River Delta Region (Revision)." It was later replaced by "The Intercity Rail Construction in the Guangdong-Hong Kong-Macao Greater Bay Area" in July 2020 [14, 15]. The total length of the plan is about 5700 km in the long-term year of 2035, which proposed a construction project of 13 intercity railways and five integrated transportation hubs in the short term from 2020 to 2025. The construction plan consists of three integrated transportation hubs in Guangzhou, Shenzhen, and the west of the Pearl River, and more than 30 intercity lines connected with nine cities and two special administrative regions of Hong Kong and Macao (see Fig. 4). In order to allow for fast construction of the HSR, many HSR stations are planned for suburban locations away from city centers.

3 China's High-speed Rail Station Site Location

In China, the 'HSR new town' model has been dominant in government planning at various levels in the site selection for HSR stations. In this model most of the new station sites are located in suburbs or exurbs distant from the large urban centers. The hope is that the HRS station will spark the development of "new towns." (In Western terms, these would be major metropolitan subcenters or districts.) The intent is to stimulate local economic development and offering an attractive alternative location to the crowded city centers.

The Beijing-Shanghai HSR line is an example. The total length of this line is 1318 km. Of the 24 cities connected by the line, 18 cities chose to build HSR stations in the suburbs. The reasons for suburban site selection included lower costs, hope of capturing rising land values, and a desire to relieve pressure on the central areas of the cities. Since there is less densely developed land in the suburbs, the cost of land acquisition can be greatly reduced compared to city centers. Suburban station development also may generate land value increment profits due to positive spillover effects of the railway station. Finally, many cities are interested in promoting the transformation of their urban spatial structure from a single center to a polycentric form in order to alleviate the pressures of high population density and intense commercial activity in the central cities.

The railway authority also wants to locate HSR stations in the suburbs. This simplifies HSR track alignment allowing straight lines which reduce project construction cost as well as operation costs in the future. Their preference for the suburbs may be compounded by the fact that the railway authority is not responsible for the connecting transport for passengers accessing the rail station.

It also should be noted that the location of HSR stations varies according to the influential capacity of local government in the cities through which rail passes or provides a station. Because of China's hierarchical administrative system, large cities are more influential in controlling the discourse in the negotiations between local government and the railway authority than are smaller cities. As a result of this, most HSR sites are located in the suburban areas of large cities, while in most medium and small cities the new stations are located in the exurban fringe (see Table 1), where it is difficult to provide good public transport service. The smaller cities thus have greater local car traffic associated with the stations.

Shanghai Hongqiao HSR Station provides a typical example [16]. The station is part of Hongqiao Integrated Transport Hub (see Fig. 5), which includes a regional airport, and Hongqiao Business Plan, which is intended to guide development in the vicinity of the transport hub (see Fig. 6). The station is located 15 km from Shanghai city center and links the Beijing-Shanghai HSR with the Beijing-Shanghai railway and Shanghai-Nanjing intercity railway to the north, and the Shanghai-Kunming railway, Shanghai-Hangzhou-Ningbo passenger dedicated line and Shanghai-Hangzhou intercity railway to the south. It began operation on July 1, 2010 with a yearly passenger dispatch of 51,076,000 in 2021 and a planned capacity of 78,380,000 passengers.

To enhance the connection between the Shanghai city center and the Yangtze River Delta Region through the station, expressway networks and rail transit networks have been constructed. So far, rail transit lines 2 and 10, which provide seamless transfer from the Shanghai city center, have been extended to the station. The planned rail transit lines 17 and 5 will also be constructed to link Hongqiao HSR Station with another part of Shanghai (Figs. 7 and 8).

Table 1 China's major HSR station site locations and rail line plans

Names of HSR way stations	Number of urban lines in operation	Distance from the city center (km)	Operation time of high-speed railway stations
North Xi'an station	3	13	2011–01
East Zhengzhou station	2	8	2012–09
East Hangzhou station	4	13	2013–06
Shanghai Hongqiao station	3	15	2010–07
South Guangzhou station	4	18	2010–01
South Nanjing station	4	10	2011–06
South Beijing station	2	5	2008–08
Wuhan station	2	12	2009–12
Tianjin station	3	0	2008–08
South Changsha station	2	9.5	2009–12

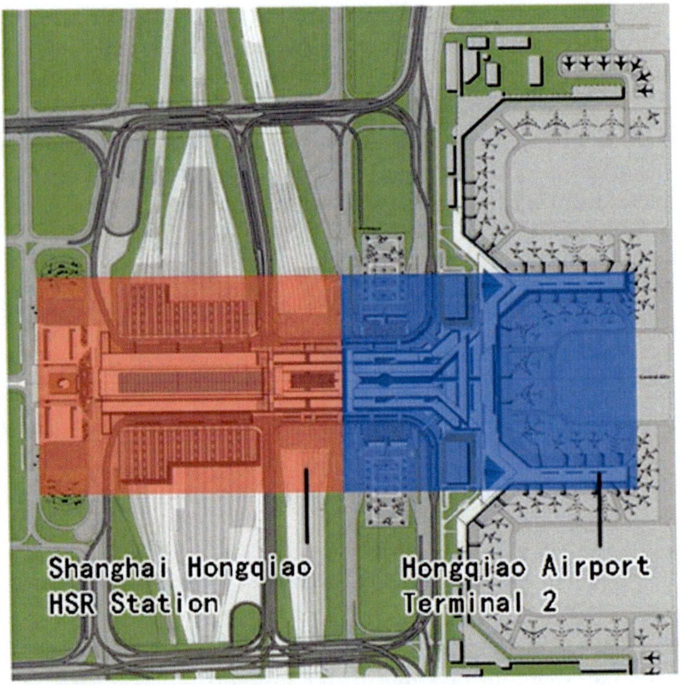

Fig. 5 Hongqiao integrated transport hub project

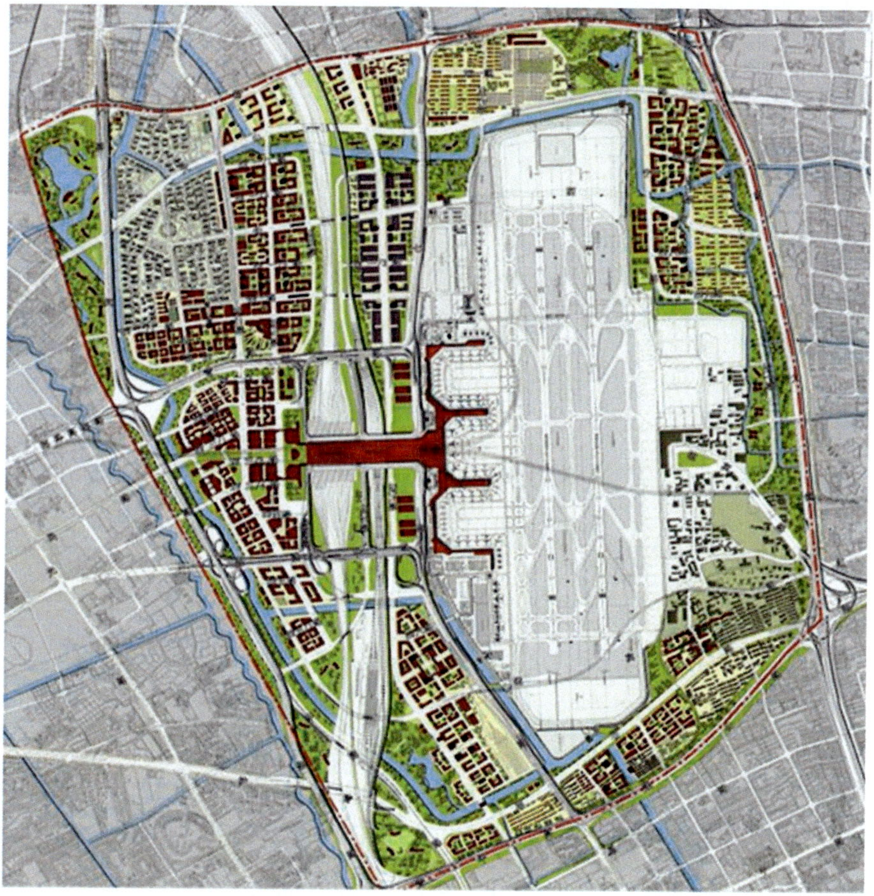

Fig. 6 Hongqiao business district plan

4 Research Questions and Survey

At present, national HSR station construction is in full swing. Relevant international and domestic research indicates that large-scale HSR construction creates new opportunities for the region and spurs urban development. Site selection for HSR stations will have an enormous impact on urban and regional spatial structure transformation, especially for stations located close to city centers. However, as noted earlier, many HSR stations are located quite far away from city centers in China, so empirical data is needed to analyze the impact of HSR station location on travel characteristics of passengers and urban spatial structure, including travel distance distribution, changes in travel time and HSR passenger distribution in different geographic locations. This information will be useful in improving the travel efficiency of intercity HSR service

Fig. 7 Location of Hongqiao HSR station and its transport connections

through station location or connecting transport service options, as well as the design of stations.

We conducted a survey of 1834 respondents from February 27, 2012 to March 3, 2012 within the Hongqiao HSR station. We randomly selected passengers in the waiting hall to conduct face-to-face interviews. We questioned passengers regarding their trip origin and destination, on-board high-speed train travel time, mode of travel to the HSR station and time station, and demographic characteristics.

From this survey we know the social and economic attributes of passengers. We also have the information for each segment of their travel from origin to destination, and travel characteristics before and after the opening of the HSR station. Based on these data, we can analyze the impact of the location of HSR stations and the connecting transport system on the travel efficiency for HSR passengers in terms of time and cost. We also can compare the findings to the forecasts that were made by transit planners in preparing for the development of the HSR service and station.

In addition, after 12 years after the opening of Hongqiao HSR Station, we have obtained the travel time from Hongqiao HSR Station on the outskirts of the city and Shanghai Railway Station in the city center by public or private transport through the network map open platform, as the accessibility of high-speed rail stations. Through this study, we compare the spatial relationship between the location and accessibility of high-speed railway stations in cities and the distribution of urban activity elements.

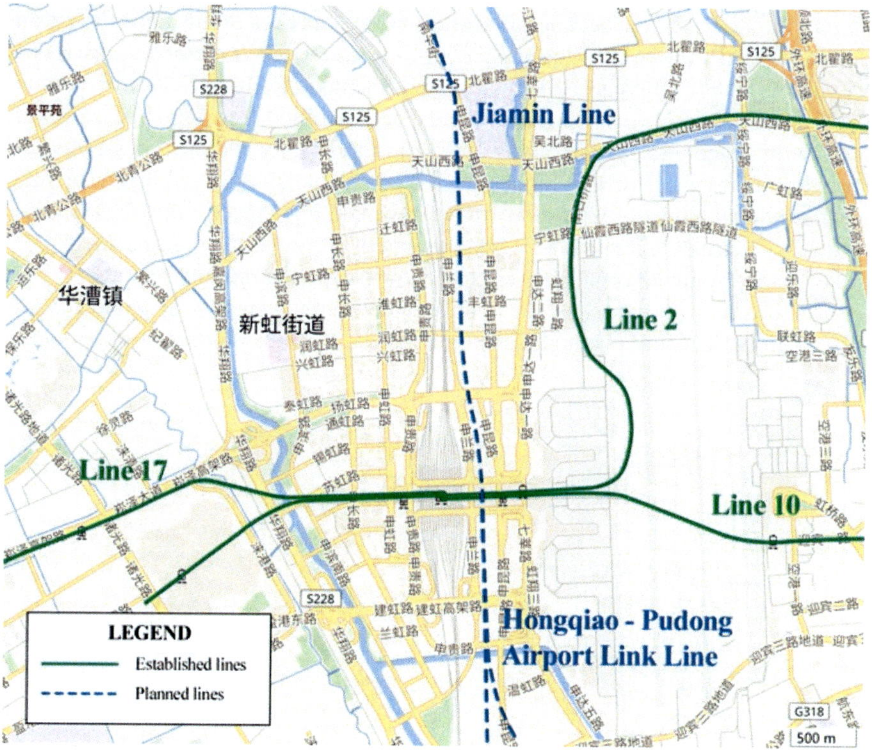

Fig. 8 Hongqiao HSR station and metro connection

The following research questions are explored in this study:

1. What are the differences in POIs accessibility of HSR stations located in different parts of Shanghai?
2. How far will the passenger travel by HSR and what factors influence passenger volume between Shanghai and a destination city?
3. Where do the passengers come from in Shanghai? As the location of the station is closer to the neighboring provinces of Zhejiang and Jiangsu, how many passengers are from regions outside of Shanghai?
4. What is the proportion of time spent in each segment of travel and how can we improve their travel efficiency?

5 The Evaluation and Comparison of Accessibility

In this study, we divide the whole of Shanghai into 0.01° * 0.01° (about 1 km * 1 km) grids, with a total of 6348 grids. We calculate the travel time with Hongqiao HSR Station and Shanghai Railway Station and use the travel time of the grid centroid point as the accessibility between the grid and the corresponding railway station point via the API of AMAP (lbs.amap.com). Figures 9 and 10 show the results of accessibility from Hongqiao HSR Station and Shanghai Railway Station by public and private transport. It is not difficult to find that the phenomenon of public transport accessibility extending outward along urban rail transit lines and the accessibility of the station area is usually much higher. On the other hand, the accessibility of private transport shows outward expansion from where the high-speed railway station is located. With an extensive road network, the accessibility at different time thresholds appears as approximate concentric circles, showing more proximity features.

Based on the above analysis, the coupling between the public transport accessibility distribution of the two high-speed rail stations and the rail transit station area

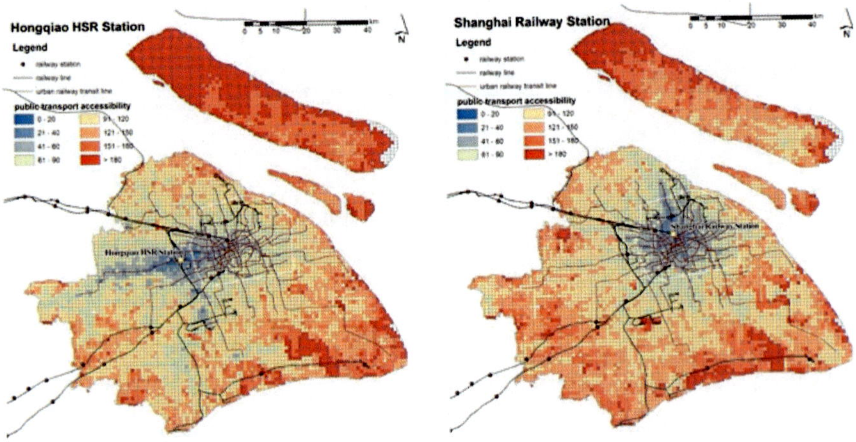

Fig. 9 Public transport accessibility of Hongqiao HSR station and Shanghai railway station

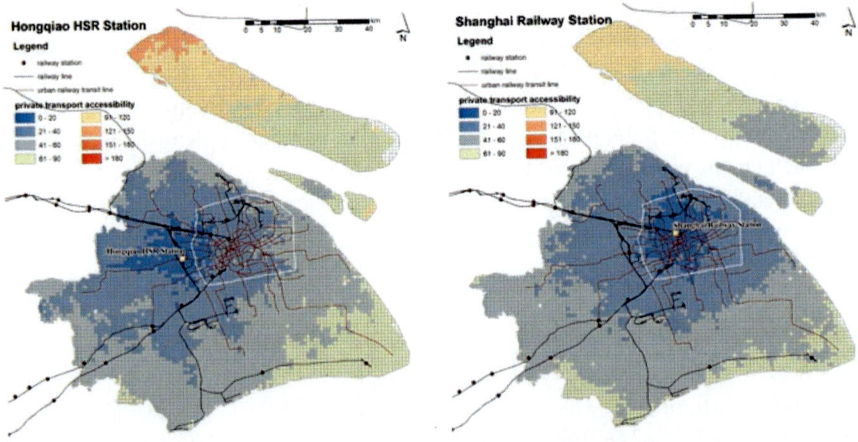

Fig. 10 Private transport accessibility of Hongqiao HSR station and Shanghai railway station

is further analyzed. According to the calculation of public transport accessibility of high-speed rail stations in Fig. 11, the cumulative accessible ranges under the time bands of 20, 40, and 60 min are defined, which are overlayed with the ranges of 800 m around rail transit stations. Then, the ratio of the overlapping area between the accessibility area and the rail transit station area to the total area of the accessibility area under the corresponding time threshold was calculated (see Table 2 and Fig. 11).

Under the time threshold of 20, 40, and 60 min, the public transport accessible area of Hongqiao HSR Station and Shanghai Railway Station is almost the same, and even within the range of 60 min, the accessible area of Hongqiao HSR Station is slightly larger than that of Shanghai Railway Station. However, no matter in the time threshold range of 20, 40, or 60 min, the coupling area and coupling proportion

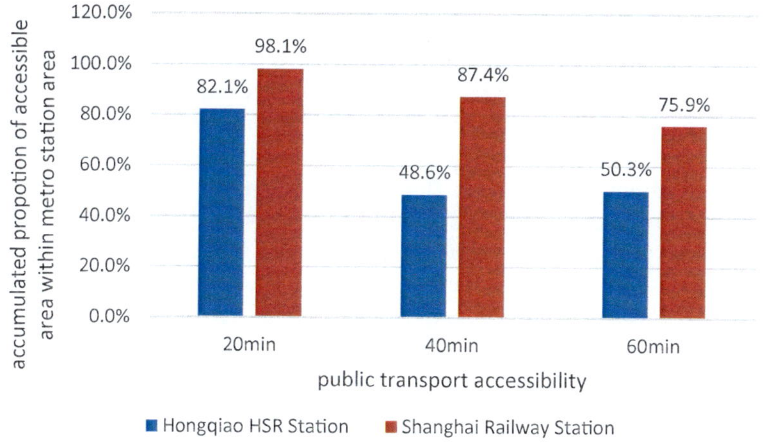

Fig. 11 The spatial coupling proportion of public transport accessibility and rail transit station area under different time thresholds

Table 2 The proportion of rail station accessibility area and coupling with urban rail transit station area under different time bands

Station	Travel time (min)			
	Area and proportion	20	40	60
Hongqiao HSR station	Accessible area (km^2)	10.6	115.2	458.4
	Accessible area within the metro station area (km^2)	8.7	56.0	230.5
	Proportion (%)	82.1	48.6	50.3
Shanghai railway station	Accessible area (km^2)	5.3	137.3	434.0
	Accessible area within the metro station area (km^2)	5.2	120.0	329.5
	Proportion (%)	98.1	87.4	75.9

between public transport accessibility and the rail transit station area of Shanghai Railway Station located in the urban center is more significant than that of Hongqiao HSR station located in the urban edge. For the business and commuter groups using the high-speed railway and intercity railway, the stations located in the downtown area can provide more convenient connection conditions for high-speed railway and intra-city travel time to/from HSR station.

On this basis, the research obtained the spatial distribution of points of interest (POI) of three types of urban activity elements within Shanghai, including commercial and life service facilities, company office facilities, and public service facilities. Furthermore, the research superimposed it with the public transport accessibility circle of the two railway stations to get the number of POI facilities in the corresponding accessibility circle layer, as shown in Figs. 12, 13 and 14. The blue represents total POI facility with accessibility in Hongqiao HSR Station, while the red represents the POI of Shanghai Railway Station.

In the corresponding accessibility circle, the number of service facilities available from Hongqiao HSR Station first increases from 0 to 90 min and then decreases from 90 to 180 min, reaching the highest level when the accessible time is 60–90 min. While for Shanghai railway stations, the number of commercial and life services and business services remains relatively balanced at a medium-to-high level in each time band with a maximum reach time of 20–120 min, but the number of public services is slightly reduced. Considering that the public transport accessible area of railway stations gradually increases from 20 to 120 min (see Table 2), it can be assumed that the density of facilities distribution is gradually reduced when the number of accessible facilities remains relatively equal. The density of facilities in the 20–40-min band to Shanghai Railway Station is the highest value, and then gradually decreases.

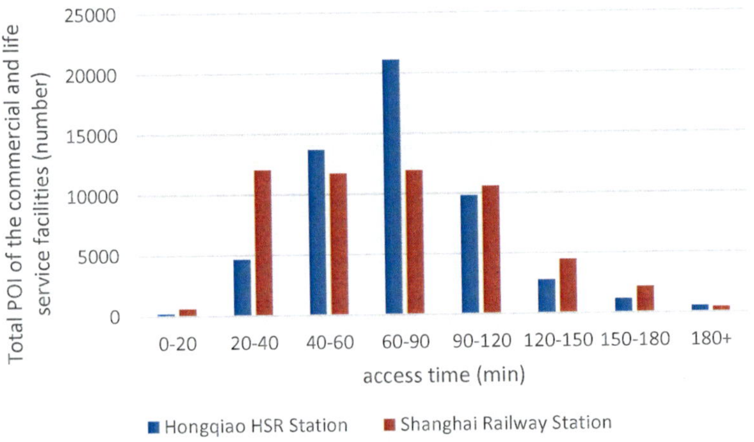

Fig. 12 Public transport accessibility and total POI of the commercial and life service facilities

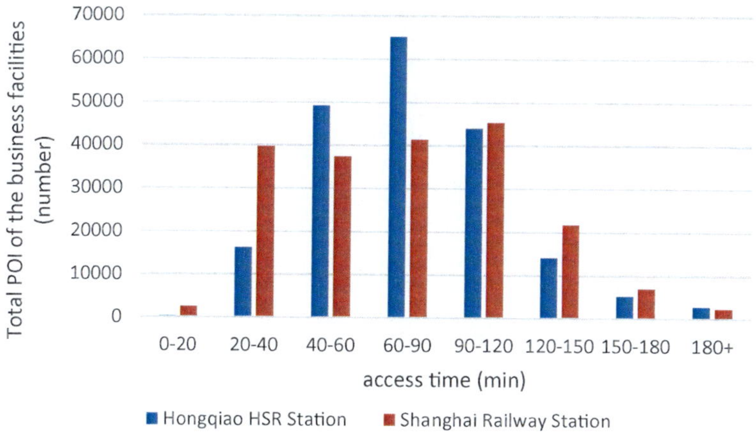

Fig. 13 Public transport accessibility and total POI of the business facilities

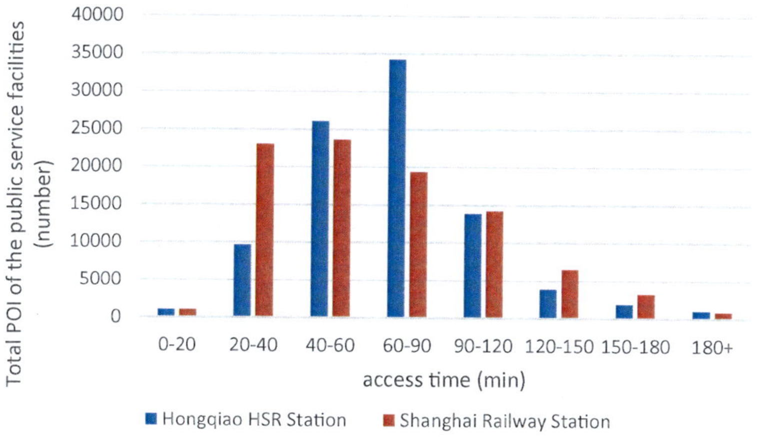

Fig. 14 Public transport accessibility and total POI of the public service facilities

6 High-speed Train Passenger Travel Distance

China covers a vast geographical territory, and the construction of a national HSR network will facilitate interregional connection. How are passengers distributed from the Hongqiao Station to their destination city? Conducting a regression analysis based on the variables of passenger number, the population of the destination city and the distance between Shanghai and the city, we reveal that the number of passengers is in direct proportion to the destination city's population, while in inverse proportion to the distance between Shanghai and the city (see Table 3). In other words, the larger the population of the destination city, the more passengers travel from Shanghai to

Table 3 Number of passengers as a function of city population and distance

Model		B	t	Sig
1	(constant)	1	2.71	0.015
	Population (0000)	0.086	2.76	0.013
	Distance(km)	−0.125	−3.01	0.008

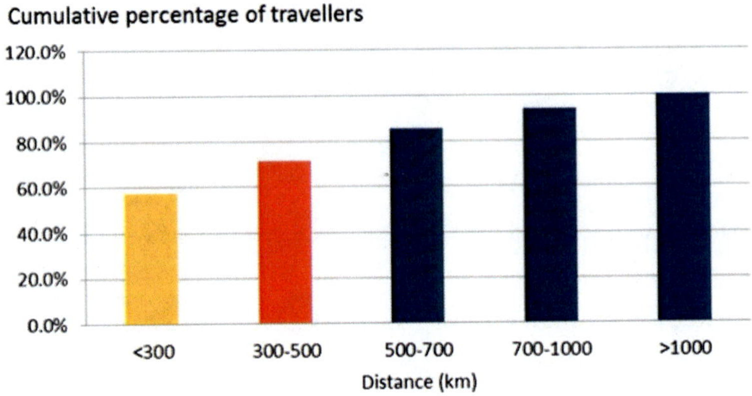

Fig. 15 Cumulative percentage of travelers

the destination city, and the shorter the distance to the destination city, the more passengers travel via HSR.

The research reveals that the average passenger travel distance is 377.4 km, which is not as far as was expected before the opening of HSR. It can be concluded from the travel distance distribution that short and medium-distance passengers still comprise a majority of all HSR travel, with 57% of them travelling less than 300 km, and 71% of all passengers travelling less than 500 km (see Fig. 15).

7 The Origination Points of High-Speed Train Passengers

From Table 4 we see that that 88% of the passengers surveyed are from different districts in Shanghai, with 4.1% of passengers transferring from the nearby Hongqiao airport and 7.8% of passengers from outside of Shanghai.

Table 4 The origin of surveyed passengers in Hongqiao HSR station

	Passengers	Percentage (%)
Hongqiao airport transfer	75	4.1
Within Shanghai	1595	88.1
Outside of Shanghai	141	7.8

Table 5 Administrative districts distribution in Shanghai

Location	District name
The central urban area	Huangpu district
	Zhabei district
	Jing'an district
	Xuhui district
	Hongkou district
	Changning district
	Yangpu district
	Putuo district
The city outskirts	Baoshan district
	Jiading district
	Pudong new area
	Minhang district
Outer suburban districts	Chongming district
	Fengxian district
	Jinshan district
	Songjiang district
	Qingpu district

To analyze the geographic location of passengers from Shanghai, we divided the 16 districts and one county in Shanghai into three major categories, the central urban area, city outskirts and outer suburban districts (Table 5 and Fig. 16).

Survey result reveals that approximately half of the HSR passengers come from the central urban area of Shanghai, 40% are from the city outskirts, and only about 10% of the passengers are from the outer suburban areas. Further analysis of passenger intensity from various administrative districts of Shanghai (where intensity is the number of passengers divided by the population of the district) shows that passenger intensity is in inverse proportion to the distance to the central urban area. In other words, the majority of passengers come primarily from the dense central urban area and adjacent districts. Thus, a HSR station located in the suburbs will require a majority of passengers to travel extra distance (Fig. 17).

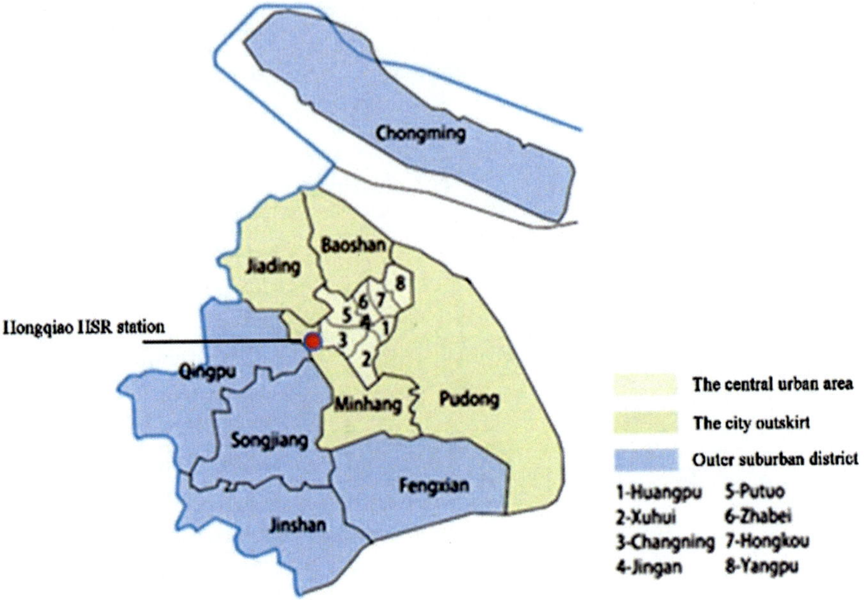

Fig. 16 Shanghai district and county administrative divisions

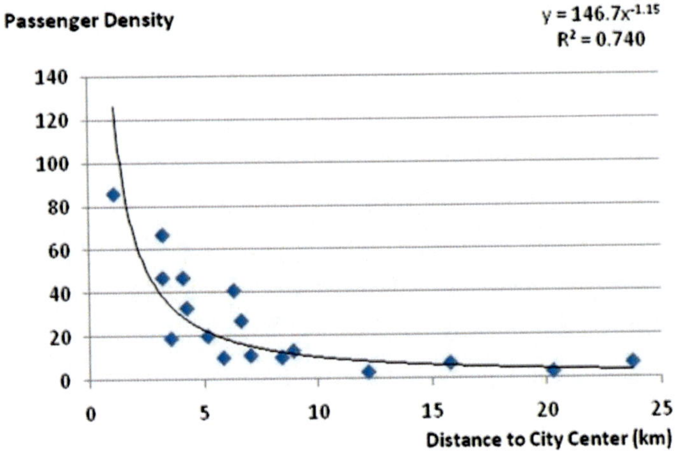

Fig. 17 The relationship between density of HSR passengers and distance to city enter

8 Mode of Travel Access to Hongqiao HSR Station

Because the location of the new station is far away from the existing city center from where there is a high demand for travel by HSR, much effort was given to improving transport between the station and the city center, including two rail transit lines, an

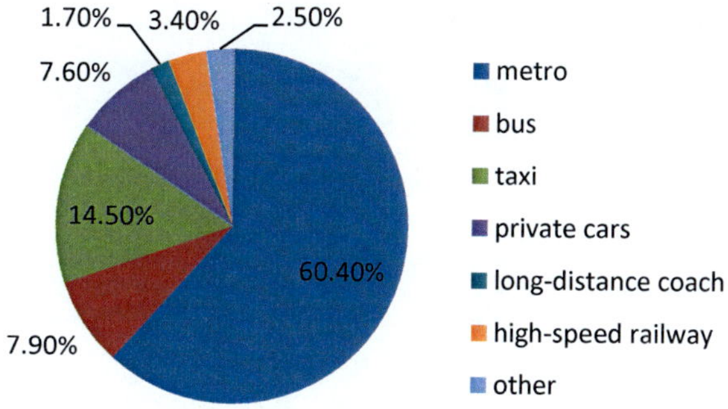

Fig. 18 Modal split of transit to Hongqiao HSR station

elevated motorway, and expansion of the conventional bus system. Figure 18 shows the mode split that we observed in the survey. We found that 60.4% of passengers accessed HSR by urban rail transit, and an additional 7.9% used the conventional bus system, bringing the total percentage using public transport to access HSR to 74.9%. This is far in excess of the planning forecast share for transit, which was 50%. Of the private transit modes, only 7.6% of passengers take a personal car to Hongqiao HSR Station, while 14.5% take a taxi. The forecasts for the project predicted considerably higher private transport access. As a result, the elevated motorways to Hongqiao station may have been over-supplied.

Figure 19 shows mode of access to the HSR station by car owners and those without a car. Even among HSR users who possess a private car, only 13.6% drove to Hongqiao Station. These private car-owning passengers showed a preference for the metro, with 53.9% of them preferring to use the metro to get to the station. However, private car-owning passengers use conventional bus transit much less than passengers who do not have cars. These results clearly show the importance of provision of high-quality public transport in connecting city centers and HSR stations in order to attract people to public transport and reduce the demand for car travel. The investment in metro to HSR stations undoubtedly reduced car mileage travelled to the HSR stations located far away from city centers—a boon from a traffic and air pollution perspective.

Comparing mode choice to the HSR station from the various districts of Shanghai, we find that 76.5% of the passengers from the central urban area take urban rail to the HSR station, while only 7% take cars to the station. For passengers from the suburban areas, which are less well served by the urban rail network, people rely more on conventional bus transit, with 38% of passengers taking buses to the station and only 27.7% taking urban rail. From suburban areas, demand for travel to the HSR station by car is 15.1%, double that of the central urban area (see Fig. 20).

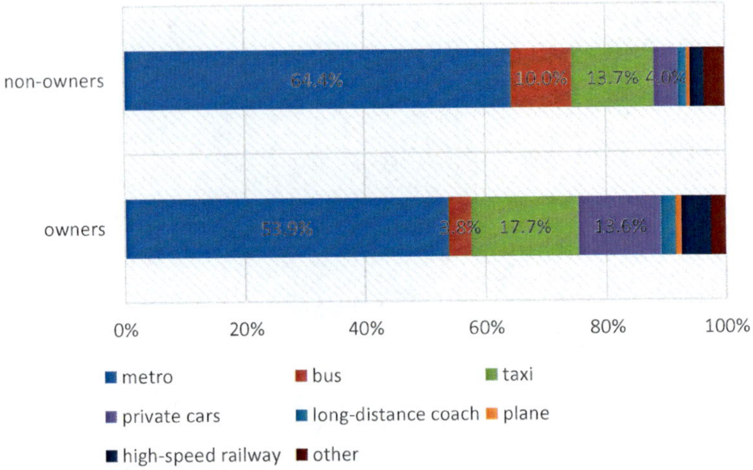

Fig. 19 Access mode choice to the station by car owners and non-owners

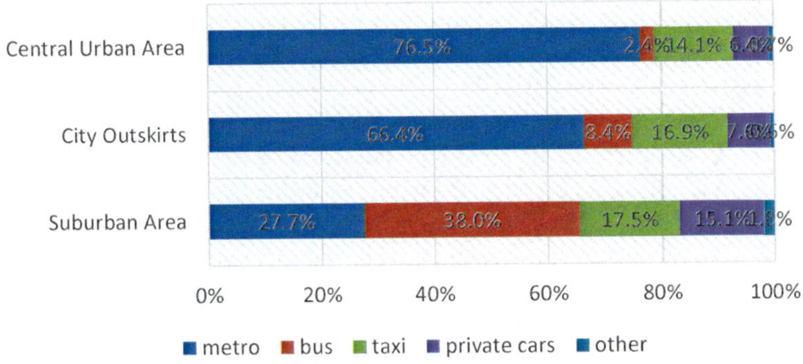

Fig. 20 Modal split of transit to the HSR station from different regions of Shanghai city

9 Analysis of Door-to-Door Transit Time for HSR Travel

As illustrated in Fig. 21, the total door-to-door travel time for HSR is comprised of four parts: time from origin to the Hongqiao HSR station, waiting time in the station, travel time on board the high-speed train, and finally time from the destination HSR station to the point of destination.

As the speed of high-speed trains is so high, the on-board time between two stations is greatly reduced from that of traditional rail or highway transit. However, if people travel a relatively short distance, the on-board time may comprise only a small portion of total travel time. In this sense, during the planning and construction of HSR, we have to pay attention to the access modes serving the HSR stations. Large improvement in travel efficiency through HSR can only be realized when the

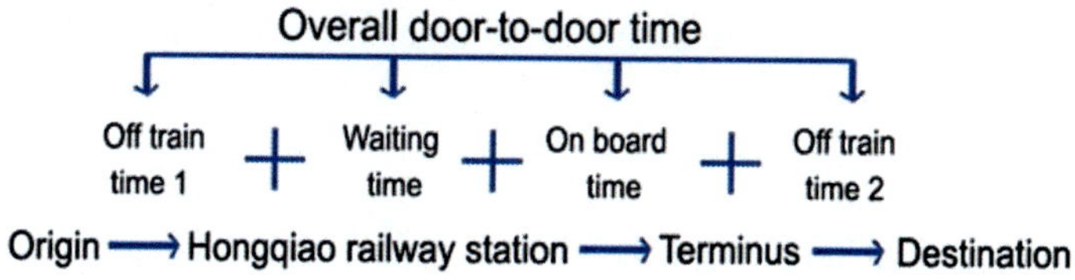

Fig. 21 The segments of door-to-door travel time consumption

connecting travel time from the origin to the HSR station and the egress time to the travel end point can be kept down in proportion to overall travel time. If this is not done, increasing train speed may make a limited contribution to total travel efficiency.

From our survey, the average on-board HSR travel time is 192 min, transit time to the HSR Hongqiao station averages 56 min, and waiting time for the HSR train averages 61 min. For the shorter trips under 300 km, on-board travel time for HSR only accounts for 25% of total travel time. Therefore, for the shorter trips, the higher train speed will produce less benefit to improved travel efficiency, and efforts during HSR planning to reduce the off-train time are key (Figs. 22 and 23).

Fig. 22 Total passengers' door-to-door time composition

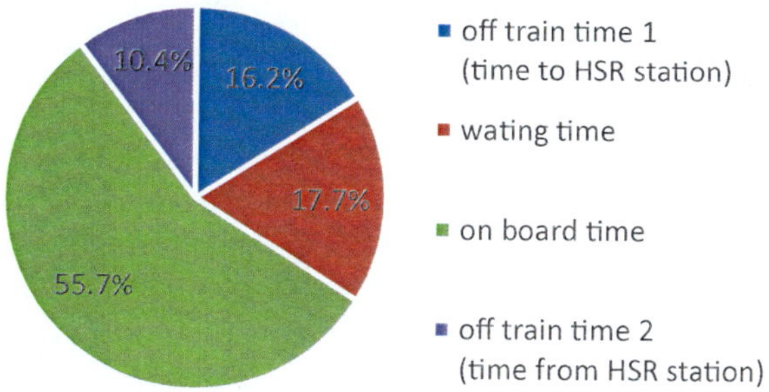

Fig. 23 Off-train time as % of total travel time compared to on-train travel times

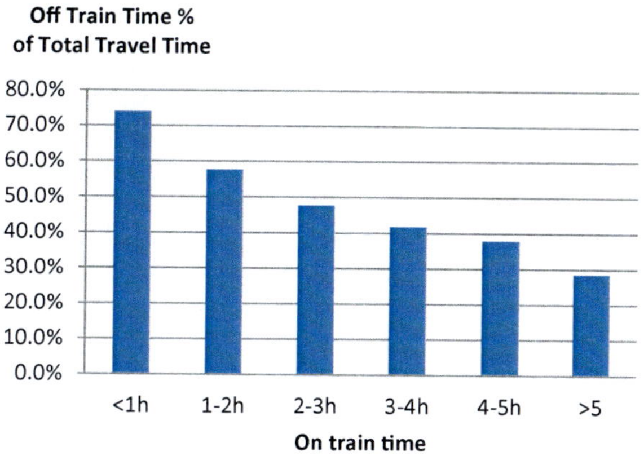

10 Changes in Access Time Before and After the Opening of the Station

Before the opening of the Hongqiao HSR Station, people could take traditional or high-speed trains in either the Shanghai Railway Station or Shanghai South Railway Station, both of which are located quite close to the city center. Comparing passenger average access time to those two stations to the average access time to the newly established Hongqiao HSR station, we find that average access time increased 18%, from 50 to 59 min (Fig. 24). The passengers whose access time increased the most are those whose origin is the central urban district. Passengers in the suburban south part of Shanghai benefit from the location of Hongqiao HSR station in the form of reduced access times to the station, but passenger intensity is relatively low there. The large-scale expansion of the Shanghai urban rail system over the past several years has partially mitigated the effects of having the HSR connection located farther away, and it appears that the travel time increase is acceptable to most passengers.

11 Conclusion

Since 2003, the China Railway Authority has been implementing the national strategic plan for a massive increase in rail construction, prioritizing high speed rail along with interregional cargo rail and rail construction in the West. The network of high-speed rail services is now the largest in the world.

HSR is a well-suited transport mode for China's rapidly growing demand for intercity passenger travel. Its high speeds, high capacity, and modest land and resource requirements are a good match for a country with a large territory and many large cities, a dense and increasingly urban population, and serious resource and environmental constraints.

Operating at speeds of up to 350 km/h, China's HSR is faster than auto travel and is competitive with domestic air travel for the majority of intercity trips, and is even competing with air travel for trips of over 1000 km. In Shanghai and many other Chinese cities, suburban and exurban HSR station locations have been chosen because they are less expensive to build, may stimulate subcenter growth, and permit straighter alignments and faster rail service.

However, our survey, conducted in Shanghai's Hongqiao Station which is fairly typical of the new HSR stations, shows that over 70% of the passengers travel on HSR for distances under 500 km. and over half travel less than 300 km. For these travelers, the amount of time it takes to access the HSR station is a major consideration; higher train speed has a decreasing contribution to total travel efficiency the shorter the total travel distance. In addition, our study found that the coupling degree between HSR stations located in urban centers and urban activity facilities is better than that of HSR stations located on the edge of cities. Therefore, the challenge is how to balance location of HSR stations with the provision of improved urban center transit

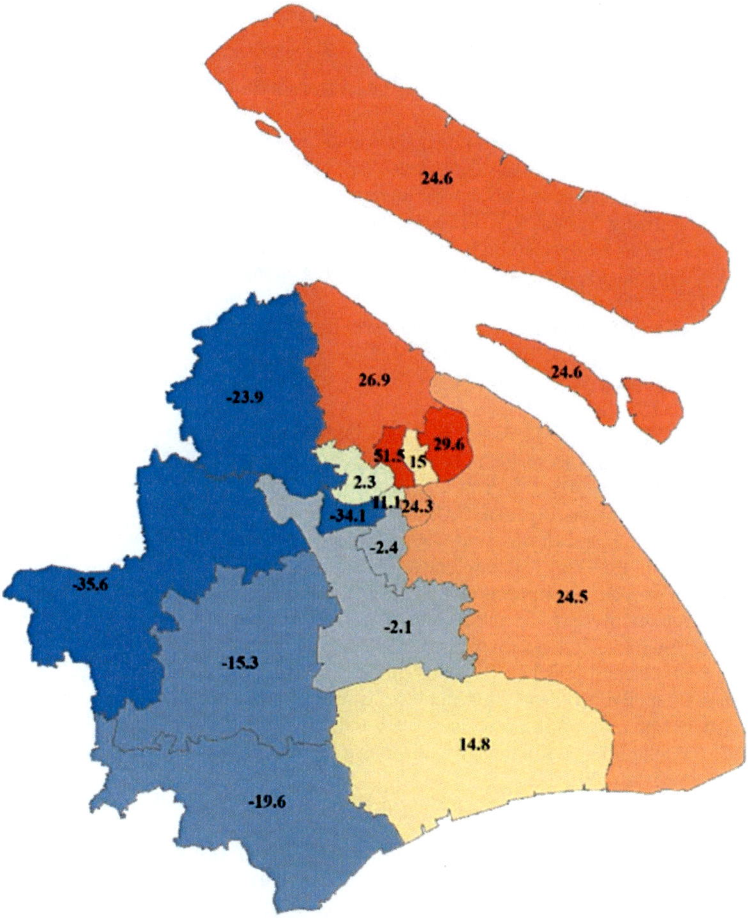

Fig. 24 Access time changes in different districts of Shanghai (unit: %)

connections where there are higher HSR passenger intensities. The Shanghai case shows that extending connections, especially rail transit connections, to the HSR station can attract many passengers. However, this comes at a high cost and HSR access still takes more time from the customer than the earlier, in-city station locations required. Instead of constructing a large HSR station in a distant suburb, planners should consider the direct connection of HSR to existing traditional rail stations, where the HSR train can be more easily accessed in the passenger-intensive urban areas. This will simultaneously decrease road transportation, and thus air pollution and road infrastructure costs. The overall efficiency and benefit to society may be greater.

References

1. Maria, B.: Forecasting demand for high speed rail. Transp. Res. Part A Policy Pract. **70**, 81–92 (2014)
2. NDRC: Mid-term and Long-term Railway Network Planning (Revision). http://www.sdpc.gov.cn/zcfb/zcfbqt/2008qita/W020090605629892360234.pdf (2008)
3. Xiao, S.S.: Impact analysis of improving integrated transport system through high-speed rail development. Integr. Transp. **9**, 9–13 (2011). (in Chinese)
4. Feng, C.A.: Impact analysis of high-speed rail construction on urban development. In: China Urban Planning Annual Conference, Tianjin 2009. Urban Planning Society of China, Beijing, pp. 175–184 (2009) (in Chinese)
5. Javier, G., Rafael, G., Gabriel, G.: The European high-speed train network: predicted effects on accessibility patterns. J. Transp. Geogr. **4**, 227–238 (1996)
6. Sean, T.: High-speed rail, the knowledge economy and the next growth wave. J. Transp. Geogr. **22**, 285–287 (2012)
7. Yu, S., João, S., Luis, M.M.: Assessing High-Speed Rail's impacts on land cover change in large urban areas based on spatial mixed logit methods: a case study of Madrid Atocha railway station from 1990 to 2006. J. Transp. Geogr. **41**, 184–196 (2014)
8. Juan, C.M., Concepción, R., Juan, C.G., Javier, G.: Spatial analysis of the competitiveness of the high-speed train and air transport: the role of access to terminals in the Madrid–Barcelona corridor. Transp. Res. Part A Policy Pract. **69**, 392–408 (2014)
9. Chen, C.L., Peter, H.: The impacts of high-speed trains on British economic geography: a study of the UK's InterCity 125/225 and its effects. J. Transp. Geogr. **19**, 689–704 (2011)
10. NDRC: Mid-term and long-term high-speed railway network plan (2016–2030). https://www.ndrc.gov.cn/xxgk/zcfb/ghwb/201607/t20160720_962188_ext.html (2016)
11. The State Council of the People's Republic of China: China to enhance global competitiveness in transport. http://www.gov.cn/zhengce/2019-09/19/content_5431432.htm (2019)
12. NDRC: Yangtze River Delta intercity rail transit planning (Revision). http://www.czdrc.gov.cn/gdwz/1820.htm (2010)
13. NDRC: Multi-Level rail transit planning in Yangtze River Delta. https://www.ndrc.gov.cn/xxgk/zcfb/ghwb/202107/t20210702_1285346_ext.html (2021)
14. NDRC: Inter-city rail transportation network planning of the Pearl River Delta Region (Revision). http://www.sdpc.gov.cn/xmsphz/t20100813_366086.htm (2009)
15. NDRC: The intercity rail construction in the Guangdong-Hong Kong-Macao Greater Bay Area. https://www.ndrc.gov.cn/xwdt/tzgg/202008/t20200804_1235524.html (2020)
16. Shanghai Construction and Management Commission: Hongqiao integrated transport hub project. http://www.shjjw.gov.cn/gb/jsjt2009/node706/node1482/userobject1ai88748.html (2014)

HSR and Land Use Impacts 2

Relationship Between the Influence Area of High-Speed Railway Stations and the Casual Effect of these Stations on the Population Change of Municipalities Alone the Railway Line in Japan

Jikang Fan, Shintaro Terabe, Hideki Yaginuma, Haruka Uno, and Yu Suzuki

Abstract The high-speed railway station has a wider influence area than the commuter railway station and is believed to impact the municipalities around it significantly. The municipalities where high-speed railway stations will be constructed might be affected similarly. It is still unclear how the size of the station's influence area affects the population of municipalities. We examine a method using a propensity score to show the effect of high-speed railway development quantitatively. Thus, we can statistically infer the causal effect of high-speed railway development while considering the influence of commuter railway lines and highways. We matched 1126 municipalities all over Japan using propensity scores and quantitatively analyzed the effect of opening Shinkansen stations using Difference-in-Differences analysis. As a result, we find that the Shinkansen station generally positively affects population changes.

Keywords Causal inference · High-speed railway · Propensity score · Difference-in-differences analysis.

J. Fan (✉) · S. Terabe · H. Yaginuma · H. Uno · Y. Suzuki
Department of Civil Engineering, Tokyo University of Science, Chiba Prefecture, Noda City 278-8510, Japan
e-mail: 7621523@ed.tus.ac.jp

S. Terabe
e-mail: terabe@rs.tus.ac.jp

H. Yaginuma
e-mail: yaginuma@rs.tus.ac.jp

H. Uno
e-mail: unoharuka@rs.tus.ac.jp

Y. Suzuki
e-mail: yusuzuki@rs.tus.ac.jp

F. Pagliara (ed.), *Socioeconomic Impacts of High-Speed Rail Systems*,
Springer Proceedings in Business and Economics,
https://doi.org/10.1007/978-3-031-26340-8_11

1 Background

The Shinkansen, Japan's high-speed rail system, has continuously developed since it opened in 1964. The Shinkansen is expected to have 9 lines connecting 92 stations through a network of more than 2700 km high-speed railways by mid 2022, which is shown in Fig. 1. These high-speed railway lines, along with the West Kyusyu Shinkansen, which was partially opened in September 2022 (from Takeo-Onsen Station to Nagasaki Station), constitute Japan's huge high-speed rail network. In the 57 years of development of the Shinkansen, it has become increasingly important as a way of replacing and/or avoiding short-haul flights and thereby reducing Green House Gas (GHG) emissions, responding to the need for strong climate action. At the same time, Shinkansen stations have a much wider influence than commuter railway line stations. The stations are considered to significantly impact the population and economy within their range of influence. Thus, the regions where new Shinkansen stations will be opened may be similarly affected.

Simultaneously, Japan has more than 12,000 km of highways and more than 2,000 interchanges, and many more highways are under construction. Due to the high cost of these traffic infrastructures and considering the population effect and economic effect of the highway on the regions along the road, it is necessary to analyze the effects of Shinkansen station development in conjunction with the effects of highways.

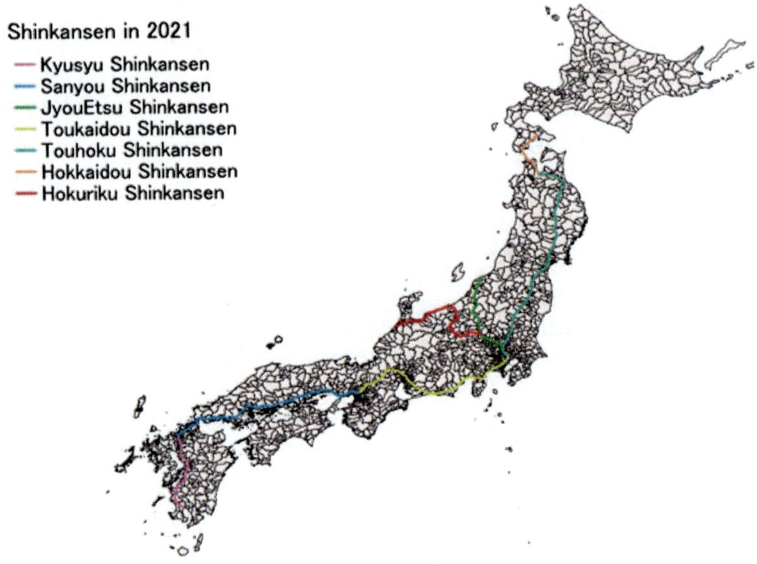

Fig. 1 Distribution of High-Speed Railway in Japan, 2022

2 Review of Existing Research and Positioning of the Research

2.1 Review of Existing Research

Many studies explore the impact of high-speed railway stations on their local area. For example, some studies [1, 2] analyzed the economic impact of stations on the region (urban and rural, coastal and non-coastal) and found that although high-speed rail stations have a positive effect on the local economy, this effect strongly depended on the attributes and classification of the stations. At the same time, Wetwitoo and Kato [3] used causal analysis to calculate the impact of high-speed railways in regional towns. The study found that compared with areas without high-speed railway stations, the economic level of areas with such stations at a certain distance (150–200 km) from large cities is higher. Okabe [4] showed some characteristics of the Shinkansen as an inter-city transportation system and also, describe the effects of railway development on business and the society.

Previous research [5] also studied the impact of the development of commuter railway line stations on population changes in municipalities. This empirically confirmed the common-sense notion that transportation improvement enhances the development of a region.

Talebian et al. [6] evaluated the impact of California's financial support for Amtrak Stations on the local population and employment using propensity score matching. They conducted a multiple regression analysis by adjusting the balance of covariates using propensity score matching between the groups—at the county and city levels—that received financial support and those that did not. The results demonstrated that cities (counties) where Amtrak Stations were developed with financial support from the state of California became attractive railway lines for people and positively impacted the population. However, the impact on local employment remained limited. The study by Jia et al. [7] utilized Difference-In-Differences (DID) analysis and propensity score matching to examine whether the construction of high-speed railways in China resulted in the development of the regional economy by comparing routes. Consequently, it was found that the effects differed between the lines but that the existence of high-speed railways in China positively impacted the economy.

The impact of high-speed rail, however, is not necessarily all positive [8]. Pagliara et al. [9] combining the construction benefits of high-speed railways with the Gini coefficient proves that the existence of high-speed railways has increase the inequity in four European countries. And in terms of passenger preference, according to the research by Cavallaro et al. [10], considering the construction cost and average travel time, passengers will prefer to choose high-grade commuter railway system (Trans-European Transport) compared to high-speed railway system. This proves that the impact of high-speed railways on short distances will be smaller than other modes of transportation, while the effect on medium and long distances will be more obvious.

2.2 Positioning and Significance of the Research

Most previous studies undertaking ex-post evaluations of the Shinkansen have taken a microscopic perspective of facility effects before and after the construction of stations, regional comparisons, or presence/absence comparisons. However, many of them were limited in the number of regions and routes analyzed.

The current research is significant as it quantitatively analyzes the causal effects of the development of the Shinkansen on the population of the area around the stations in the framework of Rubin [11].

This study focuses on a region where high-speed railways have been developed and examines how the causal effects change when the range of influence of the Shinkansen station changes. Additionally, it examines the impact of the years of existence of Shinkansen stations in their respective areas.

3 Method and Data

3.1 Scope of this Research

Considering the particularities of large urban areas, such as the three major metropolitan areas (Tokyo area, Nagoya area and Osaka area) and prefectures, where no Shinkansen were opened until 2015, the subjects of this research were 1126 municipalities across Japan. The exclusions included metropolitan areas, ordinance-designated cities, remote islands, and the Shikoku, Hokkaido, and Fukushima prefecture evacuation areas. The exclusion of the three major metropolitan areas and the ordinance-designated cities (administrative units unique to Japan) from the study was necessary as we needed to avoid the influence of metropolitan areas and analyze the actual causal effects of high-speed rail stations on local towns.

3.2 Overview of Propensity Score

If the value of the allocation variable (a variable indicating whether the municipality has the Shinkansen station) is z_i and the value of the covariate (a variable that affects both the allocation variable and the result) of the ith subject is x_i, then the municipality is in treatment group The assigned probability e_i, called the propensity score, can be expressed by Eq. (1) [12].

$$e_i = p(z_i = 1|x_i) \tag{1}$$

In situations where random assignment is not possible, the distribution of the confounding factors (the covariates that affect the dependent variable) may differ depending on the value of the explanatory variable when investigating the effect of the Shinkansen station.

Two prerequisite assumptions must be met when we use the propensity score: the conditional independence assumption (CIA) and the common support assumption (CSA).

For the CIA, when the covariate x is conditional, the joint distribution of the dependent variable y_1 that assigns the treatment group and the dependent variable y_2 that assigns the control group is independent of the disposition variable z, as shown in Eq. (2) [13].

$$(y_1, y_2) \perp z|x \quad (2)$$

In other words, the assignment by the propensity score will depend only on the observed covariate but not on the dependent variable.

For the CSA, the probability that the sample be assigned to treatment group for each possible propensity score of the covariate x is strictly within the unit interval, as is the probability of not receiving treatment. This assumption of common support ensures sufficient overlap in the characteristics of treated and untreated units to find suitable matched pairs, as shown in Eq. (3).

$$0 < \mathrm{p}(\mathrm{z} = 1|x) < 1 \quad (3)$$

When these assumptions are satisfied, the treatment assignment is strongly ignorable in the terminology of Rosenbaum and Rubin [13].

3.3 Variables Used to Calculate the Propensity Score

The variables used in this study are shown in Table 1. To calculate the existing effect of the Shinkansen stations, we used the population change ratio to calculate the ATT (Average Treatment effect for the Treated) by DID analysis. The data sources of all variables in this list are from the official statistics of the Japanese government.

For each municipality, the dummy variable of the Shinkansen station was used as the objective variable. The treatment group was the municipality group with a Shinkansen station, and the control group was the municipality group without a Shinkansen station. Due to the lack of observational data and the desire to consider the influence of the attributes of the municipalities in the propensity score, we selected the Coastal Area dummy variable, the Heavy Snowfall Area dummy variable, and the Habitable Area Ratio to explain the conditions of municipalities. To consider the influence of other transportation means (such as the commuter railway lines and stations), the construction status of that transportation infrastructure, and analyze their influence over Shinkansen stations, we chose to use the density of the commuter railway lines and the highways, the number of the commuter line stations and the interchanges of the highways. For the Population dummy variable, when we use the

Table 1 Utilized variables in the study

Variable name	Definition	Source
Shinkansen station (*)	Municipalities within a certain road distance from the Shinkansen station are set as 1	Railroad time series data
Coastal area (*)	Set the coastal municipalities to 1	Coastline data
Heavy snowfall area (*)	Set the heavy snowfall area municipalities to 1	Heavy snowfall area data
Habitable area ratio	The proportion of habitable area to the total area (%)	Habitable area ratio data
CL density	Length of the commuter railway line divided by the total area (km/km^2)	Railroad time series data
CL station	Number of commuter railway line stations in a municipality area	Railroad time series data
HW density	Length of the highway divided by total area (km/km^2)	Highway time series data
HWIC	Number of highway interchanges in a municipality area	Highway time series data
HWIC distance	The road distance from the local government to the highway interchange (km)	Highway time series data
Population (*)	The municipality with a total population of 30,000 or more is set to 1	Census
Population change ratio	Population change / the total population of the target year (%)	Census

(*) dummy variable: Discrete variable with value 1 or 0

propensity score matching, we aim to match municipality pairs of the same level. So, we distinguish the municipalities' size (large and small) by setting a population threshold (30,000). The propensity score then measures the probability that each municipality was assigned to the treatment group by performing a logistic regression analysis with covariates related to population change. The population variables are aggregated every five years from 1985 to 2015 based on the national census. Therefore, the base years of this study are 1985, 1990, 1995, 2000, 2005, 2010, and 2015.

In addition, regarding the setting of various distance-related variables, we have adopted the shortest road distance (only selected roads dedicated to cars or mixed traffic of people and vehicles above a certain size) to calculate the communication range, as shown in the Fig. 2 on the right.

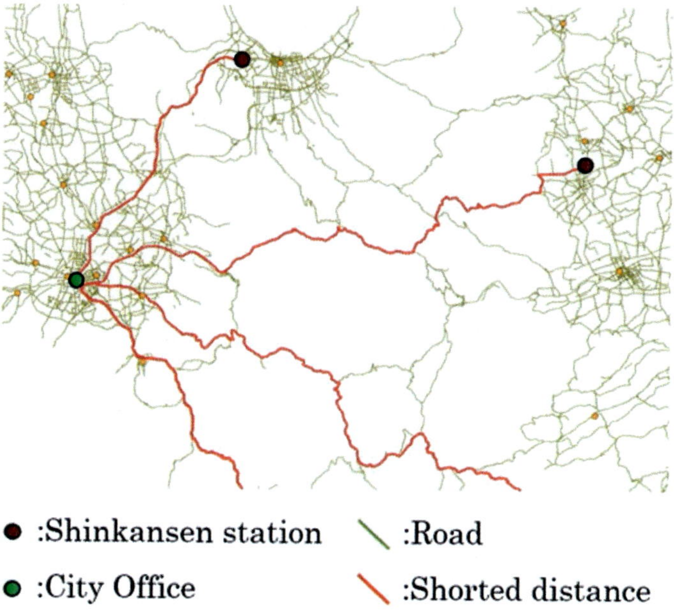

Fig. 2 The shortest road distance

3.4 Catchment Area of Shinkansen Stations

To quantify the influence of the existence of Shinkansen stations on the population changes in the municipalities around it, we need to determine the station catchment area, that is, its effect area. However, since the range of influence of the Shinkansen station cannot be defined statistically at this point, the road distance from the municipal city hall to the Shinkansen station is divided into ranges of 5 to 9 km in 1 km increment, and from the range of 10 to 20 km in 2.5 km increments. The effects of the existence of Shinkansen stations are calculated one by one for each range of influence of a Shinkansen station. By doing so, the Shinkansen station dummy variable is set to 1 for municipalities within the defined station area and 0 otherwise. Table 2 shows the number of municipalities where Shinkansen stations are defined based on the radius of their effect area.

3.5 Propensity Score Matching and Difference-In Differences Analysis [14–17]

In order to quantitative analyze the impact of high-speed railway stations on the social change rate of local population, that is the difference of building the high-speed railway station or not. We selected a set of (before and after observation) panel

Table 2 Number of municipalities with Shinkansen stations

Catchment area (Effect area)	Year						
	1985	1990	1995	2000	2005	2010	2015
5 km	33	38	38	42	50	52	63
6 km	37	44	44	48	56	57	73
7 km	46	54	54	60	68	69	87
8 km	53	62	62	70	80	81	105
9 km	62	72	72	81	91	92	120
10 km	75	87	87	96	106	108	141
12.5 km	104	121	121	132	144	146	185
15 km	153	170	170	184	198	199	251
17.5 km	197	221	221	238	252	256	316
20 km	236	260	260	280	296	300	365

data on the population social change rate of the treatment group and the control group in every five years to calculate the change trend of the social change rate over time.

From all municipalities, we extract one with a propensity score closest to the treatment and the control groups. The logistic regression and matching produce a pair of municipalities with an adjusted distribution of covariates, i.e., it conforms to the propensity score. We then use the nearest neighbor matching [14] on the treatment and the control groups.

For the Difference-in Differences analysis, it is necessary to ensure that the state of the treatment group when not receiving treatment (counterfactual state) and the state of the control group have a parallel trend to estimate the average treatment effect of the treatment group. We performed the DID analysis after ensuring that the two municipalities in a pair have the same development trend by propensity score matching. We used the propensity score for DID analysis to match the treatment and control groups. Regression analysis was performed to calculate the causal effect [13–15].

3.6 *The Method of Calculating the Average Treatment Effect of the Treatment Group*

We can estimate the average treatment effect on the treatment group using the DID regression equation by calculating the objective variable in the treated state (here, the population change ratio) and the objective variable in the untreated state, as in Eq. (4).

$$y = \mathrm{a}_1 + a_2G + a_3T + a_4GT + \varepsilon \tag{4}$$

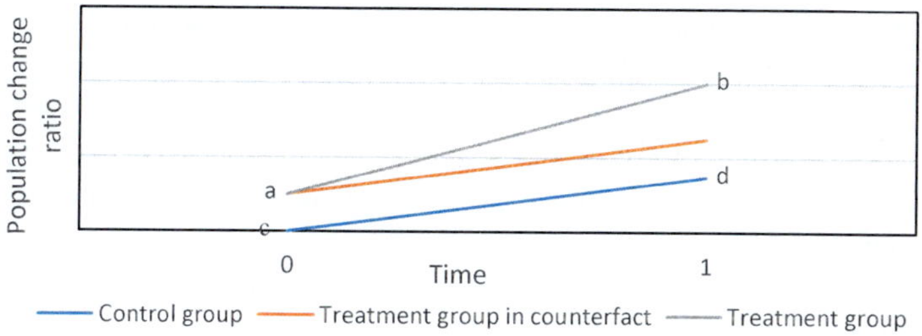

Fig. 3 Concept of DID analysis

Here, y is the population change ratio, G is the treatment dummy (whether the municipalities have a Shinkansen station or not), T is the time point dummy (that takes 0 for the reference year and 1 of 5 years after the reference year), ε is the error term, and α_1 to α_4 are coefficients; G and T are not related.

The illustration of the DID analysis is shown in Fig. 3. The population variability a for the reference year ($T = 0$) in the treatment group ($G = 1$), the population variability b for five years after the reference year ($T = 1$) in the treatment group ($G = 1$), the population variability c for the reference year ($T = 0$) in the control group ($G = 0$), and the population variability d for five years after the reference year ($T = 1$) in the control group ($G = 0$) can be calculated using Eq. (4). Then, using those four values, we can calculate the average treatment effect for the treatment group using Eq. (5).

$$\text{ATT} = (b - a) - (d - c) \tag{5}$$

This measures the effect of existing Shinkansen stations by DID analysis on municipality population changes.

4 The Average Treatment Effect on Treated (ATT) of Shinkansen Station by DID Analysis

4.1 Validation of the CIA and the CSA

Before calculating the propensity score by the logistic regression model, we need to test the CIA and CSA that must be satisfied.

There is currently no direct way to test CIA assumptions. Since it states that the distribution of every sample depends only on the covariate and not on other variables, we use the goodness-of-fit of the logistic regression model to test this assumption indirectly. The better the model's goodness-of-fit, the higher the accuracy of the prediction of the probability of each sample being assigned to the treatment group. We can show that the distribution of each sample depends mainly on the covariates.

In this study, we use ROC to test the models to ensure their goodness-of-fit meets the requirements. In Figs. 4 and 5, we select the case with the smallest number of samples (1985, 5 km) and the case with the largest number of samples (2015, 20 km) for display. We confirmed the model's goodness-of-fit by the area under the curve of the ROC; the larger the area under the curve, the more accurately the model describes the assignment.

The ROC shows that the logistic regression model has a high goodness-of-fit, so it can be inferred that CIA has been proven.

With regards to the CSA, to visually reflect the distribution of propensity scores for the treatment and control groups, we use the box plot to summarize the distribution

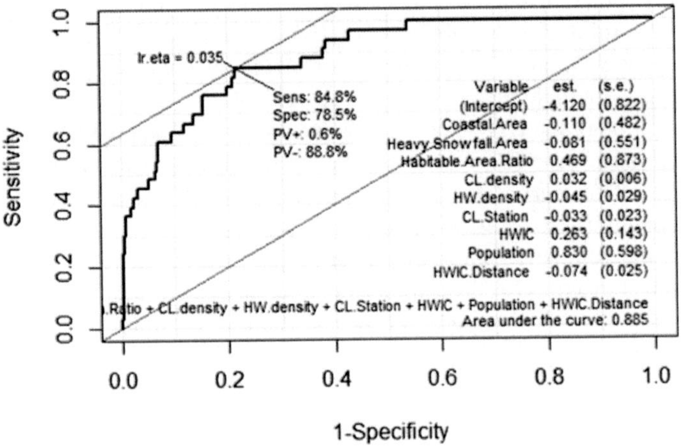

Fig. 4 ROC of 1985 when the station effect area is 5 km

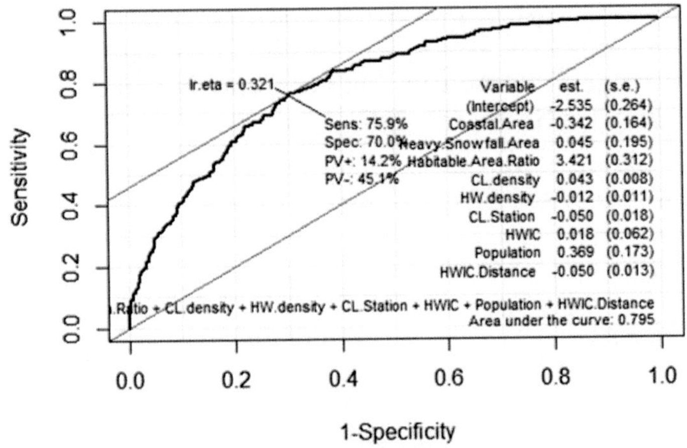

Fig. 5 ROC of 2015 when the station effect area is 20 km

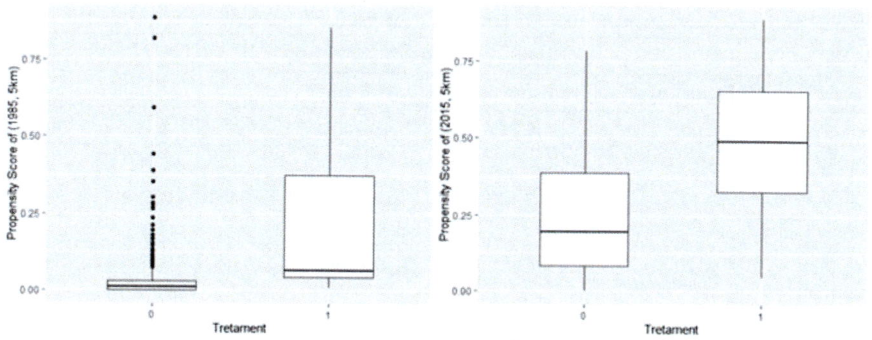

Fig. 6 The distribution of propensity scores (1985 and 2015)

of propensity scores of all samples in two cases (1985, 5 km, and 2015, 20 km, for the same reason as above), as shown in Fig. 6.

The box plot shows that the distribution ranges of propensity scores in the treatment and control groups mostly overlap. Although the distribution is not very good, we can still adjust the covariates for DID analysis by nearest neighbor matching.

4.2 *The Result of the Logistic Regression Model of Propensity Score*

The total population and the value of population change were used to calculate the IPWE. Since the variables used for DID analysis have not been weighted, it is more appropriate to use the ratio of population change rather than the value of population change. Table 3 shows the results for a station with an area of 5 km as an example.

4.3 *Matching Results*

After the logistic regression model calculates the propensity score, the propensity score needs to be used for nearest neighbor matching to adjust the covariate distribution and to form a treatment–control pair to satisfy the conditions for DID analysis. The distribution of propensity scores after matching is shown in Fig. 7.

The box plot shows that the distribution of the propensity scores of the matched samples has been greatly improved, which is a good result for this study.

Table 3 The result of propensity score (5 km) (DID analysis)

Years	1985		1990		1995		2000	
Variable	Estimate		Estimate		Estimate		Estimate	
Intercept	−4.077	***	−4.227	***	−4.616	***	−4.393	***
Coastal area	−0.137		−0.088		−0.149		−0.360	
Heavy snowfall area	−0.124		−0.347		−0.631		−0.525	
Habitable area ratio	0.611		0.811		1.249		0.931	
CL density	0.031	***	0.037	***	0.034	***	0.036	***
HW density	−0.045		−0.042		−0.002		0.003	
CL station	−0.033		−0.036		−0.034		−0.023	
HWIC	0.258		0.071		−0.092		−0.106	
Population change ratio	−0.030		−0.009		−0.022		0.013	
Population dummy	0.852		1.019		0.985		0.760	
HWIC distance	−0.075	**	−0.088	**	−0.080	*	−0.073	*
Years	2005		2010		2015			
Variable	Estimate		Estimate		Estimate			
Intercept	−4.002	***	−3.795	***	−4.129	***		
Coastal area	−0.199		−0.450		−0.128			
Heavy snowfall area	−0.651		−0.181		0.522			
Habitable area ratio	0.240		0.331		1.031			
CL density	0.048	***	0.046	***	0.060	***		
HW density	−0.005		0.024		0.002			
CL station	−0.027		−0.025		−0.018			
HWIC	−0.129		−0.264	*	−0.153			
Population change ratio	0.075		0.100	*	0.106	*		
Population dummy	0.384		0.278		0.161			
HWIC distance	−0.022		−0.013		−0.031			

Significance: *** 0.1%, ** 1%, * 5%

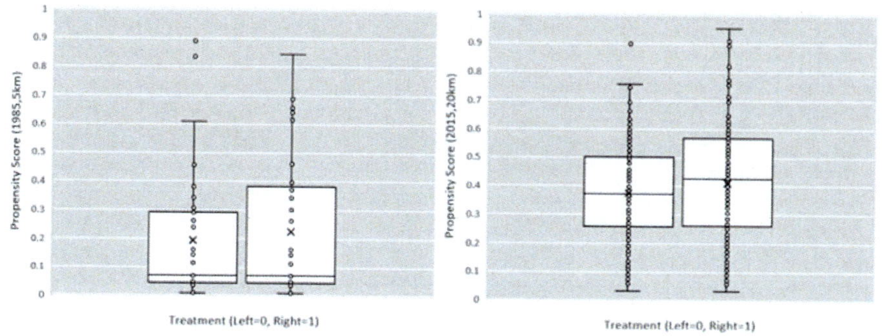

Fig. 7 Propensity score distribution after matching

4.4 The Population Change in All Target Municipalities

The total population change ratio of the municipalities with and without Shinkansen stations and the total population change ratio of municipalities without Shinkansen stations matched during DID analysis were plotted to understand the relationship between the total population change ratio of all objects in this study. The relevant graph is shown in Fig. 8 (the area of influence of Shinkansen stations is taken as 7 km as an example). Figure 8 demonstrates that the population fluctuation ratio of the municipalities with Shinkansen stations was higher than the total population change ratio of all target municipalities and that the population change ratio of the municipalities without Shinkansen stations was lower than the total population change ratio of all target municipalities. Since the population change ratio of the municipalities without Shinkansen stations is similar to that of municipalities with Shinkansen stations, it was found that the pair of treatment group and control group used for DID analysis have been appropriately matched.

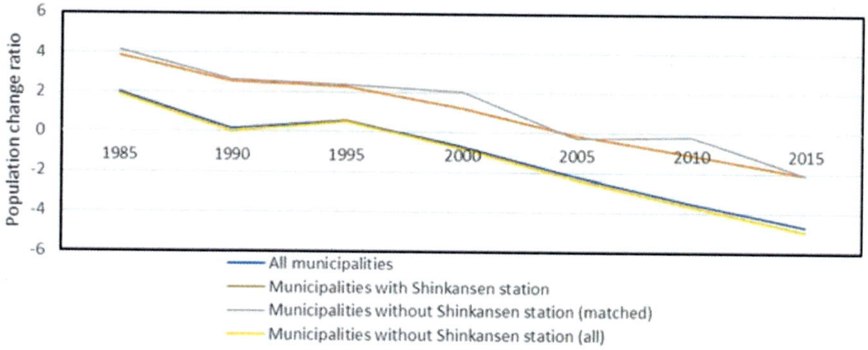

Fig. 8 The population change ratio

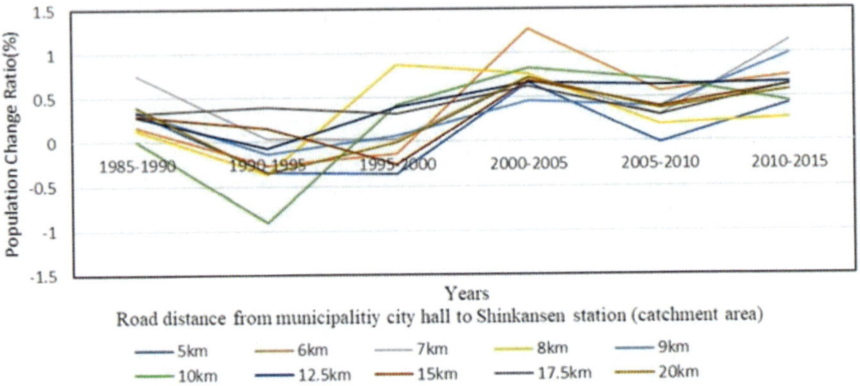

Fig. 9 Population changes in municipality due to the existence of Shinkansen stations

4.5 Causal Effect of DID Analysis

Figure 9 shows the difference in the ratio of population change between the municipalities with Shinkansen stations (treatment group) and the municipalities without Shinkansen stations (control group). It also shows the average causal effect for the treatment group, thereby demonstrating the population volatility expected to be brought about by the development of the Shinkansen each year. Similar to the analysis in the previous section, this analysis demonstrates how the causal effect changes by setting multiple stations' influence areas according to the road distance from the municipal city hall to the Shinkansen station.

The results of the DID analysis show that Shinkansen stations have a positive effect on the population of local municipalities in most years. They also have a population-increasing effect on the treatment group municipalities from 7 to 20 km. However, in the decade from 1990, the overall impact of Shinkansen stations showed a negative trend, especially from 1990, and the Shinkansen may have played a role in reducing the population. This may be due to the population movement towards the three major metropolitan areas resulting from the bursting of the economic bubble.

5 Conclusion

In this work, we utilized causal effects to empirically demonstrate the effect of the existence of Shinkansen stations on population change in their underlying municipalities. The DID analysis results showed that the Shinkansen stations in local municipalities positively affected local population change in most years in most station effect areas. However, Shinkansen stations might also have been a factor that contributed to the decline in local population between 1990 and 2000. We speculated that this was because the Shinkansen was the preferred mode of transportation as people moved to the three major metropolitan areas. Most of the population movement was via the Shinkansen during the economic bubble burst. So, it is possible that the Shinkansen had a negative impact on local population fluctuations during that period.

The results demonstrated that Shinkansen stations had a positive effect on local population change in most cases. Since other factors that had a greater impact on the population change, such as the distance from the nearest metropolis, might have been omitted, it is necessary to re-examine the regression equation to calculate the propensity score. In addition, there are still inappropriate places in propensity score matching; for example, there may be better ways of making covariate adjustment. Since population change can be divided into social increase and decrease and natural increase and decrease, future research should analyze the social increase and decrease by excluding the influence of natural increase and decrease.

References

1. Meng, X., Lin, S., Zhu, X.: The resource redistribution effect of high-speed rail stations on the economic growth of neighboring regions: evidence from China. Transp. Policy **68**(30), 178–191 (2018)
2. Kim, H., Sultana, S., Weber, J.: A geographic assessment of the economic development impact of Korean high-speed rail stations. Transp. Policy **66**, 127–137 (2018)
3. Wetwitoo, J., Kato, H.: High-speed rail and regional economic productivity through agglomeration and network externality: a case study of inter-regional transportation in Japan. Case Stud. Transp. Policy **5**(4), 549–559 (2017)
4. Okada, H.: Features and economic and social effects of the Shinkansen. Jpn. Railw. Transp. Rev. (1994)
5. Terabe, S., Maekawa, Y., Kasai, M., Kuroe, K.: Classifying high-speed rail stations based on socioeconomic changes in surrounding areas after the development of the Kyushu Shinkansen. In: Proceedings of the 95th Annual Meeting of Transportation Research Board, 16–1685 (2016)
6. Talebian, A., Zou, B., Hansen, M.: Assessing the impacts of state-supported rail services on local population, and employment: a California case study. Transp. Policy **63**, 108–121 (2018)
7. Jia, S., Zhou, C., Qin, C.: No difference in effect of high-speed rail on regional economic growth based on match effect perspective? Transp. Res. Part A **106**, 144–157 (2017)
8. Bruzzone, F., Cavallaro, F., Nocera, S.: The effects of high-speed rail on accessibility and equity: evidence from the Turin-Lyon case-study. Socio-Econ. Plann. Sci. 101379 (2022)
9. Pagliara, F., Henke, I., Russo, L., Guigon, M.: Is high speed rail a geographically inclusive system? evidence from some European countries. Appl. Spat. Anal. Policy **15**(1), 241–263 (2022)
10. Cavallaro, F., Bruzzone, F., Nocera, S.: Effects of high-speed rail on regional accessibility. Transportation (2022). https://doi.org/10.1007/s11116-022-10291-y
11. Rubin, D.B.: Estimating causal effects of treatments in randomized and nonrandomized studies. J. Educ. Psychol. **66**(5), 688 (1974)
12. Hoshino, T.: Estimation of causal effect and adjustment of survey data using propensity scores. Jpn. J. Behav. (2004)
13. Rosenbaum, P.R., Rubin, D.B.: The central role of the propensity score in observational studies for causal effects. Biometrika **70**(1), 41–55 (1983)
14. Robins, J.M., Hernan, M.A., Brumback, B.: Marginal structural models and causal inference in epidemiology. Epidemiology **11**, 550–560 (2000)
15. Dimick, J.B., Ryan, A.M.: Methods for evaluating changes in health care policy: the difference-in-differences approach. JAMA **312**(22), 2401–2402 (2014)

16. Asai, Y., Kanbayashi, R., Yamaguchi, S.: Crowding-out effect of publicly provided childcare: why maternal employment did not increase. Econpapers, ISS Discussion Paper Series (Series F), No. f177 (2015)
17. Goodman-Bacon, A.: Difference-in-differences with variation in treatment timing. J. Econ. **225**(2), 254–277 (2021)

Impacts of High Speed Rail on Residential Property Prices in Italy: A Panel-Data Set Analysis

Irina Di Ruocco, Filomena Mauriello, and Francesca Pagliara

Abstract The role played by High-Speed Rail (HSR) systems in terms of encouraging travel, tourism and boosting the labor productivity of the cities and regions served has been widely studied. Since 2008 Italy has experienced the launch of HSR in the country. Despite the great innovation in Italy, impacts of HSR on property prices remain unexplored. The added value of this contribution is to analyse the link between HSR and residential property prices for the case study of Italy. The research aims at analyzing the processes of transformation of the territory and the correlation between HSR and the real estate market, quantifying the impacts that HSR has generated. The methodology is based on the evaluation of impact assessment through a hedonic pricing model, by estimating ex post and ex ante scenarios of residential properties values close to HSR stations with a difference-in-difference approach. To demonstrate the robustness of the method the OLS estimators is applied. This study is based on a dataset of 10 cities served by HSR, considering also dummies variables to simulate the HSR cause-effect. A preliminary analysis shows a link between HSR and residential property prices.

Keywords High speed rail · Hedonic pricing model · Spatial econometrics · Italy

1 Introduction

Over the past decade, High-Speed rail (HSR) systems have attracted worldwide research interest [1–5]. Given the mutually dependence between transportation and economic development [6, 7], HSR system is functional for many activities as travelling [8, 9], tourism purposes [10, 11], and more generally it serves cities and regions

I. Di Ruocco (✉)
University of Insubria, DIECO, Varese, Italy
e-mail: idiruocco@uninsubria.it

F. Mauriello · F. Pagliara
University of Naples Federico II, DICEA, Naples, Italy

F. Pagliara (ed.), *Socioeconomic Impacts of High-Speed Rail Systems*,
Springer Proceedings in Business and Economics,
https://doi.org/10.1007/978-3-031-26340-8_12

far from each other [7, 12]. Compared to conventional rail systems [13], HSR operating over medium to long distances, reducing journey times and providing services, can also affect the real estate market.

According to the literature, the link between HSR and the property market is becoming an important research topic in many countries [13] as well in transport evaluation discussions [14] as highlighted by the several studies on cost–benefit analysis [15] that underline the focus of economic decision-making processes related to transport infrastructures and services.

The opening of high-speed stations has enabled both the development of the railway network and improved the accessibility of the territory and passengers, changing transport dynamics such as the increase in commuting for work, tourism and other purposes. As a result, the urban form of the city has changes due to the presence of HSR that has boosted the real estate market to reallocate resources near stations according to new travel demand needs.

The effects of HSR can also be observed at different spatial levels, with different endogenous factors relating to the specific economic conditions of the territory. At the micro-regional level, studies focus on smaller, connected areas, and show a change in the real estate markets and land prices [13]. HSR stations have an impact on local economies by increasing the benefits of reduced generalized transport costs and increased accessibility, resulting in higher growth and productivity. Among the various wide impacts induced by high-speed trains, another effect has been noted if a metro station is present where high-speed trains are present, inducing urban regeneration measures [7, 16].

2 Theoretical Background

2.1 HSR and Land Values Dynamics: Literary Review

With regard to the literature analyzed, the HSR positive effects on the territory do not show a uniform consensus across countries. Furthermore, diverging trends emerged when analyzing the periods of inauguration of the HSR line between the different countries, and more often in the same country there were several investments involving only certain regions. The motivations for opening a new HSR station were strongly linked to its costs and benefits, demonstrated by numerous scientific researches that have explored the different effects of HSR development in ex post analyses [17, 18].

Considering the most relevant scientific evidence, HSR induces changes in the behavior of passengers [3, 10, 19–21], on the tourism sector as it improves the attractiveness of cities, it contributes to reduce the congestion, the carbon emissions and air pollution problems due its advantages of promoting intermodality, reducing car use. The attractiveness of high-speed rail for intercity transport users has been assessed in other works, such as those by [22, 23], who analyzed the preferences and willingness to pay of Italian travellers on the Naples-Milan corridor through complex random utility models. At urban local level, HSR increases interregional accessibility, facilitating commuting and linking cities at short distances by creating one huge connected economic region [24–28]. At the macro-regional level, HSR improves long distance travelling and international trips, showing effects on the land sector [29] and on the regional economy [7, 30–33]. As suggested in Monzón et al. [34], a new HSR link could strengthen the advantageous location of cities nearby the main nodes of the network. Many studies have focused on the hedonic pricing approach [18, 35], and estimation of spillovers effects [36–39]. A slightly different analysis has been performed by [40], who analyzed the impact of the proximity of properties to metro stations on their perceived value of the properties by retailers, considering the case study of Naples, using a discrete choice experiment to model the retail location choice. However, current research is still limited in terms of investigating the HSR effect on the property market. It has also been pointed out that the impact on property values has not reached a unified consensus in the literature, as some research has found that there is a positive effect on land values [41, 42], a minor effect [43], and a negative effect [44, 45].

When a new HSR project is designed and new HSR project stations are planned, the project produces advantages to urban development, both in large cities and smaller towns. The deployment of HSR in Italy, with speeds between 280 and 300 km/h, has changed users' travelling experience, reducing car use and becoming competitive w.r.t. buses on many routes [24, 46].

2.2 *HSR and Property Values*

Many contributions on the impact of HSR on land values deal with the case studies of UK [15] and China [13, 33, 47]. This paper aims at examining the effects of HSR on the residential property prices for the Italian high speed rail network. The latter has been developed in several steps moving from high capacity to high speed, with a network still in progress. The location of stations is fundamental for the economic development of the country and enhanced travel, showing that stations planning may be a strategic decision. In this study HSR stations such as "Naples Afragola" and "Reggio Emilia AV" have been considered. Indeed, these two stations are placed not in the city centers of the cities served but in more peripheral areas, since it is becoming a real need to relocate stations outside the city to manage traffic flows. The effect of HSR on land values therefore differs when considering an urban texture such as suburbs, urban center, metropolitan city [48–50].

Preliminary economic theories of HSR effects suggest that railway infrastructure can influence property values and change the spatial distribution of these values [51]. Further benefits are evident in the connections between HSR and other infrastructure with increased intermodality, or changing the labour market in the metropolitan area by creating new opportunities, changing residential prices, constructing and/or transforming buildings for additional residential and commercial functions [7, 52, 53].

2.3 HSR and Economic Theories

From a wider perspective these previous considerations, related with concepts of accessibility and economic development, are gaining ground in the analysis on HSR and land values. A further reflection on the effects of HSR on the real estate market is related to investments and the relationships with the land [6, 54], which follows the principles of infrastructure theory proposed by Ansar et al. [55], highlighting how investments in HSR have different results on spatial development. According to the main school of thought that has dominated the mainstream discussion in economics [56, 57] the financial, social, and environmental performance of infrastructure investments is, in fact, poor [55, 58]. Starting from these fundamental considerations, evidence shows the added value of HSR on the real estate market.

With the aim of analyzing the economic impact of HSR on the real estate market, several methods have been applied from the economic theory of hedonic pricing [39] to the Difference-In-Difference (DID) method used for the HSR case studies in China and to the quasi-natural experiments approaches [59] which have been used on the spatial casual effect and on the investment efficiency. Many empirical results suggest that the impact of rail infrastructure on property prices may vary with city dimension and by the time periods of analysis [35]. The DID method has found many applications, with the aim of assessing the effects between two treatments or between different time periods. While in China the price effect is more significant in small and medium-sized cities [11], in Italy there are no confirmed results yet.

2.4 Study Motivation

The focus of this study is on the analysis of the factors influencing the real estate sector by examining the specific case of the HSR netwok in Italy, which has the potential to create relevant changes in the economic development, analyzing also the cause-effect relationship on the property market. This paper aims (1) to enrich the emerging literature on the effects of HSR on the real estate market (2) to investigate the relationship between accessibility, HSR and the residential property prices, (3) to estimate, through a quasi-experimental approach, the causal effects of HSR on property prices, showing a methodology applied to the Italian context based on the hedonic pricing approach.

3 Research Design

3.1 *The Development of the HSR Network in Italy*

Concerning the Italian context, the first high speed line was inaugurated in 1992 between Florence and Rome known as "Direttissima". The line has experienced several changes such as the introduction of a new generation of trains (travelling at 300 km/h) which started in December 2005 for the cities of Rome, Naples, Milan and Bologna. In December 2009 the project was extended with the Milan-Turin and Bologna-Florence lines. In 2010, the Italian high-speed network became fully operational and other projects are still in progress. The Italian national network and operations are all owned by FS Holdings (Ferrovie dello Stato "FS"). FS has as main operating companies Trenitalia that operates freight and passenger trains, including high-speed trains, interregional and regional services, and RFI (Rete Ferroviaria Italiana) that is responsible for the infrastructure [10, 60].

The railway line had evolved between 2005 and 2022 with different investments of upgrading from high capacity to high speed, opening important high-speed stations such as Naples, Milan and Turin in 2009. After 2009, 10 more stations have been built, included the "Grandi Stazioni - Great Stations" project with the aim of bringing the Italian railway system up to the best European standards and improving its capacity and speed. Since 2012 a new private company, Nuovo Treno Viaggiatori known as "Italo" is competing with Trenitalia on the same HSR network. Some important advantages of opening a high-speed railway line in early 2009 is the reduction in travel times along Italian cities. Investments in transportation infrastructure have often played key roles in the urban land use and land price models. The railway line is also promoted by the Italian Recovery Plan "PNRR" [61] *Mission no. 3* to reduce inequality between regions, to improve accessibility and to increase economic development. The Mission no.3 aims at developing a modern, sustainable and interconnected transport infrastructure network to boost the competitiveness of the country [62]. The case study is represented by 13 HSR stations in 10 Italian cities, with the exception of cities where HSR is under construction or where the "Great Stations" project is not yet present (Fig. 1).

3.2 *Methods and Data*

In this study, a traditional hedonic pricing model, based on properties values collected around the HSR in Italy have been considered. One limit of the hedonic pricing theory lies in the limitation of not analyzing endogenous issues in depth since this manuscript proposes an analysis from both a temporal and spatial perspectives. Indeed, the spatial extension includes Italian stations for both non-metropolitan and metropolitan cities, while the temporal extension consists of considering the 2006–2022 period of data collection. As it will be explained in the next section, it will be applied the

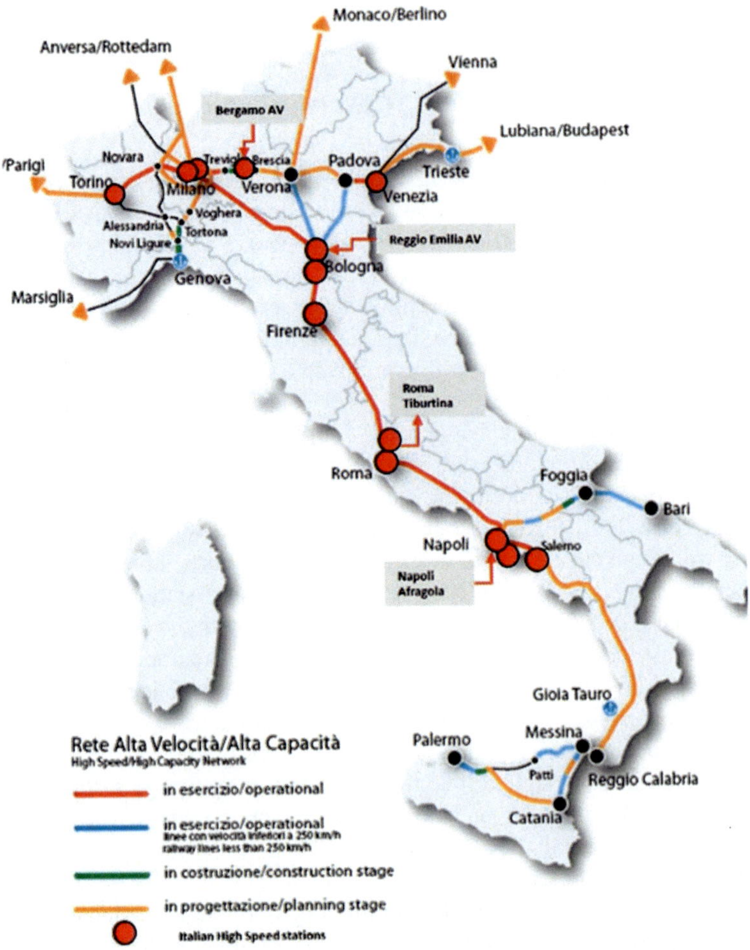

Fig. 1 Italy HSR network (*source* elaboration based RFI Source)

Generalized Estimating Equations (GEE) due to different considerations: firstly, the lack of stationary condition among stations; secondly, because of preliminary results derived from the statistical analysis of the variable. In this manuscript, the GEE model was chosen, since one of its main advantages is, as showed in Zorn [63], the quasi-likelihood methods, that differently from the standard maximum-likelihood, requires the full conditional distribution of the dependent variable.

These reasons, when considered in terms of bias, indicate the possibility of errors in price evaluation due to unobserved factors. To solve the problem of omitted variable bias in traditional hedonic pricing models, a quasi-experimental approach has been chosen. In Table 1 the characteristics of the 10 cities analyzed are reported.

Descriptive statistics are presented in the following Tables 2, 3, 4, and 5.

Table 1 Main characteristics of cities

City	Station	Italy position	Metropolitan city	GDP*per capita in Eur € (OECD, 2019)	Unemployement rate**(15–74 years, 2021)
Naples	Garibaldi	South	Yes	18,149	23.67
Naples	Afragola	South	–	18,149	23.67
Salerno	Centrale	South	–	17,479	15.12
Roma	Termini	Centre	Yes	34,625	9.76
Roma	Tiburtina	Centre	Yes	34,625	9.76
Firenze	Novella	Centre	Yes	35,642	6.18
Bologna	Centrale	Centre	Yes	38,918	4.57
Reggio E	Mediopadana AV	Centre	–	33,694	5.15
Milano	Centrale	North	Yes	50,786	6.47
Milano	Porta Garibaldi	North	Yes	50,786	6.47
Torino	Porta Susa	North	Yes	30,304	8.26
Venezia	Santa Lucia	North	Yes	30,208	6.03
Bergamo	Centrale	North	–	30,485	3.55

The GDP was considered at the regional level and the unemployment rate at a more disaggregated level [64]. The dataset deals with information regarding the residential property values for the 10 Italian cities served by HSR, observed during the years from 2006–2022 [65].

The dependent variable considered is as listed below:

- *Residential property value*: price in €/m^2 of the residential property located in the same neighbourhood of the HSR station or considered as the area nearby the same HSR neighbourhood. In this case it is the average of the total prices, if in case of several districts, the average value of the areas was considered [65].

The model is based on the logarithmic function of the dependent variable (log(Y)) and it follows a normal distribution. The outcome variable is the *Residential Price* (Log(ResPrice)) for the property location i at time t.

The independent variables are:

Transportation system variables

- *HSR opening year*: is a dummy assuming value 1 if the HSR station is present, 0 otherwise;
- *Metropolitan city*: is a dummy assuming value 1 if the HSR station is located in a metropolitan area, 0 otherwise;

Table 2 Descriptive statistics per year

Year	Residential €/mq households		Income per capita (€/inhab)		Pop density inhab./km^2		Distance HSR to main road (km)		Distance HSR to main road (min)		Presence of school		Presence of park		Presence of hospital	
	Average	St.dev	Average	St.dev	Average	St.dev	Average	St.dev	Average	St.dev	Average	St.dev	Average	St.dev	Average	St.dev
2021	3457	1106	23,017	4687	4220	2554	31	15	3	1	1	0	1	0	0	0
2020	3453	1112	19,985	4235	3701	2425	34	19	3	1	1	0	0	0	0	0
2019	3448	1120	23,067	4623	4164	2529	30	15	3	1	1	0	1	0	0	0
2018	3443	1129	23,105	4594	4130	2508	30	15	3	1	1	0	1	0	0	0
2017	3437	1137	23,143	4565	4097	2486	30	15	3	1	1	0	1	0	0	0
2016	3433	1143	23,178	4540	4065	2464	30	15	3	1	1	0	1	0	0	0
2015	3428	1150	23,203	4524	4032	2441	30	15	3	1	1	0	1	0	0	0
2014	3424	1156	23,226	4508	4001	2419	30	15	3	1	1	0	1	0	0	0
2013	3420	1161	23,249	4491	3970	2397	30	15	3	1	1	0	1	0	0	0
2012	3416	1166	23,270	4473	3940	2375	30	15	3	1	1	0	1	0	0	0
2011	3412	1171	23,292	4454	3910	2353	30	15	3	1	1	0	1	0	0	0
2010	3408	1176	23,312	4436	3880	2331	30	15	3	1	1	0	1	0	0	0
2009	3404	1180	23,332	4417	3849	2308	30	15	3	1	1	0	1	0	0	0
2008	3400	1185	23,352	4397	3819	2285	30	15	2	1	1	0	1	0	0	0
2007	3399	1186	23,374	4374	3789	2261	30	15	2	1	1	0	1	0	0	0
2006	3396	1189	23,393	4352	3758	2236	30	16	2	1	1	0	1	0	0	0

Table 3 Descriptive statistics per year

Year	Presence of shopping centre		Whether bus stops is nearby HSR		Whether metro stops is nearby HSR		Whether airport stops is nearby HSR		Whether Road/highway stops is nearby HSR		Distance from HSR (min)		Distance from HSR (km)	
	Average	St.dev	Average	St.dev	Average	St.dev	Average	St.dev	Average	St.dev	Average	St.dev	Average	St.dev
2021	1	0	1	0	1	0	0	0	0	0	3	2	11	6.69
2020	1	0	1	0	1	0	0	0	1	0	2	1	8	4.13
2019	1	0	1	0	1	0	0	0	0	0	3	2	11	6.73
2018	1	0	1	0	1	0	0	0	0	0	3	2	11	6.74
2017	1	0	1	0	1	0	0	0	0	0	3	2	11	6.76
2016	1	0	1	0	1	0	0	0	0	0	3	2	11	6.77
2015	1	0	1	0	1	0	0	0	0	0	3	2	11	6.78
2014	1	0	1	0	1	0	0	0	0	0	3	2	11	6.80
2013	1	0	1	0	1	0	0	0	0	0	3	2	11	6.81
2012	1	0	1	0	1	0	0	0	0	0	3	2	11	6.83
2011	1	0	1	0	1	0	0	0	0	0	3	2	11	6.85
2010	1	0	1	0	1	0	0	0	0	0	3	2	11	6.86
2009	1	0	1	0	1	1	0	0	0	0	3	2	11	6.87
2008	1	0	1	0	1	1	0	0	0	0	3	2	11	6.89
2007	1	0	1	0	1	1	0	0	0	0	3	2	11	6.90
2006	1	0	1	0	1	1	0	0	0	0	3	2	11	6.92

Table 4 Descriptive statistics per stations

Stations	Residential €/mq households		Income per capita (€/inhab)		Pop density inhab./km^2		Distance HSR to main road (km)		Distance HSR to main road (min)		Presence of school		Presence of park		Presence of hospital	
	Average	St.dev	Average	St.dev	Average	St.dev	Average	St.dev	Average	St.dev	Average	St.dev	Average	St.dev	Average	St.dev
Na Gar	2272	196	17,028	1021	8266	521	2	0	10	0	1	0	0	0	0	0
Na Afragola	4004	883	14,756	705	3508	79	2	0	5	0	0	0	0	0	0	0
Salerno C.	4042	955	17,413	2504	2210	39	1	0	4	0	1	0	0	0	1	0
Rome Ter.	4316	664	24,253	1310	2221	29	4	0	15	0	1	0	1	0	0	0
Rome Tib	4269	396	24,704	1318	2221	29	2	0	6	0	1	0	1	0	1	0
Firenze SMN	2443	421	22,709	1650	3584	35	3	0	12	0	1	0	0	0	0	0
Bologna C.	2863	417	24,400	1948	2715	338	4	0	12	0	1	0	0	0	0	0
Reggio R.AV	1350	0	23,581	600	2291	0	5	0	11	0	0	0	0	0	0	0
Venezia S.L.	4420	387	21,289	1853	1090	514	0	0	0	0	1	0	0	0	0	0
Torino P.S.	2523	284	28,267	3286	7132	965	20	0	1	0	1	0	1	0	1	0

(continued)

Table 4 (continued)

Stations	Residential €/mq households		Income per capita (€/inhab)		Pop density inhab./km^2		Distance HSR to main road (km)		Distance HSR to main road (min)		Presence of school		Presence of park		Presence of hospital	
	Average	St.dev	Average	St.dev	Average	St.dev	Average	St.dev	Average	St.dev	Average	St.dev	Average	St.dev	Average	St.dev
Milan C.	3521	396	28,669	2942	7416	248	6	0	23	0	1	0	1	0	0	0
Milan P.G.	4498	949	28,015	2929	7416	248	30	0	24	0	1	0	1	0	0	0
Bergamo C	1521	236	21,854	1004	2969	90	3	0	5	0	1	0	0	0	0	0

Table 5 Descriptive statistics per stations

Stations	Presence of shopping centre		Whether bus stops is nearby HSR		Whether metro stops is nearby HSR		Whether airport stops is nearby HSR		Whether Road/highway stops is nearby HSR		Distance from HSR (min)		Distance from HSR (km)	
	Average	St.dev	Average	St.dev	Average	St.dev	Average	St.dev	Average	St.dev	Average	St.dev	Average	St.dev
Na Garibaldi	1	0	1	0	1	0	0	0	1	0	10	0	2	0
Na Afragola	0	0	1	0	0	0	0	0	1	0	5	0	2	0
Salerno C	0	0	1	0	0	0	0	0	0	0	4	0	1	0
Roma Termini	1	0	1	0	1	0	0	0	0	0	15	0	4	0
Roma Tib	1	0	1	0	1	0	0	0	1	0	6	0	2	0
Firenze SMN	1	0	1	0	0	0	0	0	0	0	12	0	3	0
Bologna C	1	0	1	0	0	0	0	0	0	0	12	0	4	0
Reggio Emilia AV	0	0	1	0	0	0	0	0	1	0	11	0	5	0
Venezia SL	0	0	1	0	0	0	0	0	1	0	6	0	1	0
Torino P.S.	0	0	1	0	1	0	0	0	0	0	3	0	1	0
Milano C	1	0	1	0	1	0	0	0	0	0	6	0	23	0
Milano P.G.	1	0	1	0	1	0	0	0	0	0	6	0	23	0
Bergamo C	0	0	1	0	0	0	0	0	0	0	3	0	5	0

- *HSR distance to main road*: describes the average distance in mins and km from the HSR station to the closest infrastructure (expressed in minutes and km).

Attractiveness variables

- *School:* is a dummy assuming value 1 if a school is present, 0 otherwise;
- *Shopping centre*: is a dummy assuming value 1 if a shopping centre is present, 0 otherwise.

Socio-economic variables

- *Pop:* is the average value of the population per city [64];
- *Pop Density:* is the average value of the density €/inh. [64];
- *Income per capita:* is the income per population based on Census data [64];
- *GDP:* is the Gross Domestic Product expressed in millions [64];
- *Unemployment*: % of unemployed in the given province [64].

The choice of these independent variables is in line with the literature as highlighted in [13, 35] and new variables have been added according to the case study under analysis.

It is relevant to highlight that the unavailability of data for some cities in several periods of analysis (i.e. before 2006 for all cities and before 2013 for the station "Reggio Emilia AV") represented a limitation.

4 The Generalised Estimation Equation GEE Model

To the best of our knowledge, previous research studies have considered various empirical and methodological tools to investigate the effect of HSR on the real estate market. Few of them have analyzed the impact on residential property prices by applying the hedonic pricing model. One reason is due to complexity of modelling when using panel data, for the existence of spatial autocorrelation and the bias problem due to possible unobserved heterogeneity of the variables [66, 67]. The GEE is widely used in spatial analysis and well known in the literature [68, 69] with several results in cluster data [70]. The GEE, as reported in the literature, is a part of the Generalized Linear Models (GLM) [71]. The model is suitable for analyzing panel data made up of count data [67]. Furthermore, the GEE has a different role in the analysis of spatial correlation, as the GEE is a type of Generalized Linear Model (GLM) is used for cross-sectional analysis [72], correlated data in longitudinal study [73] and spatial analysis [74].

Some current evidence shows that the GEE has been used for clustered data analysis [75]. As mentioned in [10, 67], the relationship between the dependent variable and any independent variable was assumed to be stationary with respect to the GEE model, which, unlike other models such as the Geographically Weighted (GW) one, provides a better assessment of the temporal relationship of the variables, whereas a GW is suitable to explore spatial relationships.

To assess whether the data of this study are temporally or spatially correlated, several tests were carried out for all dependent variables as test for normality or t-student.

Concerning the autocorrelation, it is highlighted in literature [72, 76] that in order to use the GEE model with temporal and spatial data, an appropriate correlation environment compatible with spatial autocorrelation (related to panel data characteristics) should be used. The spatial autocorrelation is a concept that studies the phenomenon whereby two neighbouring regions are more correlated than distant ones [72]. For this case study, the GEE was set with a first-order autocorrelation value.

Regarding the panel dataset, 3,266 observations were collected. The choice to use panel data is due to measurements over time for the same cities, in order to avoid serial correlation leading to statistical inferences, the regression model considered is the one valid on panel data, extending the use to the GLM to account for different aspects of the correlation.

This model captures the time variation by fitting a regression model as described in the Eq. (1) proposed by Lipsitz and Fitzmaurice [10, 77, 78]:

$$\mathrm{E}\left[y_{\mathrm{it}}\right] = e^{\beta_0+\beta_1 x_{1t}+\beta_2 x_{2t}+\ldots+\beta_p x_{pt}+\Phi y_{it-1}+u_{it})} \tag{1}$$

The parameter β_0 is the intercept of regression, the terms $\beta_p, i = 0,1,\ldots,p$ are the regression coefficients, ϕ is the parameter for the autoregressive component, u_{it} is the error component, considering the disturbances correlated with the term y_{it-1}.

To solve the autocorrelation, a coefficient of autocorrelation is used of the first type, the model is fitted using population-averaged Poisson models and fixing a value of the AR process (1) in the error term, and the threshold value is < 0.1. The parameters of this model, as proposed in Pagliara and Mauriello [10], are estimated using a 'backward' elimination process, which considers all variables in the model at the beginning of modelling and at each step of the backward process one variable is eliminated.

5 Results

The impact of HSR on *residential property values* at the city level is reported in the Table 6, showing the results of the regression (only statistically significant covariates are exposed) reported above.

A first consideration arises from the GEE model since the variables are measured in logarithmic terms, the coefficients are interpreted as elastic.

The choice in the GEE model to fit the term AR (1) provides interesting results and is in line with some of the literature analysed [13, 79].

The dependent variable is mainly affected by the *distance of HSR from city centre, presence of schools* and by *metro stops nearby to HSR*, resulting both significant in terms of the P-value and positive, while *distance of HSR from city centre* shows a

Table 6 GEE results

	Coef	Std. Err.	z	P > lzl
Dummy metropolitan city	−1.382	0.481	−2.872	0.004
Distance HSR—city center	−0.030	0.014	−2.211	0.027
Presence of school	0.497	0.288	1.726	0.084
Presence of hospital	1.810	0.797	2.272	0.023
Presence of shopping center	2.513	1.022	2.459	0.014
Whether metro stops is nearby	0.224	0.126	1.777	0.076
Distance to main road (min)	0.042	0.023	1.854	0.064
Intercept	6.491	0.628	10.343	<0.001

negative coefficient. As emerged in Armstrong and Rodriguez [45], HSR has a minor effect when the distance of the property from the HSR increases. This means that the presence of a shopping center inside the HSR stations, distance to the main road, and the bus stops affect positively the property prices. According to the value of accessibility, reported by *Distance to main road (min),* it is statistically significant across all cities and all stations. This can be interpreted as an increase in accessibility to HSR stations (reduction of travel time) associated with an increase (4.2%) of property prices. The questionable result, as indicated in the literature [11], shows that the analysis has a further clustering step for the type of city classifying whether city is metropolitan or non-metropolitan, as the regression shows a negative coefficient for the dummy of the metropolitan value, suggesting a further analysis to disaggregate the cities and related stations. In this case the negative indirect effect of HSR strengthened the relationship between property values and type of cities. The negative coefficient of the variable *Dummy metropolitan city* is due to statistical estimation problems such as the omission of the spatial dependence, suggesting a further analysis of the station accessibility, linking property prices to station proximity/remoteness. Based on this consideration, this aspect also suggests a consideration about the spill over effect generated by HSR as the total effect of accessibility and, on *ceteris paribus* condition, one might expect it to be insignificant for non-metropolitan cities, while being positive and significant for metropolitan ones.

6 Conclusions

HSR can significantly improve accessibility, economic development, bringing several benefits. This study examines the effect of HSR on the residential property market in Italy around the HSR stations. This paper represents a positive correlation between HSR and property values. Furthermore, this study provides a contribution to the literature with the application of the GEE. Given the heterogeneity of the territories, a further subdivision by type of cities is fundamental. This study also proposes some social policy implications, as presented in Huang and Du [13]. Firstly, Italian government at different levels (national, regional e municipal) may benefit from these values by adapting different taxation on land in the proximity of stations (capitalisation or sale, offices rent inside the station buildings or as promoted by Rete Ferroviaria Italiana the redevelopment of buildings around stations).

Based on this study, further research will consider the enrichment of the results obtained including new stations in the pipeline.

CRediT authorship contribution statement Irina Di Ruocco: Data curation, Formal analysis, writing—review & editing.

Francesca Pagliara: Conceptualization, Supervision, review & editing.

Filomena Mauriello: Formal analysis, Supervision.

References

1. Börjesson, M.: Forecasting demand for high speed rail. Transp. Res. Part A Policy Pract. **70**, 81–92 (2014)
2. Cascetta, E., Coppola, P.: New high-speed rail lines and market competition. Transp. Res. Rec. **2475**, 15–18 (2015). https://doi.org/10.3141/2475-02
3. Chen Z., Zhou, Y., Haynes, K.E.: Change in land use structure in urban China: Does the development of high-speed rail make a difference. Land Use Policy **111** (2021)
4. Delaplace, M.: High-Speed Rail and the City. In: Vickerman, R. (eds.) International Encyclopedia of Transportation, pp. 261–265. Elsevier (2021)
5. Gao, Y., Zheng, J., Wang, X.: Does high-speed rail reduce environmental pollution? Establishment-level evidence from China. Socio-Econ. Plann. Sci. **83** (2022)
6. Chen, Z., Xue, J., Rose, A.Z., Haynes, K.E.: The impact of high-speed rail investment on economic and environmental change in China: a dynamic CGE analysis. Transp. Res. Part A: Pol. Pract. **92**, 232–245 (2016)
7. Vickerman, R.W.: Can high-speed rail have a transformative effect on the economy? Transp. Policy **62**(C), 31–37 (2018).
8. Lumsdon, L., Page, S.J. (eds.): Tourism and Transport: Issues and Agenda for the New Millennium. Elsevier, Oxford (2004)
9. Di Matteo, D.: Does high-speed rail matter for tourism? Evidence from Italy. Res. Transp. Bus. Manag. (2022)
10. Pagliara, F., Mauriello, F., Garofalo, A.: Exploring the interdependences between High Speed Rail systems and tourism: some evidence from Italy. Transp. Res. Part A Policy Pract. **106** (2017).
11. Chen, Z., Haynes, K.E.: Impact of high-speed rail on international tourism demand in China. Appl. Econ. Lett. **22**, 57–60 (2015)
12. de Palma, A., Lindsey, R., Quinet, E., Vickerman R. (ed.): A Handbook of Transport Economics. Edward Elgar Publishing, number 12679 (2011)
13. Huang, Z., Du, X.: How does high-speed rail affect land value? Evidence from China. Land Use Policy 101 (2021)
14. Vickerman, R.W.: High-speed rail and regional development: the case of intermediate stations. J. Transp. Geogr. **42**(2015), 157–165 (2015)
15. Pagliara, F., Di Ruocco, I.: How public participation could improve public decisions on rail investments? Reg. Nal Sci. Policy Pract. **10**(4), 383–403 (2018)
16. Di Ruocco, I., D'Auria, A.: The "Duomo" metropolitan station in Naples: a case of urban regeneration through public art with a focus on spillover effects on real estate market. Cities' Identity Through Architecture and Arts" 5th edition, (in progress of publication) (2022)

17. Levinson, D.M.: Accessibility impacts of high-speed rail. J. Transp. Geogr. **22**(2012), 288–291 (2012)
18. Cartenì, A., Pariota, L., Henke, I.: Hedonic value of high-speed rail services: quantitative analysis of the students' domestic tourist attractiveness of the main Italian cities (2017)
19. Pagliara, F., La Pietra, A., Gomez, J., Vassallo, J.M.: High speed rail and the tourism market: evidence from the Madrid case study. Transp. Policy **37**, 187–194 (2015)
20. Albalate, D., Fageda, X.: High speed rail and tourism: empirical evidence from Spain. Transp. Res. Part A: Policy Pract. **85**(2016), 174–185 (2016)
21. Albalate, D., Campos, J., Jiménez, J.L.: Tourism and high speed rail in Spain: does the AVE increase local visitors? Ann Tourism Res **65**(2017), 71–82 (2017)
22. Tinessa, F., Marzano, V., Papola, A.: Mixing distributions of tastes with a combination of nested logit (CoNL) kernel: formulation and performance analysis. Transp. Res. Part B Methodol. **141**, 1–23 (2020)
23. Tinessa, F.: Closed-form random utility models with mixture distributions of random utilities: exploring finite mixtures of qGEV models. Transp. Res. Part B Methodol. **146**, 262–288 (2021)
24. Martínez, H.S., Givoni, M.: The accessibility impact of a new High-speed rail line in the UK–a preliminary analysis of winners and losers. J. Transp. Geogr. **25**(2012), 105–114 (2012)
25. Ortega, E., López, E., Monzón, A.: Territorial cohesion impacts of high-speed rail at different planning levels. J Transp. Geogr **24**(2012), 130–141 (2012)
26. Cao, J., Liu, X.C., Wang, Y., Li, Q.: Accessibility impacts of China's high-speed rail network. J. Transp. Geogr. **28**(2013), 12–21 (2013)
27. Shaw, S.L., Fang, Z.X., Lu, S.W., Tao, R.: Impacts of high speed rail on railroad network accessibility in China, J. Transp. Geogr. **40**, 112 (2014); Transp. Res. Part A: Pol. Pract. **100**, Transport innovations. Environ. Plann. A **45**(8), 1977–1994.
28. Wang, L., Liu, Y., Sun, C., Liu, Y.: Accessibility impact of the present and future high-speed rail network: a case study of Jiangsu Province, China. J. Transp. Geogr. **54**, 161 (2016)
29. Jiao, J., Wang, J., Jin, F., Dunford, M.: Impacts on accessibility of China's present and future HSR network. J. Transp. Geogr. **40**(2014), 123–132 (2014)
30. Cheng, Y., Loo, B.P.Y., Vickerman, R.W.: High-speed rail networks, economic integration and regional specialisation in China and Europe. Travel Behav Soc **2** (2015)
31. Chen, Z., Haynes, K.E.: Impact of high-speed rail on regional economic disparity in China. J Transport Geogr. **65** (2017)
32. Chen, C.L., Vickerman, R.W.: Can transport infrastructure change regions' economic fortunes: some evidence from Europe and China. Reg. Stud. **51**, 144–160 (2017)
33. Yang, Z., Li, C., Jiao, J., Liu, W., Zhang, F.: On the joint impact of high-speed rail and megalopolis policy on regional economic growth in China. Transport Pol **99**, 20–30 (2020)
34. Monzón, A., Ortega, E., López, E.: Efficiency and spatial equity impacts of high-speed rail extensions in urban areas. Cities **30**(2013), 18–30 (2013)
35. Chen, Z., Haynes, K.E.: Impact of high speed rail on housing values: an observation from the Beijing-Shanghai line. J. Transp. Geogr. **43**, 91–100 (2015)
36. Dong, X., Zheng, S., Kahn, M.E.: The role of transportation speed in facilitating high skilled teamwork across cities. J Urban Econ. **115** (2020)
37. Gao, Y., Zheng, J.: The impact of high-speed rail on innovation: an empirical test of the companion innovation hypothesis of transportation improvement with China's manufacturing firms. World Dev **127** (2020)
38. Zhang, W., Tian, X., Yu, A.: Is high-speed rail a catalyst for the fourth industrial revolution in China? Story of enhanced technology spillovers from venture capital. Technol Forecast Soc Change **161** (2020)
39. Peng, W., Yin, Y., Kuang, C., Wen, Z., Kuang, J.: Spatial spillover effect of green innovation on economic development quality in China: evidence from a panel data of 270 prefecture-level and above cities. Sustain. Cities Soc. **69** (2021).
40. Tinessa, F., Pagliara, F., Biggiero, L., Delli, G.: Walkability, accessibility to metro stations and retail location choice: some evidence from the case study of Naples. Res. Transp. Bus. Manag. **40**, 100549 (2021)

41. Knaap, G.J., Ding, C., Hopkins, L.D.: Do plans matter? The effects of light rail plans on land values in station areas. J. Plan. Educ. Res. **21**(1), 32–39 (2001)
42. Debrezion, G., Pels, E., Rietveld, P.: The impact of rail transport on real estate prices: an empirical analysis of the Dutch housing market. Urban Stud **48**(5), 997–1015 (2006). https://doi.org/10.2139/ssrn.895270
43. Andersson, D.E., Shyr, O.F., Fu, J.: Does high-speed rail accessibility influence residential property prices? Hedonic estimates from southern Taiwan. J. Transp. Geogr. **18**(1) (2010)
44. Wagner, G.A., Komarek, T., Martin, J.: Is the light rail "Tide" lifting property values? evidence from Hampton Roads VA. . Reg. Sci. Urban Econ. **65**, 25–37 (2017)
45. Armstrong, R.J., Rodriguez, D.A.: An evaluation of the accessibility benefits of commuter rail in Eastern Massachusetts using spatial hedonic price functions. Transportation **33**, 21–43 (2006)
46. Nash, C.: When to invest in high speed rail. J. Rail Transp. Plann. Manag. **5**(1), 12–22 (2015)
47. Chen, Z., Haynes, K.E.: Impact of high speed rail on housing values: an observation from the Beijing-Shanghai line. J. Transp. Geogr. **65**, 80–91 (2017)
48. Ureña, J.M., Menerault, P., Garmendia, M.: The High-speed rail challenge for big intermediate cities: a national, regional and local perspective. Cities **26**, 266–279 (2009)
49. Garmendia, M., Ribalaygua, C., Ureña, J.M.: High speed rail: Implication for cities. Cities **29**(2012), S26–S31 (2012)
50. Preston, J., Wall, G.: The ex-ante and ex-post economic and social impacts of the introduction of high-speed trains in South East England. Plan. Pract. Res. **23**(3), 403–422 (2008)
51. Alonso, W.: The historic and the structural theories of urban form: their implications for urban renewal. Land Econ. **40**(2), 227–231 (1964)
52. Garmendia, M., Romero, V., Ureña, J.M., Coronado, J.M., Vickerman, R.W.: High-speed rail opportunities around metropolitan regions: the cases of Madrid and London. J. Infrastruct. Syst. **18**, 305–313 (2012)
53. Vickerman, R.W., Spiekermann, K., Wegener, M.: Accessibility and regional development in Europe. Reg. Stud. **33**, 1–15 (1999)
54. Graham, D.J.: Agglomeration, productivity and transport investment. J. Transp. Econ. Policy **41**, 317–343 (2007)
55. Ansar, A., Flyvbjerg, B., Budzier, A., Lunn, D.: Does infrastructure investment lead to economic growth or economic fragility? evidence from China. Oxf. Rev. Econ. Policy **32**, 360–390 (2016)
56. Aschauer, D.A.: Is public expenditure productive? J. Monet. Econ. **23**(2), 177–200 (1989)
57. Banister, D., Berechman, Y.: Transport investment and the promotion of economic growth. J. Transp. Geogr. **9**(3), 209–218 (2001)
58. Flyvbjerg, B., Holm, M.S., Buhl, S.: Underestimating costs in public works projects: Error or lie? J. Am. Plann. Assoc. **68**(3), 279–295 (2002)
59. Ghoul, S. El., Guedhami, O., Ni, Y., Pittman, J., Saadi, S. (2011). Does information asymmetry matter to equity pricing? evidence from firms' geographic location.
60. Cascetta, E., Papola, A., Pagliara, F., Marzano, V.: Analysis of mobility impacts of the high speed Rome-Naples rail link using withinday dynamic mode service choice models. J. Transp. Geogr. **19**(4), 635–643 (2011)
61. Italian Government: Recovery Plan Mission n.3. https://www.governo.it/sites/governo.it/files/PNRR.pdf, (2021). Last accessed 28 Oct 2022
62. RFI - Rete Ferroviaria Italiana: Piano commerciale 2022. https://www.rfi.it/it/chi-siamo/Il-rapporto-con-lo-Stato-e-con-gli-stakeholder/il-piano-commerciale.html,visited. (2022) Last accessed 28 Oct 2022
63. Zorn, C.J.W.: Generalized estimating equation models for correlated data: a review with applications. Am. J. Polit. Sci. **45**(2), 470–490 (2001)
64. Italian National Census ISTAT. www.istat.it. Last accessed 3 Nov 2022
65. Agenzia delle Entrate–Italian National Real estate market datasouce. https://agenziaentrate.gov.it/. Last accessed 3 Nov 2022
66. Brunsdon, C., Fotheringham, A.S., Charlton, M.E.: Geographically weighted regression: a method for exploring spatial nonstationarity. Geogr. Anal. **28**, 281–298 (1996)

67. Pagliara, F., Mauriello, F.: Modelling the impact of high speed rail on tourists with geographically weighted poisson regression. Transp. Res. Part A Policy Practice **132** (2020)
68. Diggle, P., Heagerty, P., Liang, K.Y., Zeger, S.L.: 2002. Oxford University Press, Oxford, UK, Analysis of Longitudinal Data (2002)
69. Hedeker, D., Gibbons, R.D.: Analysis of longitudinal data. John Wiley and Sons (2006)
70. Wang, M.: Generalized estimating equations in longitudinal data analysis: a review and recent developments. Adv. Stat. **2014** (2014)
71. Liang, K.Y., Zeger, S.: Longitudinal data analysis using generalized linear models. Biometrika **42**, 121–130 (1986)
72. Gill, J.: Generalized linear models: a unified approach. Sage QASS Series. Ken Meier and Jeff Gill. What Works: A New Approach to Program and Policy Analysis (2000)
73. Nasrin, L., Mohammad, A., Syed, H.: A generalized estimating equations approach for modeling spatially clustered data. Austrian J. Stat. **50**, 36–52 (2021)
74. Gotway, C.A., Stroup, W.W.: A generalized linear model approach to spatial data analysis and prediction. J. Agric. Biol. Environ. Stat. 157–178 (1997)
75. Luo, R., Pan, J.: Conditional generalized estimating equations of mean-variance-correlation for clustered data. Comput. Stat. Data Anal. **168**, 107386 (2022)
76. Jacquez, G.M.: Spatial Cluster Analysis: The Handbook of Geographic Information Science, vol 395, 416 (2008)
77. Lipsitz, S.R., Fitzmaurice, G.M.: Estimating equations for measures of association between repeated binary responses, Biometrics **52**(3), 903–912 (1996)
78. Fitzmaurice, G.M., Laird, N.M., Ware, J.H.: Applied Longitudinal Analysis, vol. 998. John Wiley & Sons (2012)
79. Gibbons, S., Machin, S.: Valuing rail access using transport innovations. J. Urban Econ. **57**, 148–169 (2005)

HSR: Innovation and Environment

Assessing Innovations in High-Speed Rail Infrastructure

Abderrahman Ait-Ali, Filiz Kurt, Alessa Isberner, Kristofer Odolinski, and Mats Berg

Abstract Innovations in high-speed rail (HSR) have had substantial effects on different stakeholders within and outside the railway system. As part of the European Shift2Rail research programme, several innovative solutions are developed for, among others, improving the HSR infrastructure. The Joint Undertaking behind this research program has set objectives for these innovations in terms of punctuality, capacity, and life cycle costs. With a focus on infrastructure-related innovations for HSR, this paper aims at assessing their impacts in relation to these targets. We review the relevant research literature about the effects of HSR innovations and their assessment. The paper presents a hybrid assessment methodology combing different approaches to assess capacity, punctuality, and cost effects. This contributes to reducing the existing gap that is found in the research literature. Based on a reference scenario for HSR line and collected data from different stakeholders, the results indicate that infrastructure innovations in HSR, being developed within the European Shift2Rail research programme, can contribute to reaching the target set for punctuality. Further innovations in HSR infrastructure and/or other railway assets may be needed to reach additional targets and for more accurate improvement values giving more insights into their impacts.

Keywords High-speed · Railway · Infrastructure · Innovation

A. Ait-Ali (✉) · K. Odolinski
Swedish National Road and Transport Research Institute (VTI), Malvinas väg 6, 114 28 Stockholm, Sweden
e-mail: abderrahman.ait.ali@vti.se

A. Ait-Ali
Linköping University, Bredgatan 32, 602 21 Norrköping, Sweden

F. Kurt · A. Isberner
Deutsches Zentrum für Luft-und Raumfahrt e.V. (DLR), Lilienthalplatz 7, 38108 Braunschweig, Germany

M. Berg
KTH Royal Institute of Technology (KTH), Teknikringen 8, 100 44 Stockholm, Sweden

F. Pagliara (ed.), *Socioeconomic Impacts of High-Speed Rail Systems*,
Springer Proceedings in Business and Economics,
https://doi.org/10.1007/978-3-031-26340-8_13

1 Introduction

In this section, we introduce the relevant background information of this study and briefly describe the context to which this paper is contributing. Thereafter, we present the aim of the work as well as a delimitation of the scope of the research. We conclude the section with an overview of the paper's structure.

1.1 Context

Early innovations in high-speed rail (HSR) and its infrastructure have substantially contributed to enhancing the railway system as an important means of passenger transport. With higher speed and thus shorter travel times, these innovations enabled HSR to, among others, withstand the competition from other modes such as air transport, respond to and attract more demand for train passenger traffic, and reduce the negative environmental effects from other more polluting means of transportation. Thus, to keep up with developments of other competing transport modes and to improve and further increase the modal shift to rail passenger transport, continuous research and innovations (R&I) in different assets of the rail system are needed.

In this context and as part of the Shift2Rail (S2R) research programme, the S2R Joint Undertaking (JU) defined different Innovation Programmes (IPs) focusing on several subsystems of the railway system, e.g., infrastructure (or IP3), see Fig. 1. Moreover, the JU defines various cross-cutting-activities (CCAs) including the long-term needs and socio-economic research of the different IPs, see the red box in Fig. 1.

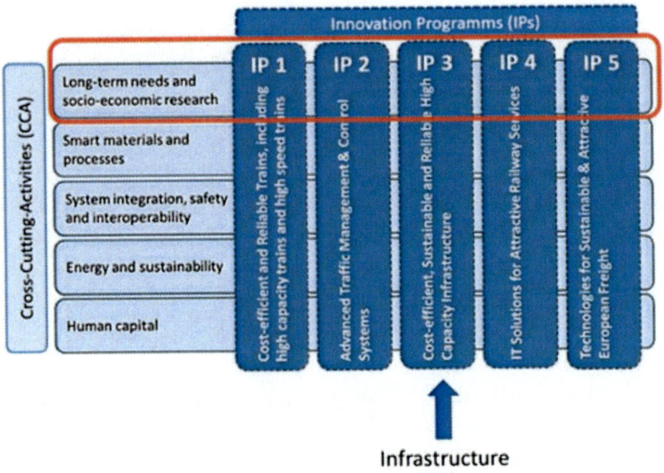

Fig. 1 Structure of the S2R IPs including infrastructure innovations [1]

As part of CCA and socio-economic research, the research project IMPACT-2 focuses on, among others, investigating the obtained and potential societal effects of R&Is within the S2R research programmes. For instance, specific key performance indicators (KPIs) are defined and monitored throughout the project, namely relating to punctuality, life cycle costs (LCC) as well as capacity.

1.2 Aim and Scope

With a focus on infrastructure-related innovations in HSR, this paper aims to quantitatively assess their effects using three different KPIs, namely capacity, punctuality, and LCC. Also, the study examines, through sensitivity analyses, how the assessed effects of such innovations in HSR infrastructure stand compared to the strategic targets set by the JU, i.e., doubling capacity (+100%), halving the life cycle costs (−50%) and increasing punctuality through improving reliability by 50% [1].

1.3 Structure of the Paper

The paper is structured as follows: Sect. 2 reviews the existing literature whereas the methodology and assessment model are described in Sect. 3. The results of the study are presented in Sect. 4. Section 5 concludes the paper.

2 Literature Review

In this section, we review the relevant literature of existing research. We first start with a historical background of R&Is in HSR and its infrastructure. Second, we present several studies on the various effects of different HSR innovations and their assessment. Finally, we identify the gap in the existing literature and present this work's contribution.

2.1 A Brief History of R&I in HSR Infrastructure

Earlier innovations in HSR took place in Japan when Shinkansen (also known as the bullet train) started operations in 1964 with train speeds reaching 210 kmph [2]. Thus, pioneering new technologies in designing and maintenance of infrastructure and rolling stocks, e.g., redesigned pantographs minimizing noise, rail welding reducing vibrations, and trains with a lower center of gravity and body weight [2].

After progressive but careful innovations, the French HSR (also known as TGV—*Train à Grande Vitesse*) started operations in 1981 with services between Paris and Lyon at 200 kmph [2]. The TGV project led to new HSR innovations such as in infrastructure (rails, bending radii, cants, switches at turnouts, catenary, pantograph, signaling system) and trains/vehicles (jointed trainsets, lower axle-load, distributed motorization, motors under engines' body, aerodynamics). Since the beginning of its operations, subsequent improvements helped achieve higher speeds, e.g., a record speed of 515 kmph in 1990. In a review of the development of HSR innovations, Walrave [2] presented the system approach, see Fig. 2, that was followed for R&Is in the French TGV project. Both estimations of the costs of infrastructure investments, rolling stock and their operations are considered alongside estimates of demand and revenues, see Fig. 2.

During the last two decades alone, China has become a major innovation actor in HSR with a total network length of more than 19 000 km, i.e., around 60% of the world's HSR network [3]. While studying the innovation evolution of the Chinese HSR industry, Chen and Mei [3] state that, after years of indigenous R&Is with support from the central government, the first milestones were reached in 2010 with a service speed reaching 350 kmph. This came out of different science and technology research projects, including different stakeholders (e.g., research institutes and laboratories, universities, academicians) working on several core technologies (e.g., EMU system assembly, car body, bogie, train control network system, and brake system) and other complementary technologies (e.g., air-conditioning system, toilet, door, window, windshield, flow receiving device, auxiliary power supply system, interior decoration materials, and seat). All of this made China one of the most active patent-filing countries in HSR technologies and the leader in many HSR technological innovations [4].

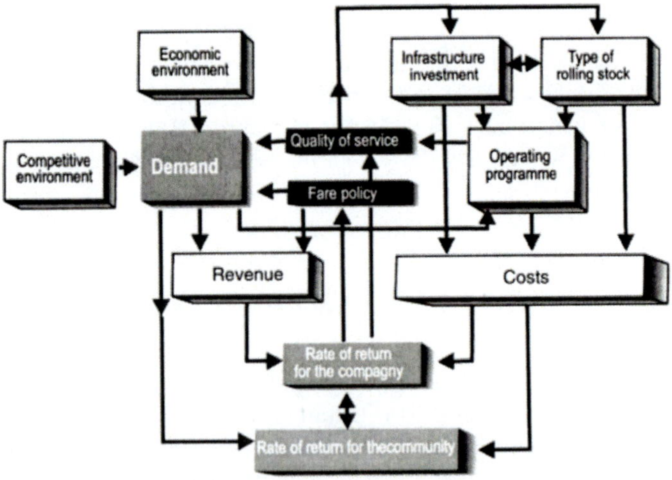

Fig. 2 System approach adopted for innovations in the TGV project [2]

2.2 Effects of HSR Infrastructure Innovations

The previously mentioned past innovations in, among others, infrastructure-related assets have led to increasing speeds on HSR lines, see Fig. 3. For speeds above 200 kmph, the infrastructure can be categorized as HSR as defined by the International Union of Railways UIC [5], also consistent with the definition by the European Commission EC [6]. These developments in operational speeds for such train services have various effects on different stakeholders in the railway market and society.

Higher speeds and shorter travel times have meant increased capacity utilization for the existing railway infrastructure capacity for both passenger and freight train services. Such infrastructure capacity can be defined as the maximum number of trains or passengers (for dedicated passenger lines) passing a specific section of the infrastructure under a given period [7]. Extensive studies show positive effects of HSR infrastructure on network capacity [8], e.g., increased infrastructure capacity [9], promotion of conventional passenger and freight rail [10], and induced travel demand [11].

Moreover, the development of HSR infrastructure worldwide has both socioeconomic and environmental effects [12]. For instance, in the case of a sustainable economy, the booming Chinese HSR has been shown to positively affect social welfare by significantly stimulating regional green innovation performance [13]. In Spain, it has been shown by Guirao, Campa [14] that the HSR has a direct positive linkage with the performance of tourism in certain regions. The authors recommend, however, further research by considering alternative explanatory indicators.

Additional effects have been associated with HSR infrastructures such as facilitating economic growth [15], reduced regional disparity [16], equity [16] and accessibility [17]. Moreover, Komikado, Morikawa [18] have recently found a positive influence of the existence of HSR on induced regional innovation.

Fig. 3 Commercial and record speeds for HSR worldwide [5]

2.3 Assessment of HSR Innovation Effects

To assess the effects of interest when studying HSR, different methods and approaches have been used such as optimization and simulation methods (for capacity assessments), econometric analysis, or monetary studies such as LCC and cost–benefit analysis (CBA), see Table 1 for an overview summary of some references.

When large sets of data exist such as empirical or historical data, econometric methods are commonly used, e.g., to analyze and compare differences (in one or several aspects) before and after the investments in HSR infrastructure. For instance, Multivariate Panel Data Analysis (MPDA) was used to analyze the significance of the HSR impacts on Spanish Tourism by Guirao, Campa [14]. Difference-in-differences (DD) is another econometric method which was used for example to study the effects of HSR infrastructure on social welfare in the case of a green economy [13].

When focusing on the monetary aspects of the effects, methodologies such as CBA and/or LCC analysis are often adopted to assess the impact of HSR infrastructure. CBA has been recently increasingly used as a decision support tool for assessing large infrastructure investment projects. Several examples of economic appraisal applications and methodology are summarized in the EU CBA guide, i.e., [19]. In a study attempting to assess the effectiveness of the HSR project Turin-Lyon, CBA was shown to fail to account for equity implications [16]. CBA has been therefore extended with strategic approaches to include the so-called wider economic impacts. Such extensions have been used, e.g., to measure the impacts of HSR in Europe [20].

LCC or life cycle costing is a monetary analysis that focuses more on the life cycle of one or more assets. It allows for estimating the costs of building, operating and maintenance, e.g., of HSR lines [21]. LCC can also be combined with a reliability assessment, for instance, to evaluate the optimal safety standards for HSR bridges

Table 1 Examples of research on assessing the effects of HSR infrastructure innovations

References	Studied effects of HSR	Assessment approach
[14]	The output of the tourist sector	Econometric analysis (MPDA)
[13]	Social welfare in the green economy	Econometric analysis (DD)
[9]	Infrastructure/network capacity	Analytical and optimization
[11]	Demand and social equity	Surveys and econometrics (logit)
[20]	Strategic impacts in Europe	CBA and wider economic impacts
[23]	Environmental and economic impacts	LCC and LCA
This paper	LCC, capacity and punctuality	Hybrid (LCC, CBA and analytical)

[22]. By focusing on the environmental impacts, such analysis is also called life cycle assessment/analysis (LCA) and can, for instance, be used with LCC to assess both the economic and sustainability impacts of HSR over their life cycle [23].

2.4 Research Gap

The literature review has revealed that most studies focus on isolating and assessing specific effects of HSR infrastructure, e.g., capacity, equity, economic growth, etc. Moreover, most of these studies adopt a single assessment approach, e.g., simulation, and econometric analysis. Table 1 gives an overview of some studies from the literature and the methods that are adopted to assess the HSR impacts.

In this paper, we present an assessment framework that adopts a hybrid approach combining different methods such as LCC, CBA and analytical analysis. Furthermore, several impacts of HSR infrastructure are assessed, namely punctuality, LCC and capacity effects.

Note that the presented studies include the effects of innovations in all HSR technologies. This paper focuses, however, on the infrastructural assets of HSR.

3 Assessment Methodology and Model

In this section, we present an overview of the methodology and describe the different sub-models that constitute the assessment model.

3.1 Overview of the Assessment Methodology

To quantitively assess the impact of innovations in HSR infrastructure, the S2R project IMPACT-2 focuses on three quantitative KPIs, namely capacity, punctuality and LCC [24]. The assessment is based on a reference scenario for HSR that could be found anywhere in Europe, also called "System Platform Demonstrator" (SPD) within S2R [25]. Thereby, the data of the SPD characteristics for HSR in the reference scenario is provided by different stakeholders in the railway market [26].

For assessing the potential impact of innovations in HSR infrastructure, various segments of the railway infrastructure system are analyzed in S2R and categorized into specific so-called "Technical Demonstrators" (TDs). Through the assessment methodology developed within IMPACT-2, it becomes possible to estimate the impacts of individual infrastructure-related innovations on the infrastructure system as well as on the whole railways. For these individual innovations in infrastructure, the relative improvements (in %) between the baseline and the future scenario are

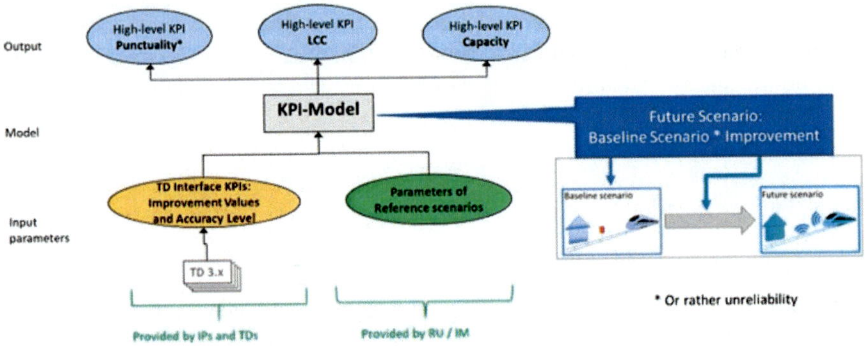

Fig. 4 Overview of the adopted methodology for assessing innovations in HSR infrastructure [28]

provided by the TDs [27]. See Fig. 4 for an overview of the assessment methodology that is adopted in this study.

3.2 *Assessment Model*

In this subsection, we briefly describe each of the sub-models forming the assessment model, namely the LCC, punctuality and capacity sub-model for assessing different effects of S2R innovations in HSR infrastructure.

Life Cycle Cost (LCC)

Based on an assessment period of 30 years, the life cycle costs are calculated by summing up both the capital and maintenance costs of all the infrastructure-related assets, i.e., switches, track, bridges, tunnels, passenger stations, power supply, and infrastructure management.

To convert the total life cycle costs to an equivalent value in the present (also called Present Net Value or PNV), we use the discounting formula in (1) where LCC_y are the costs during year y and i is the discounting factor/rate, set to 3%.

$$PNV = \sum_{y=0}^{30} \frac{LCC_y}{(1+i)^y} \tag{1}$$

Other parameters are also assumed in the sub-model such as the life service of certain assets (e.g., 20 years for switches & crossings, 100 years for bridges/tunnels). Moreover, the costs are also assumed to have a certain distribution among the different assets.

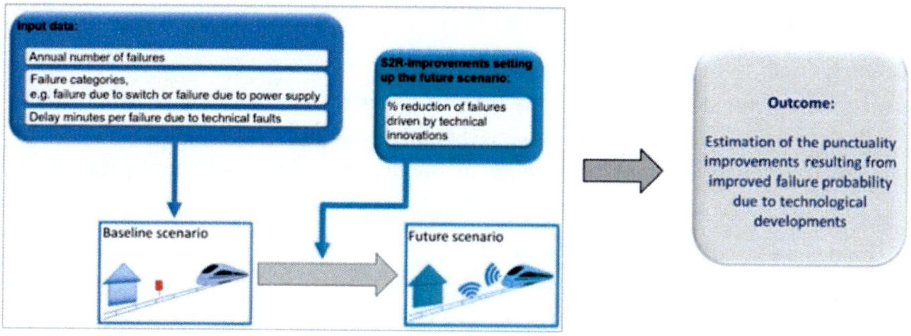

Fig. 5 Overview of the structure of the punctuality model [29]

Punctuality

Some technical innovations in HSR infrastructure, e.g., improved predictive maintenance, can provide better service reliability and hence reduced downtime and delays. Such effects are captured in the punctuality sub-model using the methodology that is illustrated in Fig. 5.

Based on input data about historical infrastructure-related failures, it is possible to estimate future on-time performances using improvement values about technical failures in future HSR infrastructure assets.

Capacity

With a focus on peak hours where capacity is most needed, the calculation of the capacity for HSR is based on the definition given in the literature review, i.e., the number of passengers passing a specific section of the infrastructure under a given period [7]. The analytical expression in Eq. (2) is hence used to calculate the maximum capacity usage (in passengers per peak hour) during peak hours with/without the innovations in HSR infrastructure.

$$Cap[pax/h] = Line[trains/h] \times Train[pax/unit] \times Coupling[unit/train] \tag{2}$$

The line capacity (*Line*) is the number of trains per peak hour and day whereas the train capacity (*Train*) accounts for the maximum number of passengers that are transported during peak hours on a train unit of the line. The coupling ability (*Coupling*) captures the number of coupled units per train on the line during peak hours.

4 Assessment Results

In this section, we describe the data collection process and present the main KPI-assessment results of S2R innovations in HSR infrastructure. We conclude the section with sensitivity analyses and discussions on the accuracy levels of some parameters.

4.1 Data Collection

Various types of data were collected from the railway stakeholders and were used in several components of the model such as reference scenarios, innovation improvements, cost distributions and accuracy levels, see Table 2 for an overview.

To characterize the reference scenario of the studied HSR line, data was collected from railway undertakings (RUs) and infrastructure managers (IMs) actively involved in S2R as well as publicly available data, e.g., national transport ministries and data from European authorities such as the European Union Agency for Railways (ERA).

For assessing the potential impact of innovations in HSR infrastructure, various segments of infrastructure-related innovations are analyzed in S2R and categorized into specific so-called "Technical Demonstrators" (TDs). Through the assessment methodology developed in IMPACT-2, it is possible to estimate the impacts of individual infrastructure-related innovations. The overall relative improvements (in %) between the baseline and the future scenario are provided by the TDs.

Additional data were also collected. For instance, the distribution of costs is used for innovations where the KPIs cannot be captured at the level of details of the TDs. Accuracy levels (and their maximal value) are used to understand the precision of the provided improvement values. Such levels are further analyzed and discussed in the sensitivity analyses later in this section.

Table 2 Data collection for assessing S2R innovations in HSR infrastructure

	Baseline	Future
Collection process	Specification of the HSR reference scenario (infrastructure characteristics)	Periodic improvement values from different technical demonstrators
Example of parameters	Maintenance/capital costs of infrastructure assets, life span, delay (due to infrastructure failure)	(Updated) improvement values corresponding to, e.g., a reduction in maintenance costs of an asset
References	Industry (RUs, IMs), research (S2R research projects)	Leaders of different TDs of the S2R innovations in infrastructure

4.2 Results

Based on the collected data for both baseline and future scenarios, the assessment methodology allows calculating the impact of innovations in HSR infrastructure elements on the total railway system. Figure 6 summarizes the assessment results (in blue) in comparison with EU targets (in red). The assessment results present the estimated percentage improvement in the KPIs (except for capacity which is lower than 1%) for the whole railway system when the different S2R innovations in HSR infrastructure are implemented.

The results in Fig. 6 indicate that innovations in HSR infrastructure have the highest impact on punctuality/unreliability, i.e., 19% which accounts for around 38% of the whole EU target of 50%. This is thanks to, among others, improved asset monitoring and condition-based maintenance which decreases the occurrences of infrastructure-related failures and delays.

However, although not presented in the figure, the impact of HSR infrastructure innovations in terms of capacity is negligible. In fact, the main gains in capacity (from infrastructure-related innovations) are thanks to the reduction in downtime time for planned maintenance. Such gains are negligible since the assessment of capacity effects is done during peak hours when the need for capacity is the highest. Innovations in other assets of HSR such as train design/capacity and/or signaling systems can be more important but are out of the scope of this paper.

Although below the EU target of −50% for LCCs, S2R innovations in HSR infrastructure allow for a reduction of around 8% in LCCs. This relative reduction in costs is mainly due to optimized maintenance, i.e., reduced costs for corrective maintenance activities.

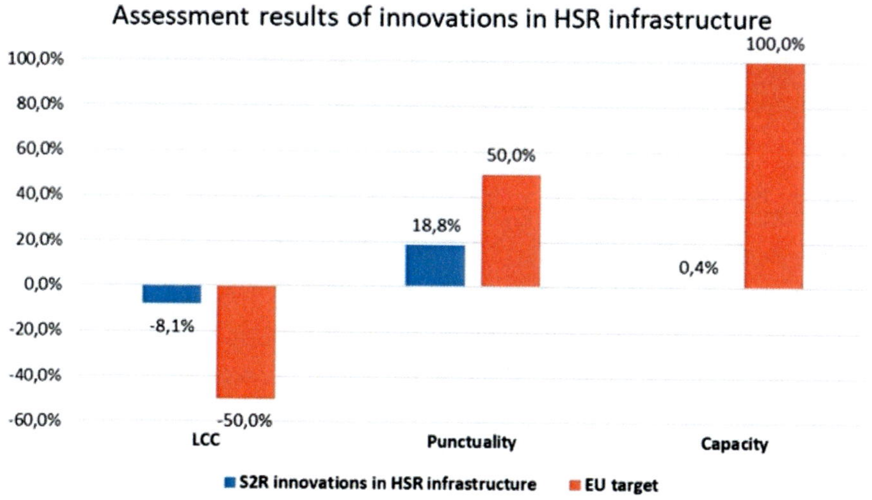

Fig. 6 KPI assessment results (in blue) of S2R innovations in HSR infrastructure (on the whole railway system) in comparison to the EU targets (in red)

4.3 Sensitivity Analyses and Accuracy Levels

To validate the assessment results, the leaders of the technical demonstrators (TDs) also provide so-called accuracy levels (AL), allowing to identify how the delivered improvement values were determined, see appendix B for detailed definitions of the ALs. The AL data are provided for two categories, namely technical and cost values, and can be assigned to one of four ALs, i.e., from highest (based on test results from laboratory/field) to lowest (e.g., expert knowledge). See Fig. 7 for the distribution of such AL for the different technical as well as cost improvements.

Figure 7 shows that most of the improvement values are based on simulation (if technical) or on a prototype (if monetary). Physical/market-based values are however absent since many of the innovations have not yet been used in the railway market. Note that, although not shown here, TD leaders may also indicate the maximum achievable AL from their demonstrators.

To further analyze the assessment results, we focus in this subsection on the improvements relative to the baseline infrastructure system, in contrast to the whole railway system as previously presented in Fig. 6. The assessment results (relative to the baseline infrastructure system) are as follows: −20,1% in LCCs and 55,2% in punctuality. The results for capacity remained as negligeable as for the whole system, i.e., 0,4%.

We perform sensitivity analyses on LCCs and punctuality. We therefore study two improvement values corresponding to different HSR infrastructure assets, namely bridge capital costs and power supply-related failures. The variations of the corresponding improvements in LCCs (for bridge capital costs) and unreliability (for power supply failure) are shown in Figs. 8 and 9, respectively.

With a focus on the effects of HSR bridge capital costs on the LCC, we these costs (on the horizontal axis in Fig. 8) around the reference value, i.e., 100%. The corresponding improvements in LCCs (on the vertical axis) are decreasing (almost

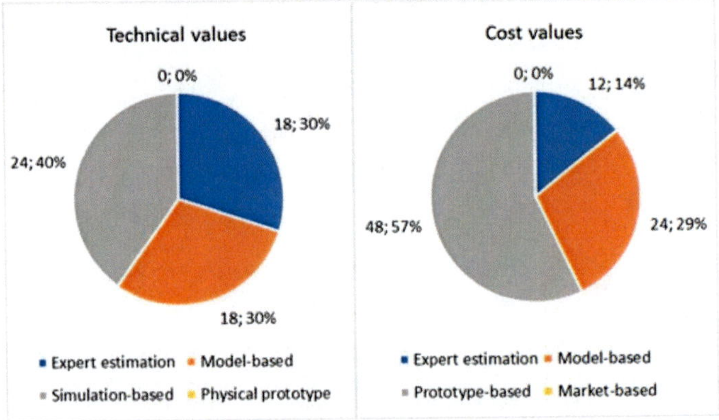

Fig. 7 Distribution of the accuracy levels of technical and cost improvement values for S2R infrastructure innovations in HSR

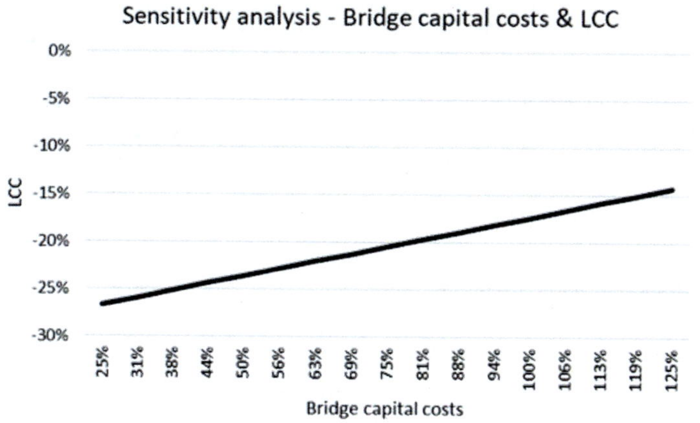

Fig. 8 Improvements in LCCs (relative to the baseline infrastructure system) when varying the capital costs of bridges

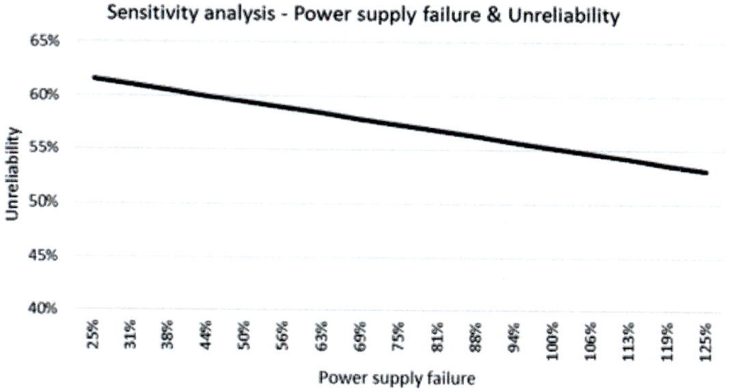

Fig. 9 Variation of the unreliability when varying the percentage improvement in power supply failures

linearly) with decreasing the capital costs for the HSR bridges. However, the results are still robust since the variation in the resulting LCCs is substantially smaller (around 13%, from −27% to −14%) in relation to the variation in bridge capital costs (around 100%, from 25 to 125%).

Another sensitivity analysis is performed on the effect of power supply failures on punctuality and unreliability. Like the LCC analysis in Fig. 8, the unreliability measure (on the vertical axis in Fig. 9) increases almost linearly with decreasing power supply failures. Both analyses (in Figs. 8 and 9) indicate that even with high variance in the improvement values (bridge capital costs and power supply failure, respectively), the measures are rather robust in terms of punctuality and LCCs (relative to the baseline infrastructure system).

5 Concluding Remarks

This section ends the paper with the main concluding remarks. We mention important conclusions as well as some of the limitations before briefly presenting some of the most relevant possible future works.

5.1 Highlights and Conclusions

In this paper, we have presented a methodology to assess innovations being developed within the Shift2Rail research programme for improving HSR infrastructures. The presented methodology combines different existing assessment approaches in the literature such as capacity analysis, life cycle costing and cost–benefit analysis.

Based on the baseline scenario of an HSR line and collected data from different railway stakeholders, we show that infrastructural innovations have the greatest effect on punctuality and can substantially contribute to the corresponding target that is set by the Joint Undertaking. Furthermore, we investigated the robustness of the assessment results using sensitivity analyses (and accuracy levels) and discussed them in relation to achieving the three different targets, namely punctuality, costs, and capacity.

5.2 Limitations and Future Works

During the development of the assessment methodology, we made a few assumptions. For instance, the case study has been parametrized based on the specifications within the Shift2Rail research programme. Therefore, the results can be considered tailored to the requirements of the programme. General conclusions as to the need for innovations in HSR infrastructure and their impacts can still be drawn.

Certain limitations are also relevant to mention in this context. In particular, the different effects (on capacity, life cycle costs and punctuality) are assessed separately, i.e., when one is assessed, the others are kept constant. Some assessed effects may, in reality, have additional contributions to another (separately) assessed effect. Moreover, several parameters, such as cost improvement values, are found to be difficult to accurately estimate as illustrated by the (maximum achievable) accuracy levels. Such limitation is beyond the scope of this work but is important to the assessment results.

The mentioned assumptions and limitations leave room for several possible future works. For instance, it is possible to continue the assessment of innovations in HSR infrastructure (and other railway assets) for future development of new ideas and innovative improvements under the European Rail Joint Undertaking. Furthermore, more market-based insights from the implementation of ongoing innovations (e.g., moving blocks, automatic/virtual train coupling, condition-based railway maintenance) would increase the accuracy of the assessment results.

Acknowledgements This project has received funding from the Shift2Rail Joint Undertaking (JU) under grant agreement No 777513. The JU receives support from the European Union's Horizon 2020 research and innovation program and the Shift2Rail JU members other than the Union.

Appendix 1—Abbreviations and Acronyms

AL	Accuracy levels
CBA	Cost-benefit analysis
CCA	Cross-cutting-activities
DD	Difference-in-differences
EC	European Commission
EU	European Union
HSR	High-speed rail
IP	Innovation programmes
JU	Joint Undertaking
Kmph	Kilometer per hour
KPI	Key performance indicator
LCA	Life-cycle analysis
LCC	Life-cycle cost
MPDA	Multivariate Panel Data Analysis
PNV	Present Net Value
R&I	Research and innovation
SPD	System Platform Demonstrator
S2R	Shift2Rail
TD	Technical demonstrator
TGV	high-speed train (Train à grande vitesse)
UIC	International Union of Railways (Union internationale des chemins de fer)

Appendix 2—Definitions of the Accuracy Levels

Accuracy levels	Cost value	Technical value
Based on the market (for costs), physical prototype (for technical)	Improvement values based on prototypes and estimation of scale effect for series production as well as acceptable market prices based on first discussions among involved stakeholders	Improvement values based on results of a test in field or laboratory conditions (foreseen to be similar in the field)
Based on a prototype (for costs), simulation/labs (for technical)	Improvement values are based on the evaluation of prototype cost without consideration of economies of scale	Improvement values based on results of the test under laboratory conditions (requiring further testing in the field), or simulations of the technology
Model-based	Improvement values are based on prototype drawings or based on calculations for similar technologies or comparable methods	
Expert estimation	Improvement values based on knowledge of experts	

References

1. Ju, S.R.: Multi-annual Action Plan. https://rail-research.europa.eu/wp-content/uploads/2020/09/MAAP-Part-A-and-B.pdf (2019)
2. Walrave, M.: The development of high speed rail innovation and tradition: prospects for the future. Rivista Internazionale di Scienze Sociali **101**(3), 375–397 (1993)
3. Chen, J., Mei, L.: Innovation evolution of China's high-speed rail industry. Front. Eng. Manage. **5**(4), 548–552 (2018)
4. Liu, X., Yang, X.: Identifying technological innovation capability of high-speed rail industry based on patent analysis. In: 2019 8th International Conference on Industrial Technology and Management (ICITM) (2019)
5. UIC: High Speed Rail: Fast Track to Sustainable Mobility (2018)
6. EC: 96/48/EC on the interoperability of the trans-European high-speed rail system. Commission of the European Communities, Brussels 23 (1996)
7. UIC: UIC Code 406-Capacity, I.U.o. Railways, Editor (2004)
8. Bracaglia, V., et al.: High-speed rail networks, capacity investments and social welfare. Transp. Res. Part A Policy Pract. **132**, 308–323 (2020)
9. Abril, M., et al.: An assessment of railway capacity. Transp. Res. Part E-Log Transp. Rev. **44**(5), 774–806 (2008)
10. Cheng, J., Chen, Z.: Impact of high-speed rail on the operational capacity of conventional rail in China. Transp. Policy **110**, 354–367 (2021)
11. Ren, X., et al.: Impact of high-speed rail on social equity in China: evidence from a mode choice survey. Transp. Res. Part A Policy Pract. **138**, 422–441 (2020)
12. Momenitabar, M., et al.: Literature review of socioeconomic and environmental impacts of high-speed rail in the world. Sustainability **13**(21), 12231 (2021)

13. Zhu, Z., Lin, X., Yang, H.: Booming with speed: high-speed rail and regional green innovation. J. Adv. Transp. **2021**, 9705982 (2021)
14. Guirao, B., Campa, J.L., López-Lambas, M.E.: The assessment of the HSR impacts on Spanish tourism: an approach based on multivariate panel data analysis. Transp. Res. Proc. **18**, 197–204 (2016)
15. Chong, Z., Chen, Z., Qin, C.: Estimating the economic benefits of high-speed rail in China: a new perspective from the connectivity improvement. J. Transp. Land Use **12**(1), 287–302 (2019)
16. Chen, Z., Haynes, K.E.: Impact of high-speed rail on regional economic disparity in China. J. Transp. Geogr. **65**, 80–91 (2017)
17. Bruzzone, F., Cavallaro, F., Nocera, S.: The effects of high-speed rail on accessibility and equity: evidence from the Turin-Lyon case-study. Socio-Econ. Plann. Sci. 101379 (2022)
18. Gutiérrez, J., Monzon, A., Piñero, J.M.: Accessibility, network efficiency, and transport infrastructure planning. Environ Plan A **30**(8), 1337–1350 (1998)
19. Komikado, H., et al.: High-speed rail, inter-regional accessibility, and regional innovation: evidence from Japan. Technol. Forecast. Soc. Chang. **167**, 120697 (2021)
20. Sartori, D., et al.: Guide to cost-benefit analysis of investment projects: economic appraisal tool for Cohesion Policy, 2014 (2020)
21. Rothengatter, W.: Approaches to measure the wider economic impacts of high-speed rail and experiences from Europe (2019)
22. Maout, E.L., Kato, H.: Life cycle cost-estimation model for building, operating, and maintaining high-speed rail systems. Asian Transp. Stud. **4**(1), 245–260 (2016)
23. Han, S.-H.: Optimal safety valuation of high-speed railway bridges based on reliability assessment and life-cycle cost concept. Int. J. Steel Struct. **17**(1), 339–349 (2017)
24. Rungskunroch, P., Shen, Z.-J., Kaewunruen, S.: Benchmarking environmental and economic impacts from the HSR networks considering life cycle perspectives. Environ. Impact Assess. Rev. **90**, 106608 (2021)
25. IMPACT-2: D4.1-Initial quantitative KPI model. Shift2Rail (2018)
26. IMPACT-2: D4.4-First SPD integrated assessment. Shift2Rail (2020)
27. IMPACT-2: D4.5-Validated quantitative KPI model. Shift2Rail (2020)
28. Kurt, F., et al.: The KPI-Model–an integrated KPI assessment methodology to estimate the impact of different innovations in the railway sector. In: SmartRaCon Workshop (2022)
29. IMPACT-2: D4.2-Initial estimation of the KPIs. Shift2Rail (2018)

High-Speed Trains Experience in Poland

Agata Pomykala

Abstract The article presents issues related to the placing in service of ED250 high-speed (HS) trainsets—on the Polish network. ED250 trains started running on the Polish railway network in December 2014. Several years of their operation allow drawing conclusions taking the right decisions regarding the purchase of these trainsets and a significant improvement in the quality of travel on the indicated, popular routes as well as to reflect on the wider aspect of using high-speed trains, i.e. the impact on reducing the negative influence of transport on the environment. This article addresses the issue of reduced carbon dioxide emissions by recalling the simulation carried out to compare the energy consumption between ED250 trains and locomotive plus carriages that, before the purchase of ED250 vehicles, served the Warsaw–Katowice and Warsaw–Gdańsk routes. The results of this simulation clearly indicate the advantages of HS trains in the ecological context. The article presents:

- technical information on high-speed trains operated in Poland,
- changes to the offer within 6 years of the ED250 trainsets in years 2015–2020,
- results of the simulation on energy consumption carried out within 2015–2020,
- the volume of reduced CO_2 emissions in 2015–2020,
- savings in the EU Emissions Trading System (EU ETS).

Keywords High-speed rail · Energy consumption · CO_2 emission

A. Pomykala (✉)
Railway Research Institute, 50 Chlopickiego Str, 04275 Warsaw, PL, Poland
e-mail: apomykala@ikolej.pl

F. Pagliara (ed.), *Socioeconomic Impacts of High-Speed Rail Systems*,
Springer Proceedings in Business and Economics,
https://doi.org/10.1007/978-3-031-26340-8_14

1 Introduction

1.1 Concepts of HSR Network in Poland

The beginnings of plans to build high-speed lines appeared in Poland in the 1970s. The Polish rail network is arranged in a shape of a cross and the main passenger flows focus in the north–south axis and the east–west axis, running to, from or through Warsaw. It is exactly for this reason that the constructors of the Central Rail Line (CMK), currently being a section of a core network line of TEN-T marked as E65,[1] already at the designing stage factored in parameters allowing its future adaptation to the speed of 200–250 km/h (sufficiently large radiuses of curves—4,000 m—and a sufficiently large distance between track axes –4.5 m) and predicted the extension of the line from Warsaw, on the eastern bank of the Vistula river, through Wyszogród, Płock and Brodnica to Gdańsk, with a branch line to Olsztyn [1]. This project was implemented only in the Silesia–Warsaw part, but the line was equipped to run trains at a maximum speed of 160 km/h[1].

In 1990–1995, the first works on the Polish high-speed rail (HSR) program were launched, under which it was assumed that two new HS lines would be built within 2010–2030:

- in the east–west corridor: the western border of Poland–Poznań–Łódź–Warsaw—the eastern border of Poland, 660 km long, adapted to a speed of 300 km/h, dedicated only to passenger traffic,
- and in the north–south corridor: northern CMK Warsaw–Płock–Gdańsk; with a length of 371 km, adapted to a speed of 300 km/h, also only for passenger traffic,

and a southern extension to the border with the Czech Republic.

In 2002, following the publication of *The directional program for high speed lines in Poland, the concept of the route running through Łódź to Wrocław and a connection with the Central Trunk Line* appeared.

Next step in 2005, was *Preliminary feasibility study for HS line Warsaw-Łódź-Poznań/Wroclaw construction* (named Y line). Between 2008 and 2011, works on the *Y Feasibility study for HS line Warsaw-Łódź-Poznań/Wroclaw construction* were carried out.

In 2015 Pre-feasibility study for the extension of HS line Warsaw-Łódź-Poznań/Wrocław to the border with Germany and the Czech Republic was ready, with a proposal of 2 line locations to Prague in the Czech Republic:

- via Jelenia Góra-Liberec-Mlada Boleslav,
- via Lubawska Pass-Hradec Kralove.

For the connection with Germany, only the corridor from Poznan was proposed.

---

[1] This section, built in 1971–1977 with the starting point in Grodzisk Mazowiecki near Warsaw and the ending point in Zawiercie in Silesia, runs for 225 km.

Since 2017, the construction of the Solidarity Central Transport Hub, including the railway component, has been a declared strategic project of the Polish government. It was assumed, among others, to rebuild the entire system of the national passenger transport system, which should function according to the hub & spoke model. The assumptions for the HS rail project have been significantly changed. It was assumed that the standard speed of 250 km/h will be adopted in the first stage of construction. In the next stage, it will be possible to increase the speed to 300–350 km/h.

1.2 Assumptions of the High-Speed Train Purchase Project in Poland

The decision to purchase high-speed trains was made as part of the preparation of a program for the construction and placing in service of HSR in Poland [2].

These trains are planned to run on modernized lines from Warsaw to Kraków and Katowice, and from Warsaw to Gdańsk. It was assumed that the maximum speed on the CMK—south line from Warsaw to Krakow and Katowice will increase to 250 km/h, and on part of the length of the line from Warsaw to Gdańsk to 200 km/h. Thanks to the new trains, a new category was to be created, called EIP (Express Intercity Premium).

The investment project *Purchase of railway passenger rolling stock for long-distance connections* entered the *List of Individual Projects for the Infrastructure and Environment Programme* [3].

After a multi-month competitive dialogue, in which five rolling stock manufacturers took part, ended with the submission of only one offer for the delivery of 20 vehicles. an agreement was signed between Alstom and PKP Intercity, on May 30, 2011. In addition to the delivery of 20 trains (at a price of EUR 20 million each; with EU funding), it included the maintenance of trains—the level of 1,2,3,4 for 17 years (cost of EUR 1.95 per 1 km) and the training of PKP IC employees.

The first train delivered in 2013 was tested for certification at a speed of 250 km/h + 10% (275 km/h). On November 24, a speed of 293 km/h was reached on the CMK section between Góra Włodowska and Psary, which was:

1. Speed record on the Polish network;
2. World record for 3 kV DC traction system;
3. Record for Pendolino trains family.

2 Regular Service and Offer Development

The introduction of ED250 trainsets into regular operation in Poland took place on December 14, 2014.

In the first stage, ED250 trains ran on the following routes (marked in red in Fig. 1).

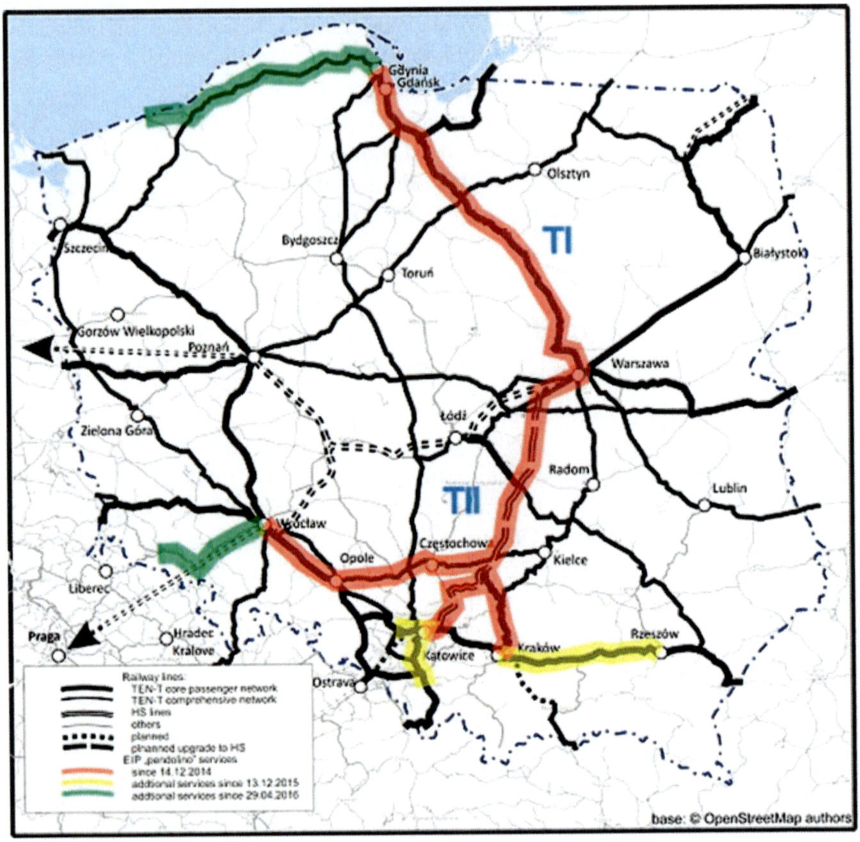

Fig. 1 EIP (ED250) services since 14.12.2014. *Source* T. Bużałek, based on: © OpenStreetMap authors

- Warsaw–Gdańsk—Gdynia (with intermediate stations: Sopot, Gdańsk Oliwa, Gdańsk Wrzeszcz, Gdańsk Główny, Tczew, Malbork, Iława, Warszawa Wschodnia);
- Warsaw–Krakow (with intermediate stations: Warszawa Zachodnia);
- Warsaw–Katowice (with intermediate stations Warszawa Zachodnia, Sosnowiec Główny, Zawiercie
- Warsaw-Wrocław.

Since December 2015 the service was extended to the southern and the south-east regions (to Rzeszów and Bielsko Biała, Gliwice), marked in yellow in Fig. 1.

In December 2016 the service was extended to the south-west and the north-west regions (Jelenia Góra and Kołobrzeg)—in green colour in Fig. 1.

The changes in the running of ED250 trains were not only related to changes in the marketing concept, but also to the planned modernization works carried out on the network:

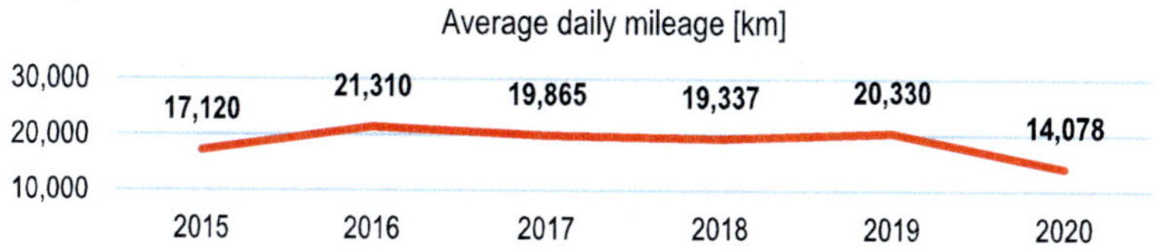

Fig. 2 ED250 average daily mileage in 2015–2020. *Source* Alstom

1. On 13.03.2016, some of the connections to Wrocław were temporarily redirected from line No. 4 to line No. 1, which was associated with an extension of travel time and a stop at the Częstochowa Main Station instead of Częstochowa Stradom. The reason for the change was the renovation works on line 4 [4].
2. 2.6.08.2016–11.08.2016, due to renovation works at the Biała Rawska station, the movement of trains on the Grodzisk Mazowiecki—Idzikowice section was completely suspended, and trains were directed to detour routes: to Krakow via Koluszki, Tomaszów Mazowiecki and Idzikowice, and to Katowice via Skierniewice, Koluszki and Częstochowa [5].
3. 10.07.2017–10.08.2017 on the CMK section Opoczno Południe–Włoszczowa Północ/Knapówka there was a round-the-clock double-track closure, as a result of which most trains ran through Skierniewice and Częstochowa. In addition, 2 EIP Warsaw—Jelenia Góra courses took place through the Łódź Widzew station [6].

In addition, in March 2020, due to the conditions resulting from COVID-19 pandemic, the operation of these trains was suspended [7].

Changes in the offer and limitations related to infrastructural works impacted the average daily mileage, as shown in Fig. 2.

3 ED250 Technical Parameters

ED250 trains are electric multiple units consisting of 7 coaches connected in a semi-permanent way. These trains based on the ETR610 Pendolino design are not equipped with a tilting mechanism. The setup can operate at different voltages.

Max speed is 250 km/h but in Poland is run no more than 200 km/h on modernized lines. Each train can seat 402 passengers. The details of technical parameters are specified in Table 1.

4 Energy Efficiency Analysis

Until the introduction of the ED250 trains, traffic at a speed of Vmax of 200 km/h was possible by trains composed of coaches and a 6 MW locomotive. It was interesting

Table 1 ED250 technical parameters

Parameter	Data
Class	ED250
Maximum speed	250 km/h
Number of coaches	7
Number of seats	402 (57 × 1 class including 12 × mother&child comp, 345 × 1 class with 2 PRM)
Axles arrangement	1A'A1' + 1A'A1' + 2'2' + 2'2' + 2'2' + 1A'A1' + 1A'A1'
Length of train	187,4 m
Axle load	14 643 kg (static ready to service) 15 793 kg (static normal load)) 17 600 kg(static maximum load))
Mass	414 t / 445 t
Traction power	5500 kW (8 × 708 kW)
Traction supply	25 kV 50 Hz & 15 kV 16,7 Hz & 3 kV DC
On-board control command system	ETCS L1 + L2, SHP, SIFA, PZB/LZB, Mirel

Source Alstom

to analyze to what extent the replacement of locomotive-driven trains with ED trains achieved energy and environmental effects in line with the Green Deal policy.

In 2021, professor Adam Szeląg from the Warsaw University of Technology conducted a simulation based on the usage of simulation methods of running ED250 and equivalent locomotive passenger trains to compare parameters on running on to basic railway lines where ED trains are operating. This allows comparing energy effectiveness of ED250 trains with locomotive-driven ones [8].

Simulations of the movement were carried out for:

- ED250 trains with parameters shown in Table 1
- Locomotive-driven trains (PAS) setup as in Table 2
- 3 kV DC network
- 2 lines & 2 directions of movement (See Fig. 1)
  - TI Warsaw-Gdańsk (W-Gd), 328,10 km

Table 2 Assumed parameters of a passenger train PAS with a BoBo locomotive

Parameter	Value
Nominal power	6 MW
Locomotive mass	90 t
Mass of one wagon	45 t
Number of wagons	10
Number of passengers	450
Maximum speed	200 km/h

  - TII section Grodzisk Maz.-Zawiercie (G-Z), 224.05 km on line Warsaw-Kraków/Katowice

- Number of seats: ED250—402, locomotive-driven—450
- Mass: ED250—445 ton, PAS—540 ton
- Occupancy of seats—100% & 60%
- Max speed 200 km/h
- CO_2 emissions data Poland: 0.719 t/1MWh of electricity.

The simulation results and comparison of traction and motion parameters are summarized in Table 3.

A comparison of the unitary energy consumption for traction per ton km between the two types of trains indicates lower consumption of nearly 3.5 Wh/t for ED250 over PAS trains on the TI route and about 3 Wh/tkm on the TII route. Figure 3 compares the unitary energy consumption for traction used by ED250 and PAS trains, with occupation 60% and 100%. For TI rout: trains energy consumption is lower for ED about 10 kWh/1000paskm (with 60% occupation of seats) and about 6 kWh/1000 paskm for 100% occupation. For TII rout: ED consumes less about 9 kWh/1000 paskm energy than locomotive-driven train—in case of 60% occupation and 7 kWh/1000 paskm less for full occupation—shown in Fig. 3.

5 Environmental Outcomes of ED250 Trains

Sustainable transport has been of interest to international institutions and governments of individual countries for many years. The needs related to protecting the natural environment and carrying out activities considering the needs of the present and the future generations are the basis of sustainable development.

One of the significant emitters of greenhouse gases (GHG) is transport, whose historical CO_2 emissions dynamics in Poland is different from the European Union's average. In Poland, in 2005–2017, a significant increase in emissions was observed (by 76%), where in the EU in the same period, a 3% decrease in emissions was visible. The trend rising emissions in the transport sector in Poland may persist in the coming years, which will have a significant impact on the achievable reduction of CO_2 emissions [9].

The simulation conducted by prof. A. Szeląg also enabled to determine carbon dioxide emissions for both types of trains. Data on CO_2 emissions for Poland were adopted in accordance with the publication of 2020 [10]. The CO_2 emissions results are shown in Table 4.

A comparison of CO_2 emissions for both types of trains indicates, as a consequence of lower energy consumption, lower CO_2 emissions from high-speed trains compared to locomotive-driven trains. Total CO_2 reductions over the six-year period was more than 95 tons. Figure 4 show the volume of reduction in subsequent years.

Table 3 Results of simulations ED250 and PAS trains and calculated energy

Train & direction	vav [km]	vavs [km/h]	Ec [kWh]	Ecut[kWh/trainkm]	Ecrek [kWh]	RS [%]	Eup[kWh/1000paskm]100%	Eup[kWh/1000paskm]60%	Ecu[Wh/tkm]
ED250 TI W-Gd	153.4	132.6	4249.15	12.95	3805.84	0.10	32.22	53.69	29.10
ED250 TI Gd-W	153.1	132.0	4421.03	13.47	3977.36	0.10	33.52	55.87	30.28
PAS TI W-Gd	152.3	131.5	5655.44	17.24	5319.55	0.06	38.30	63.84	31.92
PAS TI Gd-W	152.0	131.2	5900.25	17.98	5565.01	0.06	39.96	66.60	33.30
ED250 TII G-Z	179.4	179.4	2716.79	12.13	2606.07	0.04	30.16	50.27	27.25
ED250 TII Z-G	179.3	179.3	2187.48	9.76	2067.04	0.06	24.29	40.48	21.94
PAS TII G-Z	178.8	178.8	3573.11	15.95	3491.30	0.02	35.44	59.07	29.53
PAS TII Z-G	178.7	178.7	2891.54	12.91	2807.22	0.03	28.68	47.80	23.90

vav—average velocity without time of stops [km/h]
vavs—average velocity with time of stops [km/h]
Ec—energy consumption for traction purposes only on a route (energy for auxiliary needs is not taken into account) [kWh]
Ecut—unitary energy consumption for traction purposes per train km (energy for auxiliary needs is not taken into account) [kWh/ train km]
Ecu—unitary energy consumption for traction purposes per ton km (energy for auxiliary needs is not taken into account) [Wh/ t km]
Ecrek—energy consumption for traction purposes only with recuperation on a route [kWh]
%RS—savings in energy consumption due to recuperation [%]
Eup—energy consumption for traction per 1000 pas km (assumed 100% seats occupied—normal payload) [kWh/1000 pas km].
Eup60%—energy consumption for traction per 1000 pas km (assumed 60% seats occupied—60% of normal payload) [kWh/1000pas km]
Source A. Szeląg's elaboration based on simulations

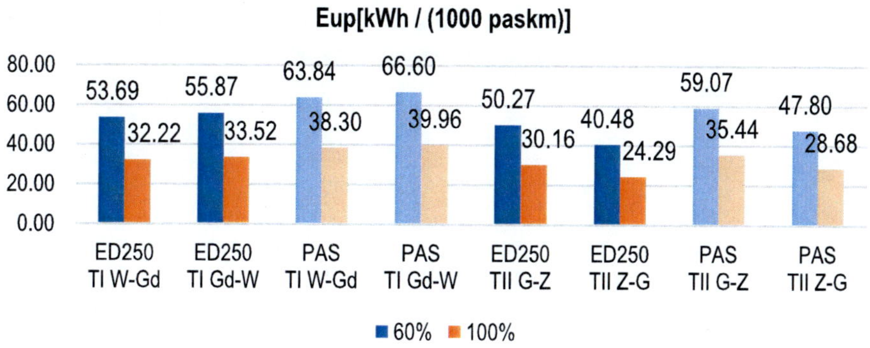

Fig. 3 Comparison of energy consumption, with 100% and 60% occupation of seats (2015–2020) *Source* [8]

Table 4 CO_2 unitary emission in [t/1000 passenger km]

Train & direction	CO_2 emission [t/1MWh]	CO_2 unitary emission [t/1000paskm]
ED250 TI W-Gd	0.7190	0.0232
ED250 TI Gd-W	0.7190	0.0241
PAS TI W-Gd	0.7190	0.0275
PAS TI Gd-W	0.7190	0.0287
ED250 TII G-Z	0.7190	0.0217
ED250 TII Z-G	0.7190	0.0175
PAS TII G-Z	0.7190	0.0255
PAS TII Z-G	0.7190	0.0206

CO_2 emission during generation of 1 MWh of electrical energy-average in Poland [t/1MWh]
CO_2 unitary emission in [t/1000 passenger km]
Source A. Szeląg elaboration based on simulations

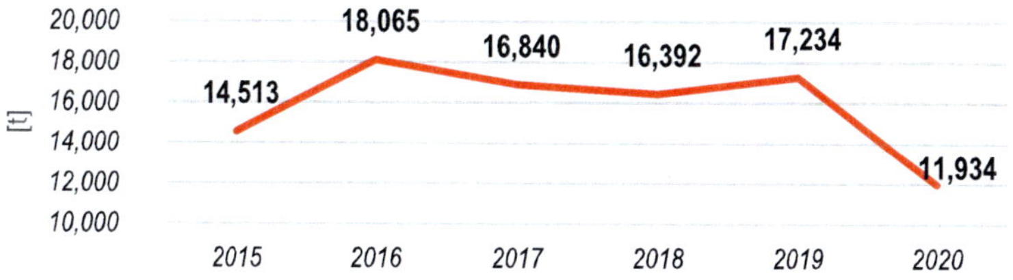

Fig. 4 CO_2 emission reduction by ED250 service in 2015–2020

These reductions in the following years depended mainly on the number of kilometers traveled, resulting from infrastructural upgrade works.

6 Conclusion

The new trains were to form a new category called Express Intercity Premium (EIP), offering the highest quality in terms of comfort and travel time.

Benefits related to the operation of the first high-speed trains in Poland are the following:

- changing the image of the railways to more modern ones
- higher transport efficiency
- lower negative impact on the environment, lower energy consumption, lower CO_2 emissions:
  - lower unitary energy consumption for traction purposes per ton km (TI 3.5 Wh/tkm and TII 3 Wh/tkm for ED250 over PAS),
  - lower energy consumption for traction per 1000 pas km (TI 60% occupation 10 kWh/1000paskm and 100% occupation 6 kWh/1000paskm; TII 60% occupation 9 kWh/1000paskm and 100% occupation 7 kWh/1000paskm for ED250 over PAS),
  - over 95 thousand ton CO_2 emission reduction associated with the operation of the ED250 in the period 2015–2020.

The introduction of the ED250 class trains into service to a greater extent will have a significant impact on reducing the negative impact on the environment. Moreover, it will contribute to the implementation of the climate goals set out in the EU Green Deal and it will also affect the competitiveness of railways in relation to road transport.

References

1. Pomykała, A., Engelhardt, J.: Concepts of construction of high-speed rail in Poland in context to the European high-speed rail networks. Socioecon. Plann. Sci. 101421 (2022). https://doi.org/10.1016/j.seps.2022.101421
2. Uchwała Nr 276/2008 Rady Ministrów z dnia 19 grudnia 2008 r. w sprawie przyjęcia strategii ponadregionalnej „Programu budowy i uruchomienia przewozów kolejami dużych prędkości w Polsce". Warszawa (2008)
3. CUPT: 'Projekt indywidualny: Zakup kolejowego taboru pasażerskiego do obsługi połączeń dalekobieżnych—Etap I', CUPT. www.cupt.gov.pl/archiwum/indexc68c.html?id=1025 (2011)
4. Pendolino pojedzie wiedenką do Wrocławia: *Kurier Kolejowy*, 2016. https://kurier-kolejowy.pl/aktualnosci/27396/pendolino-pojedzie--wiedenka--do-wroclawia.html
5. 'Szybkie prace na CMK: Kurier Kolejowy. https://kurier-kolejowy.pl/aktualnosci/28408/szybkie-prace-na-cmk--by-podroze-byly-krotsze.html (2016)

6. 'Czasowe zmiany na CMK: Rynek Kolejowy. https://www.rynek-kolejowy.pl/wiadomosci/czasowe-zmiany-na-cmk-82465.html (2017)
7. Pomykała, A.: Polish regional and long-distance railway transport during the Covid-19 Pandemic. Probl. Kolejnictwa—Railw. Rep. **65**(192), 185–194 (2021). https://doi.org/10.36137/1925e
8. Pomykala, A., Szelag, A.: Reduction of power consumption and co_2 emissions as a result of putting into service high-speed trains: polish case. Energies **15**(12) (2022). https://doi.org/10.3390/en15124206
9. GUS, Raport 2020: Polska na drodze zrównoważonego rozwoju, Warszawa. https://raportsdg.stat.gov.pl/2020/cel13.html (2020)
10. 'www.elektrowoz.pl (2020)

Spatio-Temporal Distribution of Chinese Cities' Air Quality and the Impact of High-Speed Rail to Promote Carbon Neutrality

Qingchen Liu, Hongchang Li, Wen-long Shang, and Kun Wang

Abstract With China's continuous urbanization and industrialization, the emissions in Chinese cities have become increasingly serious. Chinese government has set a goal to reach carbon peak and achieve carbon neutrality, endeavoring to gradually realize net-zero carbon dioxide (CO_2) emission. Moreover, the new transport technology has also been fast developed, represented by the quick expansion of the high-speed rail (HSR) system. Based on the Air Quality Index (AQI) of 286 Chinese cities over the 2016–2019 period, this paper first adopts the spatial auto-correlation analysis to quantify the spatio-temporal characteristics of Chinese cities' AQI. Then, a spatial difference-in-differences (SDID) model is estimated to shed light on how Chinese cities' air quality can be affected by HSR. Our paper identifies apparent spatio-temporal distribution patterns in Chinese cities' air quality. Our empirical results that the HSR opening can help reduce emissions to improve the city's air quality. Moreover, HSR opening in the adjacent city can also improve one city's air quality (i.e., the neighboring effect). We also highlight and verify the mechanism of such a positive HSR impact on the city's air quality. First, as a cleaner transport mode, HSR helps divert traffic from other more polluting modes (i.e., positive direct "transport substitution effect"). HSR also helps promote the city's tertiary industry, leading to fewer emissions (i.e., positive indirect "industrial structure effect"). Our heterogeneous analyses further demonstrate that HSR is more effective to improve

Q. Liu
School of Economics and Management, Beijing Jiaotong University, Beijing 100044, China

H. Li
College of Metropolitan Transport, Beijing Key Laboratory of Traffic Engineering, Beijing University of Technology, Beijing 100124, China

W. Shang
School of Traffic and Transportation, Beijing Jiaotong University, Beijing 100008, China

Centre for Transport Studies, Imperial College London, London SW7 2AZ, UK

W. Shang · K. Wang (✉)
School of International Trade and Economics, Business and Economics, University of International, Beijing 100029, China
e-mail: kunwang@uibe.edu.cn

F. Pagliara (ed.), *Socioeconomic Impacts of High-Speed Rail Systems*,
Springer Proceedings in Business and Economics,
https://doi.org/10.1007/978-3-031-26340-8_15

the air quality in eastern and western regions. But the neighboring effect is only obvious in eastern China as the cities are closer to each other in terms of economic relations and geographic locations.

Keywords Air quality · Emissions · High-speed rail · Spatio-temporal analysis · Spatial difference-in-differences (SDID)

1 Introduction

Since its economic reform and opening up in 1980s, China has experienced rapid economic development, accompanied by the fast industrialization and urbanization. Although the rapid economic development improved people's living standards, China's energy consumption has also been increased rapidly, especially from fossil energy sources (e.g., coal, natural gas, oil). The carbon dioxide (CO_2) emission and other related air pollutants have risen accordingly, seriously damaging the climate and people's health threatening the sustainable development of the whole country. Chinese government thus formulated the carbon neutral and carbon peak target to control the emissions and air pollution.

The Technical Specification for Ambient Air Quality Index (Trial) formulated by China puts forward the standard of Air Quality Index (AQI) to carry out quantitative monitoring of air pollution in various regions [1]. Since 2016, the Chinese government has adopted GB 3095–2012 air quality standards, replacing the original air pollution index (API) with the air quality index (AQI). The pollutant indicators add fine particulate matter ($PM_{2.5}$), carbon monoxide (CO) and ozone (O_3) to the original inhalable particulate matter (PM_{10}), nitrogen dioxide(NO_2), and sulfur dioxide (SO_2) to reflect the overall situation of air quality. With the frequent occurrences of air pollution problems, the characteristics of regional distribution of air pollution attracted more attention from public and academic research. The analysis of spatio-temporal distributions and patterns of the air quality in among regions and its influencing factors have also become a research hotspot.

Among different sources of emissions and air pollution, transport is an dominant one in China. Fuel powered vehicles emit a large number of organic compounds, nitrogen oxides and other chemical substances, leading to the greenhouse effect and increasing ground-level ozone concentration [2]. According to the China Mobile Source Environmental Management Annual Report (2020) published by the Ministry of Ecology and Environment of China, private cars are the main source of pollutants, and the policy of limiting the purchase and driving of private cars in cities cannot effectively solve the urban air pollution problem [3]. In addition, China's transport composition is not reasonable enough. For the inter-city travel, road transport, dominated by diesel vehicles, contributes about 73.0% of goods transport and 74% of passenger transport. The advantages of low energy consumption and low emission of railway and water transport have not been fully exerted [4].

Transport infrastructure is one of the key factors in promoting national economic growth. Zheng, Kahn and Zhang found that the opening of the high-speed rail (HSR) is conducive to integrating regional resources, promoting coordinated economic development, and promoting economic growth in cities along the way [5, 6]. Air Pollution Prevention and Control Briefing published by China's Ministry of Ecology and Environment suggests that the country should focus on optimizing the structure of transport, shifting from road transport to rail transport. Compared with normal-speed railway, HSR has obvious advantages in terms of economic and environmental benefits [7, 8]. Therefore, based on the development of green transport, promoting regional development and other factors, the environmental benefit of developing HSR is obvious.

Based on such background, this paper first quantifies the spatio-temporal evolution of Chinese cities' air quality over the 2016–2019 period. Then, the spatial difference-in-differences (SDID) model is employed to study the HSR impact on Chinese city's air quality. The contributions of this paper are multi-fold. First, this is one of the first studies to adopt spatial auto-correlation approach to quantify the spatio-temporal distribution and characteristics of Chinese cities' air quality. Second and more importantly, this paper provides one of the first rigorous empirical examinations on the impact of HSR opening on city's air quality. The spatial spillover effect of HSR impact is also identified. In addition to the overall effect, we investigate the mechanism of such HSR impact. Specifically, we consider the direct effect brought by the change of transport traffic volume and the modal splits, and the indirect effect caused by HSR on the city's industrial structure and production scale. The heterogeneous analysis is also done to shed light on the HSR impacts on different kinds of cities that can be categorized by the geographic location and the degree of economic development.

The spatio-temporal analysis of AQI shows clear inter-annual and seasonal patterns of air quality among Chinese cities. Moreover, the spatial pattern suggests the southern and coastal regions have overall better air quality than northern China and inland regions. This is mainly related to the climate type, terrain, economic development levels. The northern cities are low-lying, where pollutants are difficult to diffuse, and where economic development and industrialization progress are rapid, causing more serious environmental damage. Most coastal cities in southern China have a tropical or subtropical monsoon climate, and the terrain is relatively flat. Air quality in China has obvious positive spatial auto-correlation, and also demonstrates obvious aggregation trend over time. The cities with severe air pollution could have a negative impact on the air quality of neighboring cities, suggesting strong negative spatial spillover effect.

In addition, our SDID estimations suggest that the HSR opening in one city can reduce the emission and improve its air quality (i.e., defined as the the local effect), while the HSR opening in the adjacent city can also improve the city's air quality (i.e., defined as the neighboring effect). We also highlight and verify the mechanism of such positive HSR environmental impact. First, as a cleaner transport mode, HSR helps divert traffic from other more polluting modes (i.e., positive direct transport substitution effect). Moreover, HSR helps promotes the city's tertiary industry development (i.e., the service sectors, such as tourism, high-tech industries), leading to fewer emissions (i.e., positive indirect "industrial structure effect"). Our heterogeneous analyses further demonstrate that HSR is more effective to improve the air quality in the eastern and western regions of China, given the more prevailing transport substitution effect and industrial structure effect, respectively. But the neighboring effect of HSR is only obvious in the eastern China as the cities are closer with each other in terms of economic relations and geographic locations within this region. Moreover, the HSR proves to improve air quality in more economically developed cities, as the HSR traffic is larger in this region and thus plays more significant role.

The rest of the paper is organized as below. Section 2 reviews the relevant literature. The detailed research design and methodology are discussed in Sect. 3. Section 4 quantifies the spatio-temporal distribution characteristics of China's AQI. The estimation results of the HSR impact on city's air quality are presented and discussed in Sect. 5. Section 6 summarizes this study.

2 Literature Review

This study is related to three streams of literature, namely the spatio-temporal distribution of city's emissions and air quality, the influencing factors of city's air quality, and the impact of HSR opening on city's air quality. In the subsequent subsections, we review the relevant literature for each stream, respectively.

2.1 The Spatio-Temporal Distribution of City's Air Quality

The temporal studies on air quality can be done at annual, quarterly or intra-day dimension. In the annual analysis, Xu, Liu, and Wang explored the changes of China's urban AQI from 2014 to 2016, and found that the AQI showed a downward trend and the number of cities with air pollution decreased [9]. Guo, Lin, and Bian discussed the spatial and temporal distribution characteristics of air quality in China from 2015 to 2017, and found that air quality was improving year by year [10]. In the quarterly analysis, most studies concluded that the temporal variation characteristics of city's air quality in China are seasonal. Niu et al. found that the seasonal distribution of air pollution in China is U-shaped, and the main pollutants change with the seasons. This is because China has a northwest monsoon in winter and a southeast monsoon in summer. Meanwhile, man-made emissions lead to a sharp rise in the level of particulate matter in winter, so the proportion of particulate matter in the air in winter is the highest, and the pollution of fine particulate matter is the most serious [11]. Yan et al. and Chen et al. believed that the poor air quality in winter is mainly due to seasonal straw burning and coal burning for heating in winter, which produces a large amount of PM2.5 [12, 13].

Many scholars believe that there is also spatial correlation between air quality among Chinese cities. Fang et al., Zhang and Luo found that the Moran Index showed that the AQI measure values of Chinese cities had positive spatial auto-correlation, and the spatial distribution of high AQI values tended to aggregate rather than disperse [14, 15]. Dai and Zhou adopted PMFG network method and found that the AQI in the sample period presented obvious spatial positive correlation [16]. Fang based on the STIRPAT model and the environment Kuznets curve (EKC) hypothesis, using a spatial difference-in-differences approach, concluded that city's smog pollution exhibits strong spatial correlations [17, 18]. Lin and Wang used LISA aggregation map to assert that the spatial aggregation trend of city's air quality in China has been increasing year by year, showing the characteristics of pollution diffusion from key cities to overall regional pollution. Therefore, air pollution control should also be initiated by key cities, and regional pollution should be jointly controlled [19].

Li et al., Zhang and Luo concluded that China's topography is high in the south and low in the north, high in the west and low in the east, which has a significant impact on the distribution of air pollutants. Therefore, the spatial distribution of air pollutants has a significant difference between the east and the west. High-AQI urban agglomerations are mainly distributed in the north and northwest, while low-AQI urban agglomerations are mainly distributed in the south and southwest [15, 20]. Particularly, Jia found that the CO_2 emission reduction effect of HSR is more significant in eastern China, large cities, and resource-based cities. Higher levels of HSR service intensity in large cities and resource-based cities are not conducive to reducing CO_2 emissions in neighboring cities [21].

2.2 *Air Quality Influencing Factors*

Many scholars have summarized the socioeconomic factors that influence air quality in Chinese cities. Some scholars believed that population gathering and traffic congestion caused by industrialization and urbanization are the main causes of air pollution. Zhang and Griffin argued that city' air quality is affected by both environmental and socioeconomic factors, especially the urbanization development and industrial enterprise agglomeration [22, 23]. Wang believed that population density, highway passenger volume and vehicle ownership are all important factors that aggravate city's air pollution [24]. Fang et al. believed that the increase in the proportion of secondary industry and population density is the main reason for frequent air pollution incidents [14]. Therefore, the vigorous development of public transport can alleviate air pollution problems which caused by excessive urban population density and traffic congestion [25]. Fang found that the relationship between per capita GDP, urban population and urban smog pollution all follow n-shaped curve, and smog is proved to reduce to a certain extent as per capita GDP increases [18].

Other scholars believe that the energy consumption accompanied by economic growth is closely related to the atmospheric environment. Chi et al. found that although the level of urban economic development is positively correlated with air quality, the level of economic development does not play a decisive role in the change of air quality. This is because the environmental Kuznets curve is inverted U-shaped, and economic growth will lead to environmental quality deterioration and then improvement [26]. Bi et al. and Grey et al. proposed that total energy consumption and economic growth have a serious impact on air quality, and the improvement of scientific and technological level will improve the air quality [27, 28]. According to the research of Ying, the average labor input, average fixed asset input, average energy consumption, average GDP and average CO_2 emission efficiency of high-income cities are significantly higher than those of upper-middle income cities [29]. Lin and Wang believed that energy consumption and industrialization are important factors of causing air pollution in cities, but by the regional natural environment and the limitation of social and economic development stage, all kinds of social and economic factors on the air quality in different cities have different degrees of influence, so there is a need for subregional research [19].

However, scholars have different views on the effect of urban greening degree on air quality. The study of Jiang et al. showed that the increase of the proportion of tertiary industry and urban greening coverage rate is conducive to improve air quality in the Yangtze River Economic Belt, supporting the theory of "pollution sanctuary" [25, 30]. Another view is that urban greening does little to improve air quality. Wang and Wang constructed the panel model of 31 provincial capital cities in China to explore the main factors of air quality, and found that the level of urban greening could not significantly improve city's air quality [31].

2.3 The Impact of HSR on City's Air Quality

With fast HSR development, especially in China, more scholars have paid attention on HSR's impact on the air pollution. To quantify the HSR development, many studies developed indexes to measure the HSR connectivity or accessibility. For example, Zhang et al. and Jiao et al. defined the HSR connectivity as the number of HSR lines or the train frequency passing through the city and the weighted centrality in the passenger train network [32, 33]. Liu et al. used HSR accessibility measured by the shortest HSR travel time between two cities [34]. Zhang et al. and Zhu et al. used the "with or without comparison method" to reflect the actual effects of policies by comparing the data in the two states of "with HSR line" and "without HSR line". This method is adapted to the difference-in-differences (DID) method, which is applicable to measure the impact of HSR on environment [35, 36].

Existing research did not reach a consensus on the impact of railway or HSR on city's air quality. Some studies found that the development of rail transport can effectively improve the city's air quality. For example, Qin and Chen studied the impact of Beijing-Shanghai the high-speed railway on the air quality of cities along the route by using the breakpoint regression method, and found that the operation of Beijing-Shanghai the high-speed railway significantly improved the air quality of cities along the route by replacing private cars and reducing the number of ordinary railway trains with high energy consumption [37]. Yu et al. found through augmented gravity model research that railway services are improved under the drive of speed, and air passengers turn to railway transport, leading to a reduction in carbon dioxide emissions, and optimizing the traffic structure to reduce carbon dioxide [38]. The traffic substitution effect is due to the fact that the development of rail transit encourages travelers to reduce the original ground transport mode, to reduce the exhaust emissions of private cars, and to improve city's air quality [39]. Zhang and Feng used the DID model to study the significant improvement of haze pollution from the perspectives of scale effect, structure effect and technology effect [40]. Fang used causal mediation analysis method to test the two mechanisms related to the HSR, and found that industrial structure upgrading can reduce haze pollution, while real estate market development can increase haze pollution [18]. Jia adopt a continuous spatial difference-in-differences (SDID) model to investigate the effect and its mechanism of HSR service intensity on CO_2 emissions, and the results show that due to the influence of transport substitution, market integration, industrial structure and technological innovation, the intensity of the high-speed railway service greatly reduces city's carbon dioxide emission [21]. Another view is that the development of rail transit does not have a significant positive impact on city's air quality: Beaudoin and Lin found that rail transit did not significantly reduce the emission concentration of air pollutants, because of the traffic creation effect that the development of rail transit will promote urban population aggregation and economic development, aggravating the air pollution problem [41].

3 Research Design

The research framework and methodology are specified in this section. First, we highlight the theoretical mechanisms of HSR impact on city's air quality (see Sect. 3.1). Then, the detailed empirical research methods are introduced in Sect. 3.2. Last, Sect. 3.3 introduces our data sources.

3.1 *Theoretical Mechanism*

Figure 1 illustrates the mechanism of the HSR's impact on the city's air quality, which can be divided into the direct and indirect impacts. First, the direct impact includes the transport substitution effect and inducing effect. HSR is a cleaner and more emission efficient transport mode compared to the highway and aviation [36]. HSR is attractive to passengers with short and medium distance mobility demand. When HSR attracts more than 10 million passengers from other transport modes, net emission reduction can be achieved, improving the overall air quality [33, 42]. However, HSR could also stimulate new traffic demand, thus aggravating the greenhouse effect (i.e., the inducing effect) [43].

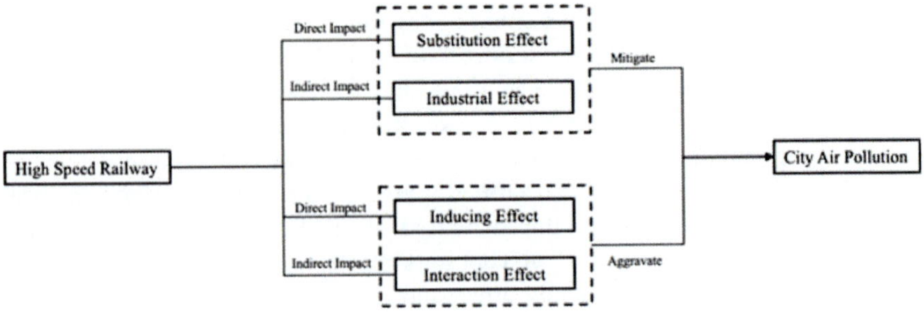

Fig. 1 The mechanism of HSR's impact on city's air quality

The indirect impact of HSR on air quality is achieved by the impact on other urban economic variables, such as GDP, population, urban area and industrial compositions, and then to affect city's air quality indirectly. For example, HSR is conducive to improving the accessibility of cities along the railway line, promoting the accumulation of technology and capital in these cities. HSR could help reduce costs of transporting passengers and goods, thus stimulating the service industries and facilitating the industrial upgrading to high tech but low pollution industries [35]. This can reduce the emissions from fossil energy consumption, thus improves the city's air quality. However, the transport cost reduction brought by HSR can also enhance the industrial agglomeration, enlarging the production and population size of the city. Especially, some large cities will further utilize the resources from the nearby cities with HSR to quickly expand the city size (i.e., the so-called Siphon effect). This would increase the energy consumption and possibly deteriorate the air quality.

Given the existence of both direct and indirect impacts and the complex mechanisms, the current empirical studies have not yet clearly identified the impact of HSR on city's air quality. Thus, in this paper, we not only identify the overall impact of HSR on China's city's air quality, but also try to distinguish the different causes.

3.2 *Research Methods*

In this paper, we first adopt the spatial auto-correlation analysis to quantify the spatio-temporal patterns of Chinese cities' air quality over recent years (see Sect. 3.2.1). Then, the spatial difference-in-differences (SDID) model is proposed in Sect. 3.2.2 to examine the HSR impact on city's air quality, while accounting for the possible spatial spillover effects as well.

3.2.1 Spatial Auto-correlation Analysis

First, our quantitative study is conducted to quantify and describe the spatio-temporal characteristics of Chinese cities' AQI. In particular, the spatial autocorrelation method is used to account for the spatial spillover effect of neighboring region's AQI on one city's AQI. This method detects whether the observed value of a certain point in space is correlated with the observed value of an adjacent (or neighboring) point. Global Moran index and local Moran index are the most commonly used spatial auto-correlation functions, to represent the correlation degree and clustering mode of geographical variables. The positive and negative values of statistics represent the positive and negative values of spatial auto-correlation. The larger the statistic is, the stronger the spatial auto-correlation is. The global Moran index measures global spatial correlation, as shown below:

$$I = \frac{n}{\sum_{i=1}^{n}\sum_{j=1}^{n} w_{ij}} \frac{\sum_{i=1}^{n}\sum_{j=1}^{n} w_{ij}(x_i - \overline{x})(x_j - \overline{x})}{\sum_{i=1}^{n}(x_i - \overline{x})^2} \tag{1}$$

where n is the total number of cities; x_i and x_j are the observed values of urban environment monitoring points in cities i and j, respectively; $\overline{x}$ is the mean of x; w_{ij} is the entry of the spatial weight matrix, if the city i is adjacent to the city j, $w_{ij} = 1$ ($w_{ij} = 0$, otherwise). The range of the global Moran index is $[-1, 1]$. $I>0$ indicates that city's air quality has an aggregation trend, and $I<0$ implies that the city's air quality has a discretization trend. When I is equal to 0, the city's air quality is spatially independent and distributed irregularly and randomly.

Local spatial auto-correlation captures the spatial relationships among cities, so it can effectively measure the degree of spatial correlation between observed cities and other cities. In the local spatial auto-correlation, we can compute the local Moran index as [44, 45]:

$$I_i = \frac{n(x_i - \overline{x})\sum_{j=1, j\neq i}^{n} w_{ij}(x_j - \overline{x})}{\sum_{i=1}^{n}(x_i - \overline{x})^2} \tag{2}$$

where n is the total number of cities; x_i and x_j are the observed values of urban environment monitoring points in city i and j, respectively; $\overline{x}$ is the mean of x; w_{ij} is the element of the spatial weight matrix, when the city i is adjacent to the city j, $w_{ij} = 1$, otherwise $w_{ij} = 0$. I_i is used to compared the degree of spatial agglomeration between region i and the surrounding areas, and we can classify the association modes into four types: high–high (HH), high–low (HL), low–high (LH), and low–low (LL). HH and LL groups indicate that the air quality of the city i is consistent with that of the neighboring cities. On the other hand, HL and LH groups indicate that the city i's air quality is a high (or low) value in a low (or high) air quality neighborhood.

3.2.2 Spatial Econometric Model

In addition to the spatial auto-correlation analysis of AQI, we examine the impact of HSR on AQI. Specifically, a spatial difference-in-differences (SDID) method is used. SDID analysis takes into account that air pollution often spreads to surrounding areas through atmospheric circulation and spatial spillover effects measured by spatial weight matrix W_{ij}. Given that not all the cities have access to the HSR service, the sample cities with HSR operation are classified into the experimental group, while the rest cities without HSR operation are the control group [46, 47]. Considering different years of HSR operation in different cities, this paper adopts the multi-phase DID model on the experimental group and the control group [48]:

$$
\begin{aligned}
AQI_{it} = & \alpha + \beta_1 Open_{it} \times Year_{it} + \beta_2 \sum_j w_{ij} Open_{it} \times Year_{it} \\
& + \boldsymbol{\gamma} \boldsymbol{X}_{it} + \boldsymbol{\theta} \sum_j w_{ij} \boldsymbol{X}_{it} + \rho \sum_j w_{ij} AQI_{it} \\
& + \lambda \sum_j w_{ij} \varepsilon_{it} + A_i + B_t + \varepsilon_{it}
\end{aligned} \tag{3}
$$

where i represents the index for cities, t represents the index for periods; AQI_{it} represents the air quality index of city i in period t, $Open_{it}$ is a dummy variable, which equals to 1 if the HSR is available in the period t in city i, otherwise, this variable equals 0. $Year_{it}$ is the dummy variable of the year in which HSR is opened. w_{ij} is the spatial adjacency weight matrix. $Open_{it} \times Year_{it}$ is the core explanatory variable, which represents the effect after the opening of HSR between the experiment group and the control group. X are several control variables selected in this paper (as shown in Table 1). A_i, B_t is a dummy variable to control for the fixed effect of city and time, respectively. ε_{it} is the residual term. α is a constant term. β_1, β_2, $\boldsymbol{\gamma}$, $\boldsymbol{\theta}$, ρ, and λ are the coefficients.

3.3 *Data Sources*

This paper selects the data of cities in China from 2016 to 2019 as the research object. As a dimensionless index, AQI data can quantitatively describe the air quality within certain region in a certain period, and the value ranges from 0 to 500. The higher the AQI, the worse the air quality. All AQI data are obtained from https://www.AQIstudy.cn.We use the 24-h average AQI value of the day as the daily data, the monthly average AQI as the monthly data, and the quarterly AQI and annual AQI definitions are similar. The data of cities are mainly from "Statistical Yearbook of Chinese Cities". After removing the cities with serious missing data, a total of 286 sample cities are obtained [49–51]. There are a few data missing problems in the sample cities, and this paper uses SPSS to make interpolation calculation for supplement.

Table 1 Definitions of the control variables

Variable types	Variable name	Explanations	Measurements
City variables	Area	Urban size	The administrative area of land
	Urban	Urbanization level	Proportion of urban construction land in the area of the district
	Pop	The urban population	Average annual population
Urban environmental protection variables	Green	Development of urban green spaces	The area of urban green space
	Bus	Degree of public transport development	The actual number of bus operation at the end of the year
Economic variables	GDP	GDP	Gross regional product
	Growth	Economic growth rate	GDP growth rate

The data of opening time of the high-speed railway stations are from China High-speed Railway Route Database (CRAD) in China Research Data Service Platform (CNRDS).

4 Spatio-Temporal Distribution Characteristics of AQI

This section reports the calculated spatio-temporal distribution of Chinese cities' air quality for the study period. The Sect. 4.1 focuses on the temporal distribution characteristics of AQI, while the Sect. 4.2 discusses the spatial distribution. Then, in the Sect. 4.3, we present the Moran index to quantitatively analyze the spatio-temporal characteristics of Chinese cities' air quality.

4.1 Temporal Distribution Characteristics of AQI

In 2016–2019 period, the evolution of Chinese cities' AQI is depicted in Fig. 2. It is observed that the AQI values in China evolves over time but demonstrates a decreasing trend. This reflects the significant improvement of the air quality in Chinese cities in recent years. In terms of the degree of air quality improvement, the number of cities that experienced air quality improvement accounted for 82.1% of the sample cities, among which the cities with more than 20% improvement accounted for 31.9%. In terms of air pollution degree, the proportion of cities with air quality reaching light pollution level (AQI annual mean > 100) decreased year by year, from 15% in 2016 to only 4.9% in 2019.

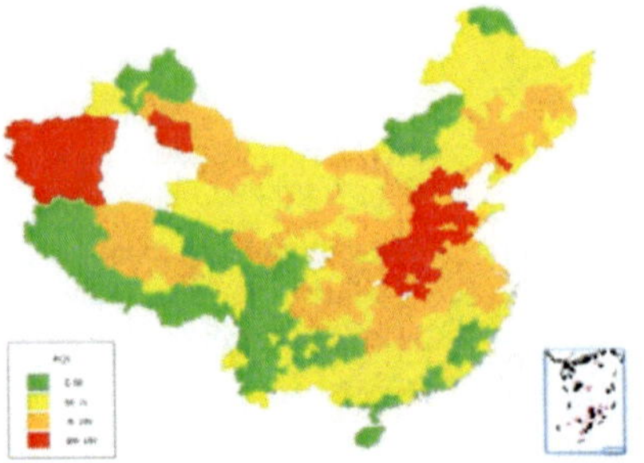

(a) Distribution of annual average AQI values in China in 2016

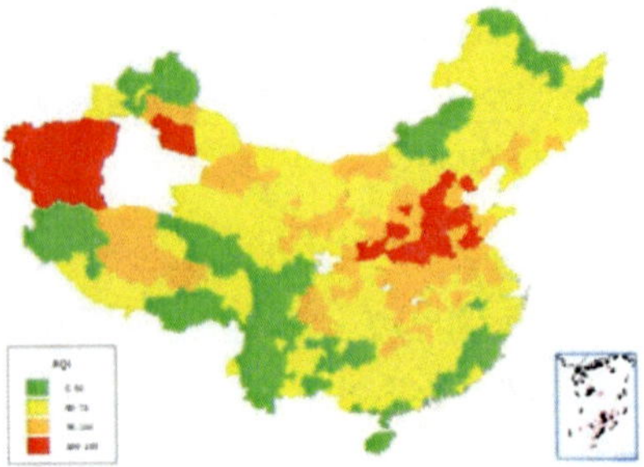

(b) Distribution of annual average AQI values in China in 2017

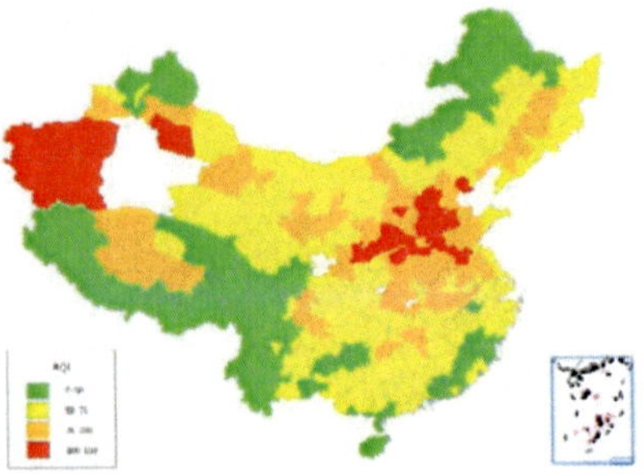

(c) Distribution of annual average AQI values in China in 2018

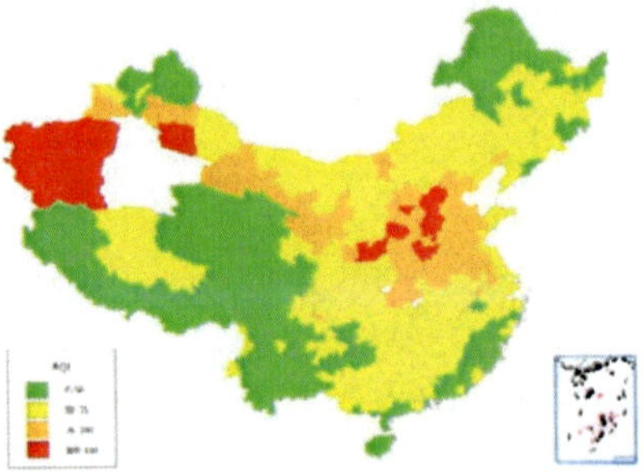

(d) Distribution of annual average AQI values in China in 2019

Fig. 2 Annual average AQI distribution in China from 2016 to 2019

As can be seen from Fig. 3, the seasonal average value of AQI is characterized by winter, spring, autumn, and summer, with the highest AQI index and the most serious pollution degree in winter, the lowest AQI index and the best air quality in summer. The main reason for the quarterly variation characteristics of AQI is that the winter temperature in the north is low, and the demand for heating has led to a substantial increase in the amount of coal burned. At the same time, it also increases the pollutants released by fuel combustion. In addition, the winter climate in the north is dry, and the reduction of precipitation is not conducive to diluting the pollutants in the atmosphere, making the AQI value reaches the highest level of the year. Precipitation increases in summer, and frequent rainfall and strong winds are conducive to disperse pollutants in the air, so the AQI value is the lowest in summer. Spring is windy and sandy, and dust storms bring a lot of dust, so the air quality in many areas is still not optimistic in spring. Autumn is drier than summer, and the burning of plant stalks after autumn harvest releases a large number of pollutants,

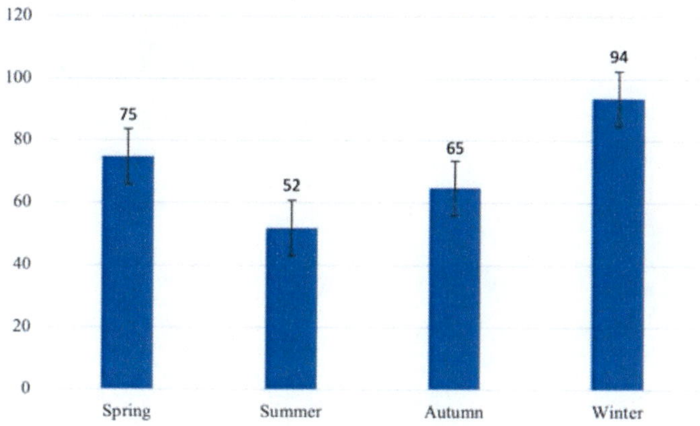

Fig. 3 Average AQI of four seasons in China from 2016 to 2019

leading to a decline in air quality in autumn, but it is slightly better than winter and spring.

4.2 *Spatial Distribution Characteristics of AQI*

To focus on the discussion of the spatial distribution of AQI, this subsection selects one particular year, 2019, for the analysis. As shown in Fig. 4, the air quality of coastal cities in Hainan and Fujian is relatively better. This is because the southeast part of China has a tropical or subtropical monsoon climate, and the terrain is relatively flat. The humid monsoon from the ocean is conducive to the diffusion of urban pollutants in summer. The cities' air quality in the high-altitude areas represented by Yunnan and Xizang also at a relatively good level. This is because there are few human activities in the high-altitude west southern region, which causes little damage to the ecological environment.

The cities with poor air quality are concentrated in north and northwest parts of China. The air quality and topography of the north China are potentially related to the east Asian monsoon, leading to the accumulation of regional air pollutants. Northern China, represented by Shijiazhuang, has the problem of low vegetation coverage, which is unable to resist the dust from the northwest. In addition, because of the low terrain in north China, the conditions for the diffusion of air pollutants are poor. In north China, the problem of high population density, rapid economic development and industrialization and urbanization still exist, so that the damage to the environment is quite serious. In cities in northwest China, such as Urumqi, water vapor is difficult to reach far away from the sea, so there is less precipitation, scarce vegetation and bad natural environment. All these reasons lead to poor air quality in northwest China.

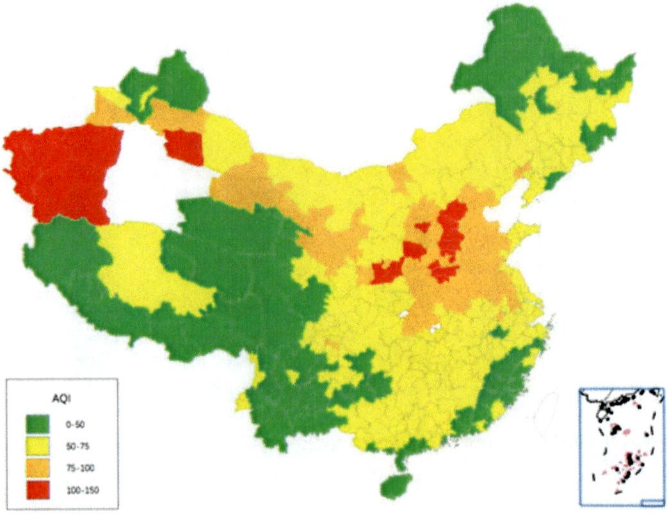

Fig. 4 Air quality distribution in China in 2019

4.3 Spatial Auto-Correlation Analysis Based on Moran Index

In this paper, daily AQI data of the sample cities from 2016 to 2019 are used to calculate the annual average. The global Moran index and local Moran index of each city are also calculated with the equations Eqs. (1–2). The results are reported in Table 2. Moran index calculation results passed the statistics test at a significance level of 5%. Moran index statistics are all positive, ranging between 0.5 and 0.6, The highest value in 2017 is 0.572 and the lowest in 2018 is 0.530, which shows that Chinese cities' air quality demonstrates an obvious positive space correlation. It can be concluded that the air quality of Chinese cities is also significantly affected by their surrounding cities and even provinces.

In this paper, the local spatial auto-correlation analysis is carried out through the Moran scatter plot in Fig. 5. The AQI samples in China are mostly concentrated in the first quadrant (H–H) and the third quadrant (L-L), which indicates that most provinces in China have the characteristics of spatial dependence, while other provinces have characteristics of spatial heterogeneity.

Table 2 Global Moran Index from 2016 to 2019

Year	Moran index
2016	0.533
2017	0.572
2018	0.562
2019	0.530

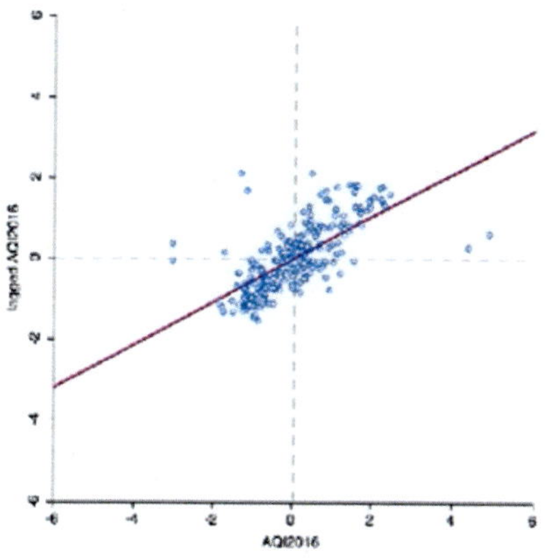

(a) The Moran Scatter Chart of the National AQI Annual Mean in 2016

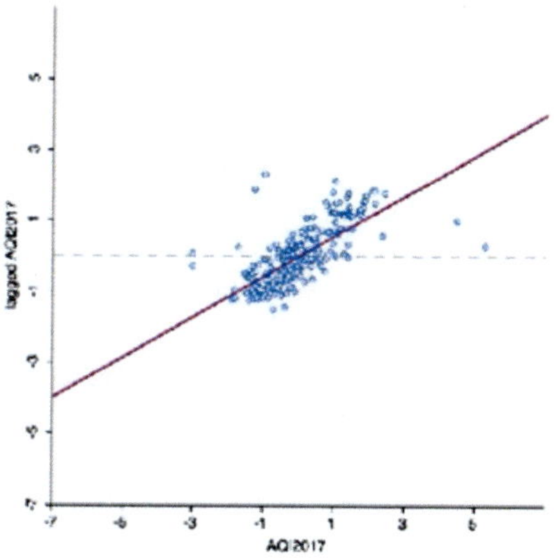

(b) The Moran Scatter Chart of the National AQI Annual Mean in 2017

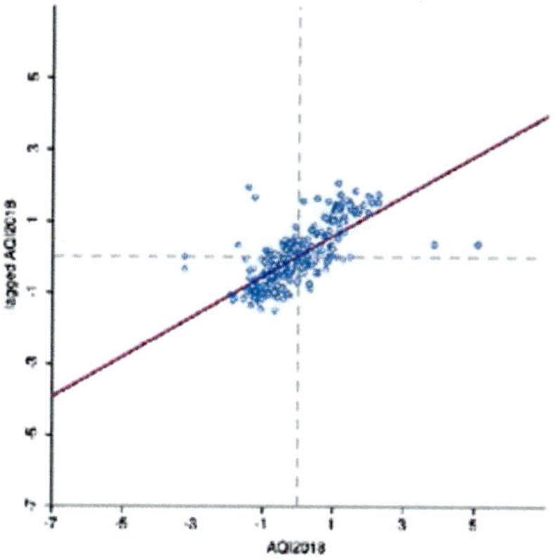

(c) The Moran Scatter Chart of the National AQI Annual Mean in 2018

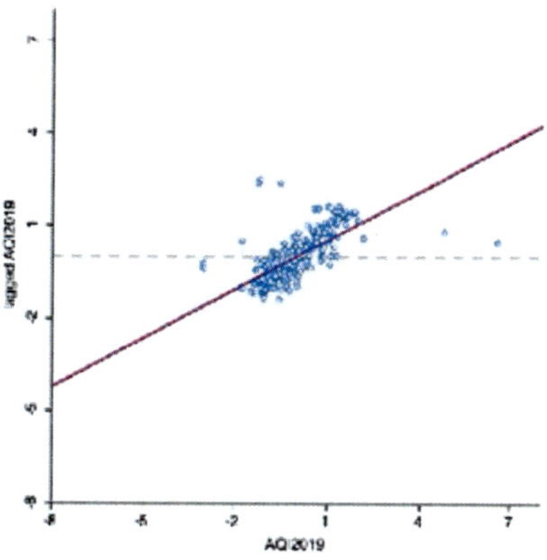

(d) The Moran Scatter Chart of the National AQI Annual Mean in 2019

Fig. 5 Scatter chart of Moran index from 2016 to 2019

It can be seen from Fig. 6 that H–H aggregation areas in China are mainly distributed in north and central China, and the AQI index of cities in these regions is high, and the surrounding cities have similarly high AQI indexes. L-L aggregation areas in China are mainly distributed in the southwest and the southern regions, and their distribution characteristics are related to the fact that these cities are located in the coastal areas, which is conducive to the diffusion of pollutants, or to the fact that they are located in the plateau areas and there are few human activities. There are fewer cities in H–L and L–H regions, which are mainly distributed in the northwest and northern China.

Through the above analysis, it can be seen that the city's air quality in China exhibits an obvious trend of aggregation. Therefore, it is necessary to focus on the

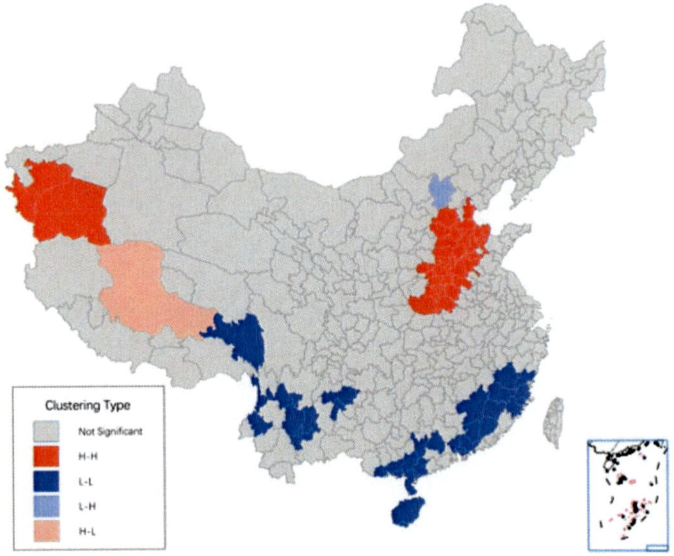

Fig. 6 LISA aggregation of city's air quality in China in 2019

control of air pollution in H–H aggregation areas to prevent its diffusion effect from continuing to reduce the air quality of surrounding cities and maintain good air quality in L–L aggregation areas.

5 Estimations of HSR Impact on City's Air Quality

This section presents the SDID estimation results to investigate the HSR impacts on city's air quality. Specifically, the Sect. 5.1 summarizes and discusses the benchmark estimation results. Then, Sect. 5.2 empirically verifies the mechanisms of such HSR impact by referring to our proposed theoretical mechanism in Sect. 3.1. In the Sect. 5.3, heterogeneous analysis are conducted on sub-sampled cities, categorized by the geographic locations and the degree of economic development.

5.1 Benchmark Estimation Results

This paper selects the panel data of 286 cities in China from 2016 to 2019 to build a multi-phase DID model to study the impact of HSR on city's air quality. The natural logarithm has been taken for all the variables (except dummy variables), including urbanization level, proportion of secondary industry and economic growth rate as the control variables.

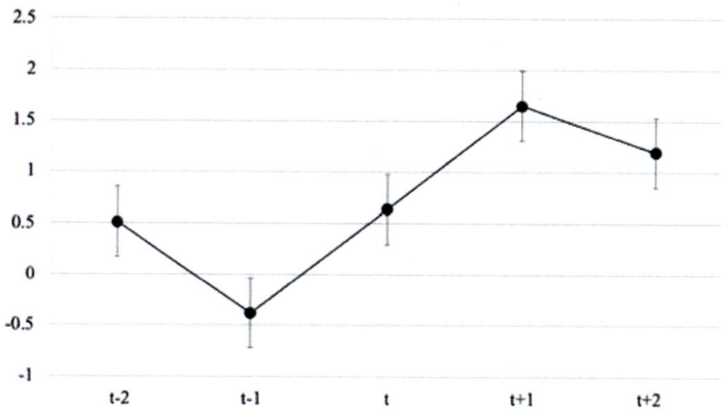

Fig. 7 Parallel trend test result

$$\begin{aligned}\ln(AQI_{it}) =& \alpha + \beta_1 Open_{it} \times Year_{it} + \beta_2 \sum_j w_{ij} Open_{it} \times Year_{it} \\ &+ \gamma_1 \ln(Area_{it}) + \gamma_2 Urban_{it} + \gamma_3 \ln(Pop_{it}) + \gamma_4 \ln(Green_{it}) \\ &+ \gamma_5 \ln(Bus_{it}) + \gamma_6 \ln(GDP_{it}) + \gamma_7 Growth_{it} + \theta \sum_j w_{ij} X_{it} \\ &+ \rho \sum_j w_{ij} AQI_{it} + \delta \sum_j w_{ij} \varepsilon_{it} + A_i + B_t + \varepsilon_{it} \end{aligned} \tag{4}$$

Since HSR opening time are different for different cities. Thus, we adapt the conventional DID model by using the SDID model. As a premise for valid DID analysis, it is still necessary to test whether the air quality index of the control cities and the experimental cities can meet the parallel trend. As can be seen from Fig. 7, there is no statistically significant time trend differences between the control and experimental cities in the two periods before HSR opening. Therefore, this model can satisfy the assumption of parallel trend to justify the use of DID method.

The spatial correlation of spatial econometric model can be caused by dependent variable, independent variable or error term. Spatial Dubin Model (SDM) is general because it can not only represent the spatial correlation of the above three aspects, but also can be transformed into the Spatial Lag Model (SLM) or the Spatial Error Model (SEM) under different coefficient settings. The following Table 3 shows the benchmark regression results of this paper. Models 1–4 are SLM, SEM, Spatial Lagged X(SLX) and SDM, respectively.

According to the estimation results in Table 3, the basic correlation of the benchmark model comes from dependent variables, independent variables and error terms, such that SDM cannot be decomposed into SLM and SEM. Therefore, we shall restrict our attention to the Model 4 (SDM). The core variable $Open \times Year$ is significantly negative in all models. According to Model 4, the opening of HSR in Chinese cities significantly reduces AQI by about 4%. This shows that HSR opening in one city can help improve this city's air quality. In the Model 4, the spatial autocorrelation coefficient ρ is 0.0897 and significant, suggesting that the air quality of a

Table 3 Benchmark regression results

Explanatory variables	Model 1 SLM	Model 2 SEM	Model 3 SLX	Model 4 SDM
Open*Year	−0.0392*** (0.0027)	−0.0491*** (0.0007)	−0.0402** (0.0408)	−0.0417** (0.0335)
lnArea	−0.0923 (0.561)	−0.0823 (0.426)	−0.0904 (0.2369)	−0.0891 (0.628)
Urban	0.0028 (0.2711)	0.0109 (0.3386)	0.0012 (0.1248)	0.0047 (0.5628)
lnPop	0.0158 (0.3192)	0.0026 (0.7832)	0.003 (0.8233)	0.0384 (0.6738)
lnGreen	−0.0201*** (0.0032)	−0.0159*** (0.0007)	−0.0144*** (0.0016)	−0.0311*** (0.0047)
lnBus	−0.034 (0.4716)	−0.0294 (0.7948)	−0.016 (0.335)	−0.067 (0.578)
lnGDP	−0.0308** (0.0227)	−0.0197*** (0.0051)	−0.0242*** (0.0064)	−0.0494*** (0.0085)
Growth	0.0031* (0.0668)	0.0022** (0.0289)	0.0042** (0.0466)	0.0056** (0.0359)
*w**Open*Year			−0.0127* (0.0623)	−0.0136* (0.0742)
*w**lnArea			0.2547 (0.3661)	0.436 (0.7531)
*w**Urban			0.9276 (0.2855)	0.8355 (0.5321)
*w**lnPop			0.1214* (0.0747)	0.2545* (0.0512)
*w**lnGreen			−0.3731** (0.0265)	−0.5433* (0.0429)
*w**lnBus			0.0807 (0.1921)	0.0754 (0.5643)
*w**lnGDP			0.1214 (0.7471)	0.4537 (0.845)
*w**Growth			0.3996 (0.1209)	0.4002 (0.5783)
$w*lnAQI(\rho)$	0.0975*** (0.0000)			0.0897*** (0.0000)
$w*\varepsilon(\lambda)$		0.5378** (0.0246)		
N	1144	1144	1144	1144
R-squared	0.6417	0.6437	0.6532	0.6586

Note 1. The values in parenthesis are P values of the estimated coefficients
2. ***, ** and * indicate significant levels of 1%, 5% and 10%, respectively
3. Without specification, the above two notes apply to all the following tables of estimation results

city will be significantly affected by surrounding cities. In addition, the HSR also has a spatial spillover effect due to the high level of ρ. The opening of HSR in one city's neighboring cities can further reduce this city's AQI by 1.3%. Such findings calls for cautions when evaluating the HSR's environmental impact because the neglect of such spatial spillover (i.e., the neighboring effect) will lead to an underestimated benefit of HSR opening to improve city's air quality.

Moreover, it is interesting to notice that several socioeconomic factors are found to affect city's air quality insignificantly, such as city's area size, urbanization degree, and population size. On the other hand, the city's economic development is found to downgrade the city's air quality. Such result shows the existence of considerable heterogeneous patterns of air quality among Chinese cities, and the economic development is more vital to determine the city's air quality, compared to other socioeconomic factors. Thus, we would examine such heterogeneous patterns in more details (see Sect. 5.3).

5.2 Mechanism Analysis

According to the previous analysis, HSR can affect the emissions and city's air quality through transport substitution and industrial structure (see our research frameworkighlighted in Fig. 1). The following two-stage mediating effect model is thus constructed on the basis of SDM to verify and disentangle such mechanism. In the first stage, we test the impact of HSR opening on potential mechanism variables M_{it} (i.e., the mediator) based on the following equations.

$$
\begin{aligned}
M_{it} =& \tau_0 + \tau_1 Open_{it} \times Year_{it} + \tau_2 \sum_j w_{ij} Open_{it} \times Year_{it} + \phi \boldsymbol{X}'_{it} \\
& + \boldsymbol{\psi} \sum_j w_{ij} \boldsymbol{X}'_{it} + \tau_3 \sum_j w_{ij} M_{it} + H_i + V_t + \kappa_{it}
\end{aligned} \tag{5}
$$

Second, this paper tests the influence of mechanism variables on AQI based on the following equation:

$$
\begin{aligned}
\ln(AQI_{it}) =& \sigma_0 + \sigma_1 M_{it} + \boldsymbol{\omega X}_{it} + \boldsymbol{\lambda} \sum_j w_{ij} \boldsymbol{X}_{it} + \sigma_2 \sum_j w_{ij} M_{it} \\
& + \sigma_3 \sum_j w_{ij} \ln(AQI_{it}) + P_i + Q_t + \mu_{it}
\end{aligned} \tag{6}
$$

Here, our mediator variables M_{it} refer to transport substitution (TS) and industrial structure (IS). TS variable is the ratio of HSR passenger traffic to airline passenger traffic for one city. IS variable is the ratio of the city's tertiary industry to the secondary industry. The control variables remain the same as before. According to the theoretical framework shown in Fig. 1, the transport substitution effect (TS) is the direct effect to

Table 4 Estimation of the positive mediating effect of HSR on city's air quality

	Model 1 (TS)	Model 2 (AQI)	Model 3 (IS)	Model 4 (AQI)
Open*Year	0.0252* (0.0717)		0.0131*** (0.0035)	
TS		−0.0623*** (0.0043)		
IS				-0.0422*** (0.0027)
Control variables	Yes	Yes	Yes	Yes
N	1144	1144	1144	1144
R-squared	0.0178	0.652	0.096	0.6497

help improve city's air quality, while the industrial structure effect (IS) is the indirect effect to help improve city's air quality.

As shown in Table 4, in model 1 (TS), the coefficient of $Open_{it} \times Year_{it}$ is significantly positive, indicating that the opening of HSR can significantly increase the market share of railway compared to airline traffic. This suggests an obvious transport substitution effect. Then model 2 (AQI) confirms the significant positive impact of the transport substitution to improve city's air quality. Then, as shown by model 3 (IS), the opening of HSR also brought apparent industrial structure effect, promoting the shares of tertiary industry relative to other sectors. Then, the estimation results of model 4(AQI) show that such industrial upgrading also helps improve the city's air quality. Although, as also shown in Fig. 1, HSR could also aggravate emissions and damage air quality through other direct effect (i.e., traffic inducing effect) or indirect effect (i.e., industry interaction effect), our estimation results in the benchmark and this mediation regressions suggest that the positive emission mitigation brought by the HSR is more dominant and lead to an overall lower emissions and better air quality.

5.3 Heterogeneity Analysis

This subsection explores the possible regional heterogeneity in HSR impact on the city's air quality. Specifically, we categorize our 286 sample cities based on geographic locations and the economic development. Our SDID regressions are then conducted again for different sub-sampled cities to distinguish the heterogeneous HSR impacts. First, according to the geographic locations and economic characteristics, sample cities are divided into the "eastern region", "northeast region", "western region" and "central region". Table 5 summarizes the regions and the corresponding cities in each region.

Table 6 collects our SDID estimation results for each region. We focus on the local and spatial effects of HSR impacts on city's air quality. In terms of the local

Table 5 Regional division of China

Region	Included provinces and cities
The eastern region	Beijing, Tianjin, Hebei, Shanghai, Jiangsu, Zhejiang, Fujian, Shandong, Guangdong and Hainan
The northeast region	Liaoning, Jilin and Heilongjiang
The western region	Inner Mongolia, Chongqing, Guangxi, Sichuan, Guizhou, Yunnan, Tibet, Shaanxi, Qinghai, Gansu, Ningxia and Xinjiang
The central region	Shanxi, Henan, Hubei, Hunan, Anhui and Jiangxi

effects, it is found that the opening of HSR significantly improved air quality of western and eastern China, while the impacts on the northeast and central region are not statistically significant. Such results are sensible in that HSR can help upgrade the industrial structure, especially promoting the tertiary industries that are environmentally friendly. Western China cities benefit more in such industrial upgrading, reflected by better air quality. On the other hand, the transport substitution effect could be more dominant in eastern China, where highway and normal-speed railway systems are the most developed. Thus, HSR, as a cleaner mode, can attract more traffic than other modes, which improves the air quality. For the neighboring effect, it only significantly exists in eastern region. That is, HSR opening in one neighboring city can also improve this city's air quality. The city density in eastern China is high such that their economy and transport correlations are closer. However, in other regions, especially western China, the cities are distributed sparsely and the economic linkage is looser, so that the neighboring effect is not significant.

In addition to above categorization based on geographic locations, the sample cities can also be grouped by their economic development degrees. Although the overall economy and people's income have grown dramatically over the past decades, China has a large population base and exists huge income disparity among different cities. The impact of HSR opening could also depend on the city's economic development, which can be divided by its per capita GDP as shown in Table 7.

According to Table 8, in terms of both local and neighboring effects, HSR opening is shown to significantly improve the air quality of those developed cities, especially those relatively developed cities. Many of these cities traditionally rely on the manufacturing and other pollutant industries to develop the economy. HSR can help them to

Table 6 Estimation results of each region

	Eastern Region	Northeast	Western Region	Central Region
Local effects	−0.0588* (0.0837)	−0.0519 (0.6943)	−0.0752** (0.027)	−0.0631 (0.1312)
Neighboring effects	−0.0137** (0.0321)	−0.0015 (0.5511)	−0.055 (0.1047)	−0.0478 (0.5829)
N	300	136	348	360
R-squared	0.752	0.793	0.766	0.719

Table 7 Classification of GDP per capita

Economy development level	Number of cities	Range
Developed	16	US$20,000 ≤ per capita GDP
Relatively developed	29	The world average (US$14,660) ≤ per capita GDP < US$20,000
Medium level	93	US$8,000 ≤ per capita GDP < the world average (US$14,660)
Low poverty	129	US$4,000 ≤ per capita GDP < US$8,000
Medium to high poverty	19	per capita GDP < US$4,000

Table 8 Estimated results of HSR opening effect on cities classified by economy development

	Developed	Relatively developed	Medium level	Low poverty	Medium to high poverty
Local effect	−0.0323* (0.0911)	−0.0983* (0.0606)	−0.0488 (0.1171)	−0.056* (0.0648)	0.084 (0.8849)
Neighboring effect	−0.0227* (0.0244)	−0.0362* (0.0921)	−0.0835 (0.4267)	−0.0352 (0.3922)	0.0572 (0.4632)
N	64	116	372	516	76
R-squared	0.8368	0.6534	0.6368	0.5935	0.7581

upgrade industrial structure to cleaner production and service industries. Moreover, the traffic volume is higher in these cities, such that the HSR transport substitution effect could be stronger as more residents have relatively high income and can afford HSR service. However, for those less developed cities, HSR demand is lower and its impact on industrial structure upgrading and transport substitution effect is limited.

6 Conclusions

This paper first quantified the spatio-temporal evolution of city's air quality in China. Then, a SDID model was employed to study the HSR impact on city's air quality. The contributions of this paper are multi-fold. First, this is one of the first studies to adopt spatial auto-correlation approach to quantify the spatio-temporal distribution and characteristics of Chinese cities' air quality. Second and more importantly, this paper provides one of the first rigorous empirical examinations on the local and spatial impact of HSR opening on city's air quality. In addition to the overall effect, we investigate the mechanism of the HSR impact. Specifically, we consider the direct effect brought by the change of traffic volume and the modal splits, and the indirect effect caused by HSR on the city's industrial structure and production scale. The heterogeneous analysis is also done to shed light on the HSR impacts on different kinds of cities categorized by the geographic location and the degree of economy development.

The spatio-temporal analysis of air quality first shows clear inter-annual and seasonal patterns of air quality among Chinese cities. Moreover, the spatial pattern demonstrates the southern China and coastal region have overall better air quality than northern China and inland areas. This is mainly related to climate type, terrain, economic development level and other factors. The northern cities are low-lying, where pollutants are difficult to diffuse, and economic development and industrialization progress are rapid, causing serious environmental damage. Most coastal cities in southern China have a tropical or subtropical monsoon climate, and the terrain is relatively flat. Air quality in China has obvious positive spatial auto-correlation, and also demonstrate obvious aggregation trend over time. Cities with severe air pollution will have a negative impact on the air quality of neighboring cities.

Our SDID estimations suggested that the HSR opening can reduce the emissions and improve the air quality (i.e., the local effect), while the HSR opening in the adjacent city can also improve the city's air quality (i.e., the neighboring effect). We also highlight and verify the mechanism of such positive HSR impact on city's air quality. First, as a cleaner transport mode, HSR helps reduce traffic in other more pollutant modes (i.e., positive direct transport substitution effect). In addition, HSR helps promote the city's tertiary industry, leading to fewer emissions (i.e., positive indirect industrial structure effect). Our heterogeneous analysis further exhibits that HSR is more effective to improve the air quality in eastern and western regions of China, given the more prevailing transport substitution effect and industrial structure effect. But the neighboring effect of HSR is only significant in the eastern China as the cities are closer with each other in terms of economic relation and geographic locations in this region. Moreover, the HSR proves to improve air quality in more economically developed cities, as the HSR traffic is larger in this region.

References

1. The ecological environment [Internet]: Environmental air quality index of technical specifications (trial). http://www.mee.gov.cn/ywgz/fgbz/bz/bzwb/jcffbz/201203/t20120302_224166.shtml (2012). Accessed 29 Feb 2012
2. Auffhammer, M., Kellogg, R.: Clearing the air? The effects of gasoline content regulation on air quality. Am. Econ. Rev. **101**(6), 2687–2722 (2011)
3. Cao, J., Wang, X., Zhong, X.H.: Have traffic restrictions improved air quality in Beijing? Econ. Q. **13**(03), 1091–1126 (2014)
4. The ecological environment[Internet]: China mobile source environmental management report (2020). http://www.mee.gov.cn/xxgk2018/xxgk/xxgk15/202008/t20200810_793252.html (2020). Accessed 10 Aug 2020
5. Zheng, S., Kahn, M.E.: Mitigating the cost of bullet trace market integration and mitigation of megacity growth. Proc. Natl. Acad. Sci. **110**(14), E1248–E1253 (2013)
6. Zhang, J.: High-speed railway construction and county economic development – a study based on satellite light data. Econ. Q. **16**(04), 1533–1562 (2017)
7. Mohring, H.: Optimization and scale economies in urban bus transport. Am. Econ. Rev. **62**(4), 591–604 (1972)

8. Lan, B.X., Zhang, L.: Revenue Management Model of High-speed Railway Passenger Dedicated Line. Chin. J. Manage. Sci. **17**(04), 53–59 (2009)
9. Xu, Y.T., Liu, X.Z., Wang, Z.B.: Spatio-temporal distribution of air quality in Chinese cities based on AQI index. J. Guangxi Normal Univ. (Nat. Sci.). **37**(01), 187–196 (2019)
10. Guo, Y.M., Lin, X.Q., Bian, Y.: Spatio-temporal evolution of air quality in urban agglomerations in China and its influencing factors. Ecol. Econ. **35**(11), 167–175 (2019)
11. Niu, H.M., Tu, J.J., Yao, Z., Ha, L., Li, J.B.: Spatial and temporal distribution characteristics of air quality in Chinese cities. Henan Sci. **34**(08), 1317–1321 (2016)
12. Yan, S., Cao, H., Chen, Y., Wu, C., Hong, T., Fan, H.: Spatial and temporal characteristics of air quality and air pollutants in 2013 in Beijing. Environ. Sci. Pollut. Res. **23**(14), 13996–14007 (2016)
13. Chen, W.W., Liu, Y., Wu, X.W., Bao, Q.Y., Gao, Z.T., Zhang, X.L., et al.: Spatial and temporal distribution characteristics of air quality and causes of heavy pollution in Northeast China. Environ. Sci. **40**(11), 4810–4823 (2019)
14. Fang, C., Liu, H., Li, G., Sun, D., Miao, Z.: Estimating the impact of urbanization on air quality in China using spatial regression models. Sustainability **7**(11), 15570–15592 (2015)
15. Zhang, X.M., Luo, S., Li, Z.F., Jia, L., Zhai, H.M., Wang, L.M.: Study on influencing factors of spatial pattern of air quality in China. J. Xinyang Normal Univ. (Nat. Sci. Edn.). **33**(01), 83–88 (2020)
16. Dai, Y.H., Zhou, W.X.: Temporal and spatial correlation patterns of air pollutants in Chinese cities. PLoS ONE **12**(8), e0182724 (2017)
17. Zhao, L.X., Zhao, R.: Study on the relationship between economic growth, energy intensity and air pollution. Soft Sci. **33**(6), 60–66+78 (2019)
18. Fang, J.: Impacts of high-speed rail on urban smog pollution in China: a spatial difference-in-difference approach. Sci. Total Environ. **777**, 146153 (2021)
19. Lin, X.Q., Wang, D.: Spatiotemporal evolution characteristics and socio-economic driving forces of city's air quality in China. Acta Geogr. Sin. **71**(08), 1357–1371 (2016)
20. Li, M.S., Ren, X.X., Yu, Y., Zhou, L.: Spatial and temporal distribution of PM_(2.5) pollution in Chinese mainland. China Environ. Sci. **36**(03), 641–650 (2016)
21. Jia, R., Shao, S., Yang, L.: High-speed rail and CO_2 emissions in urban China: a spatial difference-in-differences approach. Energy Economics. **99**, 105271 (2021)
22. Zhang, X.P., Lin, M.H.: Analysis of the regional differences of air pollution in Chinese cities and the socio-economic influencing factors: A comparative study of two air quality indexes. J. Univ. Chin. Acad. Sci. **37**(01), 39–50 (2020)
23. Griffin, P.W., Hammond, G.P., McKenna, R.C.: Industrial energy use and decarbonisation in the glass sector: a UK perspective. Adv. Appl. Energy. **3**, 100037 (2021)
24. Wang, L.: Panel data analysis of correlation between automobile consumption and air pollution. China Popul. Resour. Environ. **24**(S2), 462–466 (2014). https://doi.org/10.1016/j.en.2016.01.004
25. Shang, W.L., Chen, J., Bi, H., Sui, Y., Chen, Y.: Impacts of COVID-19 pandemic on user behaviors and environmental benefits of bike sharing: a big data analysis. Appl. Energy 285, 116429 (2021)
26. Chi, J.Y., Zhang, Y., Yan, S.Y.: City's air quality is affected by the level of urban economic development: a case study in China. Econ. Manag. **28**(05), 26–31 (2014)
27. Bi, H., Shang, W.L., Chen, Y., Wang, K., Yu, Q., Sui, Y.: GIS aided sustainable management for urban road transport systems with a unifying queuing and neural network model. Appl. Energy 116818 (2021)
28. Gray, N., McDonagh, S., O'Shea, R., Smyth, B., Murphy, J.D.: Decarbonising ships, planes and trucks: an analysis of suitable low-carbon fuels for the maritime, aviation and haulage sectors. Adv. Appl. Energy **1**, 100008 (2021)
29. Li, Y., Chiu, Y., Lin, T.Y.: Energy and environmental efficiency in different Chinese regions. Sustainability **11**(4), 1216 (2019)
30. Jiang, L., Zhou, H.F., Bai, L., Wang, Z.J.: Dynamic change characteristics of city's air quality index (AQI) in China. Econ. Geogr. **38**(09), 87–95 (2018)

31. Wang, B.H., Wang, S.: An empirical study on influencing factors of city's air quality in China: based on the analysis of panel data of 31 major cities in China. J. Fujian Agric. For. Univ. **18**(06), 29–33 (2015)
32. Jiao, J., Wang, J., Zhang, F., Jin, F., Liu, W.: Roles of accessibility, connectivity and spatial interdependence in realizing the economic impact of high-speed rail: evidence from China. Transp. Policy **91**, 1–15 (2020)
33. Zhang, A., Wan, Y., Yang, H.: Impacts of high-speed rail on airlines, airports and regional economies: a survey of recent research. Transp. Policy **81**, A1–A19 (2019)
34. Liu, S., Wan, Y., Ha, H.K., Yoshida, Y., Zhang, A.: Impact of high-speed rail network development on airport traffic and traffic distribution: evidence from China and Japan. Transp. Res. Part A Policy Pract. **127**, 115–135 (2019)
35. Zhang, Z., Yu, D.H., Sun, T.: Impact of the opening of high-speed railway on green restructuring of urban production system. China Popul. Resour. Environ. **29**(07), 41–49 (2019)
36. Zhu, S.J., Yin, S.S., Zhong, T.L.: Does the opening of high-speed rail curb urban pollution? J. Econ. Manag. **33**(03), 52–57 (2019)
37. Qin, B.T., Xie, R.B., Ge, L.M.: Study on the relationship between FDI, economic growth and environmental pollution–based on the analysis and test of China's provincial panel data. J. Guangxi Univ. Financ. Econ. **33**(02), 81–96 (2020)
38. Yu, K., Strauss, J., Liu, S., Li, H., Kuang, X., Wu, J.: Effects of railway speed on aviation demand and CO_2 emissions in China. Transp. Res. Part D Transp. Environ. **94**, 102772 (2021)
39. Park, Y., Ha, H.K.: Analysis of the impact of high-speed railroad service on air transport demand. Transp. Res. Part E Logist. Transp. Rev. **42**(2), 95–104 (2006)
40. Zhang, H., Feng, F.: Green high-speed rail: Will it reduce smog pollution? Acta Ecol. Sin. **6**(03), 114–147 (2019)
41. Beaudoin, J., Lin, L.C.Y.C.: Is public transit's "green" reputation deserved? : Evaluating the effects of transit supply on air quality. University of California at Davis Working Paper (2016)
42. Chen, P., Lu, Y., Wan, Y., et al.: Assessing carbon dioxide emissions of high-speed rail: the case of Beijing-Shanghai corridor. Transp. Res. Part D Transp. Environ. **97**, 102949 (2021)
43. Lalive, R., Luechinger, S., Schmutzler, A.: Does expanding regional train service reduce air pollution? J. Environ. Econ. Manag. **92**, 744–764 (2018)
44. Li, H., Strauss, J., Liu, L.: A panel investigation on China's carbon footprint. Sustainability **11**(7), 2011 (2019)
45. Ren, T.X, Lin, J.Y.: Spatial spillover effect of high-speed rail on urban economic growth: evidence from 277 core cities. Math. Stat. Manag. 1–16 (2021)
46. Zheng, C.L., Zhang, J.T.: The impact of high-speed railway opening on urban innovation quality: An empirical study based on PSM-DID model. Techn. Econ. **40**(02), 28–35 (2021)
47. Liu, J., Huang, X.F., Chen, J.: High-speed railway and high quality urban economic development: an empirical study based on the data of prefectural cities. Contemp. Finance Econ. **01**, 14–26 (2021)
48. Beck, T., Levine, R., Levkov, A.: Big bad banks? The winners and losers from bank deregulation in the United States. J. Financ. **65**(5), 1637–1667 (2010)
49. Zhou, J., Zhang, Y., Zhang, Y., et al.: Parameters identification of photovoltaic models using a differential evolution algorithm based on elite and obsolete dynamic learning. Appl. Energy **314**, 118877 (2022)
50. Yang, Z., Shang, W., Zhang, H., Garg, H., Han, C.: Assessing the green distribution transformer manufacturing process using a cloud-based fuzzy multi-criteria framework. Appl. Energy 118687 (2022)
51. Zhai, G.Y.: Overview of data analysis based on python. Gansu Sci. Technol. Rev. **47**(11), 5–7+26 (2018)

HSR: Accessibility and Tourism

The Accessibility Impact of High Speed Rail in Italy: A User-Based Approach

Mario Tartaglia, Lorenzo Vannacci, and Martina Farsi

Abstract Among the impacts that a High Speed Rail network has on a country's society, economy and environment, the impact resulting from accessibility changes plays a key role. In fact, the main direct change generated by a new transportation system is the variation (generally a reduction) of travel times. This produces secondary effects as well, such as changes in accessibility, and shifts of transport demand from other modes of transport to rail. These changes generate other broader effects, which have an impact on various socio-economic and environmental issues, such as energy use, environmental structure, tourism, social dynamics, human health, safety, general economy, business clustering, labour productivity, land use, real estate, transport system resilience, technology development, innovation, and policies. While the concept of accessibility has been defined in several ways by various studies in the scientific literature, it usually refers to the ease of obtaining socioeconomic opportunities (activities, services, goods) within a geographic area or—from a more passive perspective—the ease with which such opportunities can be reached. While accessibility indicators can be distinguished by whether they focus on infrastructure, utilities, location, or even people, they are usually based on travel times between geographic locations. Although in continuous transport systems, such as road networks, travel times are highly significant for the purpose of depicting accessibility, that is not true for scheduled systems like collective ones, such as bus, rail, air, and maritime transport. In these latter cases, the onboard travel time is not fully representative of the user's actual travel time, since the time spent accessing the transport system and the waiting time for services both play a crucial role in the users' perception of accessibility. Several studies have been conducted to analyse the accessibility changes generated by the Italian High Speed Rail system built and put into service over the first two decades of this century. However, these studies have often focused on an accessibility concept based on service travel time, rather than on user travel time, including access and waiting times. This paper offers an analysis of the accessibility impact of High Speed Rail in Italy, focusing on a user-based approach. Accessibility

M. Tartaglia · L. Vannacci (✉) · M. Farsi
Ferrovie dello Stato Italiane, Roma, Italia
e-mail: l.vannacci@fsitaliane.it

F. Pagliara (ed.), *Socioeconomic Impacts of High-Speed Rail Systems*,
Springer Proceedings in Business and Economics,
https://doi.org/10.1007/978-3-031-26340-8_16

indicators are therefore set up taking into account rail service timetables, and therefore the access to the services from the rail stations themselves, which, for the users, are the actual access points to the transport system. Moreover, the results of this user-based analysis are compared to the simple service travel time approach, in the order to estimate the differences and to highlight the methodological dissimilarities.

Keywords Accessibility · High speed rail · Impact

1 Introduction

Rail transport is a strategic sector, and its importance is fully recognized by the European Green Deal: the new growth strategy which, as an integral part of the European Commission's strategy to implement the United Nation's 2030 Agenda, is aimed at transforming the EU into a fair and prosperous society, with a modern and competitive economy with zero net greenhouse gas emissions by 2050.

The European Green Deal considers rail to be the backbone of an environmentally sustainable multi-modal transport system capable of achieving a 90% reduction in transport emissions by 2050 by making the switch from road transport to rail transport. In order to make the European Union the first carbon neutral region in the world by 2050, it is necessary to work on lowering greenhouse gas emissions in the transport sector, and rail is already the most energy-efficient transport mode. The trends in greenhouse gas emissions by main sectors in the EU-28 from 1990 to 2019 show that transport is the second sector in terms of emissions, representing about 25% of total greenhouse gas emissions, after the energy supply sector. While most sectors have managed to reduce their emissions over time, in the transport sector they have generally increased, with direct emissions having declined only in the railway sector, where they have decreased by 66% and now represent just 0.5% of the emissions generated by the transport sector as a whole.

One of the key factors in determining the passenger mode choices made by users is the accessibility level offered by the transport modes. Even if rail should become the main transport system utilized for short and medium distances (<400 km) due to decarbonization efforts, many situations will still remain in which the car will be the preferred choice in terms of passenger mobility.

Conventional planning approaches tend to overlook and undervalue the concept of accessibility, since they are based on a mobility-oriented analysis, and mainly evaluate transport system performance based on quantity and quality of physical travel. More recently, however, there has been a shift towards accessibility-based approaches in transport planning; there is thus a growing demand for planning support systems to have the ability to evaluate accessibility. Railway systems are also often less accessible than other transport modes due to their highly infrastructure-based nature and their scheduled service organization, and special attention must therefore be placed on measuring the accessibility of these systems. However, a number of the available studies concerning railway accessibility have certain methodological weaknesses, as they are usually based on infrastructure data (e.g. distance, planned speed, average travel time) rather than on the actual performance of the service, which relies on timetables.

On 14 December 2008, HS service was launched in Italy along the Turin-Salerno corridor, and in 2012 a private competitor even launched its own service on the same network. In 2019, after just 10 years, the network boasted over three hundred and fifty million passengers served, 380 million kilometres travelled, and over 80 cities connected. Not to mention the environmental sustainability figures, with a reduction in carbon dioxide emissions of over 20 million tonnes between 2008 and 2018 thanks to the modal shift from private cars and planes to rail: the perfect ecological means of transport.

These are just some of the impressive results achieved by the Italian High Speed train system since its launch 10 years ago, which has had a major impact in terms of specific socio-economic and territorial characteristics, as well as different levels of service variables. This paper is not intended to provide a thorough review of the existing accessibility indicators, but rather an evaluation of the Italian HS transport system in light of the changes in nationwide accessibility over the span of a decade, focusing on the user's point of view. In fact, the examination of the accessibility effect of the introduction of the High speed rail system within Italy could explain this initiative's success on the market.

2 Literature Review

Over the past few decades, the concept of accessibility has been widely debated in the scientific literature related to transportation, geography, land use, economy and social sciences. Although accessibility has been defined in several ways by different authors [1], it generally refers to the ease of obtaining/reaching opportunities, or rather activities, services, goods or simply destinations [2]. It is commonly agreed that the notion of accessibility associated with land use and transport initially appeared in the 1950s [3], when Hansen defined accessibility as the "potential of opportunities for interaction" [4]. Several other authors later defined accessibility from slightly different perspectives. For instance, the concept of accessibility as the ease of reaching businesses starting from a defined place, and using a given transport system, is conveyed by Dalvi and Martin [5], Morris et al. [6], and Johnston et al. [7]. This introduces the need to clarify the reference spatial patterns to be considered [8], by focusing, for example, on the relative accessibility between single origin–destination pairs, or on the integral accessibility between a single point in space and all the possible destinations to be reached [9]. In the latter case, the influence of both the spatial distribution of the destinations and the intensity of their use is an essential issue to be considered [10]. In addition to the distribution of the destinations, another key factor that influences accessibility is the transport network, including its topology and levels of service [11]. Moreover, while the definition of accessibility for authors like Ben Akiva and Lerman [12] focuses on the utility provided by the transport and land use systems, others highlight the importance of accessibility in relation to people, taking into consideration both the number of people and individual behaviour and perception [13]. A suitable summary of all these concepts is provided by Geurs and Ritsema van Eck [14] and Geurs and van Wee [15], who argue that the main determinants of accessibility are the land-use system (the spatial distribution and the characteristics of both opportunities and access demand by people), the transport supply (its structure, the available modes and the levels of service), the time constraints (availability of opportunities and willingness of people to access them), and people (needs, abilities and resources of individuals).

Depending on the definition of accessibility adopted, a wide range of accessibility measures can be used for analysis purposes. In recent years, several authors conducted a review of accessibility measures based on existing studies (see, for instance, [3, 16–18]). These measures can be categorised in different ways, depending on their nature, their use, and so on. An example of a classification of accessibility indicators based on analysis goals is provided by Paez et al. [19], who draw a distinction between descriptive measures (i.e. those aimed at describing the actual situation) and prescriptive measures (i.e. those aimed at defining a desirable scenario). Several other authors have proposed classifying such indicators based on their meanings and the variables considered. For instance, Handy and Niemeier [10] focused on attraction, gravity and utility measures; Kwan [20] compared place-based and person-based accessibility indicators; Bath et al. [3] proposed a framework that includes spatial separation, opportunities, gravity, utility, and time–space indicator categories; Curtis and Scheurer [1] classified accessibility indicators into spatial separation, contour, gravity, competition, time–space, utility, and network measures. However, Geurs and Van Wee [15] have proposed a more general classification scheme consisting of four categories of accessibility measures: infrastructure-based, location-based, person-based and utility-based measures. The infrastructure-based measures, often used in transport planning, focus uniquely on the transport system performance indicators, such as network density, travel times, and congestion levels, typically overlooking land-use, economic and social components; however, they can be used as inputs for calculating consumer surplus in transport projects assessments [21]. The location-based indicators, which are typically used in transport planning and geography, evaluate accessibility in certain places, and highlight the ease of connecting various points in space. They include connectivity indicators, such as distance measures between two places [9]; isochrones or contour measures [22–24], which can also take transport services into account [25, 26]); potential accessibility measures (e.g. [4, 9, 27]). The potential accessibility indicators are relative to an origin point, and are essentially based on the measure of opportunity variables at destinations, as well as the impedance functions denoting the cost between the origin and destination ([28], p. 322). They can also incorporate competition effects, as showed by Geurs and Ritsema van Eck [29]. The person-based measures focus on the accessibility of individuals or social groups starting from the theoretical space–time geography issues introduced by Hägerstrand [30] and later developed by other authors, such as Miller [31, 32]), Kwan [20] and Recker et al. [33]. Finally, the utility-based measures, especially used in economic studies, aim to estimate the benefits that people can achieve by accessing space distributed opportunities. They mainly originate from random utility theory [34], or entropy modelling [35]. A large number of studies and applications that make use of the aforementioned classes of indicators can be found in the scientific literature (some reviews are reported in [3, 16–18]). Moreover, a number of innovative approaches that expand on the conventional accessibility concepts have been also proposed, such as the Structural Accessibility Layer approach [36] and the Space Syntax methodology [37].

Some studies are specifically concerned with rail accessibility. This topic can be addressed from different points of view: accessibility to rail systems, or simply to rail stations (e.g. [38, 39]) or accessibility to opportunities by means of rail systems (an example is given by Monzòn, [40]). Concerning the latter perspective, the researchers seem to be primarily interested in estimating the economic impacts of planned rail projects, especially those for high speed rail developments (see, among others, [41–46]). Two frequently highlighted issues are that improvements in rail networks and services are able to boost regional or national accessibility [47–49] and change travel patterns [50]. Nevertheless, the railway systems' impact on accessibility strictly depends on the configuration of the networks and services, as becomes clear when fast or high speed connections trigger hub-and-spoke patterns [51], tunnel effects [52], and corridor effects [50]. As showed by Vickerman [53], this can lead to unsolved regional disparities in accessibility within the intermediate areas between the stations connected by high-speed trains with fast and improved services. Accessibility impacts affect different socio-economic aspects, like the locations of industries and workplaces [54, 55], real estate value [56], competitiveness [57], and population change [58]. Other topics investigated in rail accessibility studies include the effects on territorial cohesion [59, 60] and on equity [16, 40, 61]. The accessibility indicators most often used in rail studies are location-based measures, such as connectivity and potential accessibility indexes (e.g. [62, 63]). Some authors have proposed synthetic global accessibility indicators using more complex techniques, like data envelopment analysis (e.g. [64]) or principal component analysis (e.g. [65]). However, despite the fact that travel time is always more frequently considered in accessibility indicators (e.g. [41, 43, 58, 62, 63, 66, 67]), most of the studies neglect to fully consider penalties due to the need to use multiple rail services, i.e. they only consider onboard time, and do not include waiting time (or transfer time between onboard times) at railway stations. Some studies take transfer times into account, but usually in a simplified way [42, 44]. When adequate computing capacity and comprehensive data are available, this approximation can be overcome by considering all the information included in the rail timetables [68–70]. Such an opportunity can be provided through data availability.

3 The Case Study: Italy's High Speed Rail Network

The Italian High Speed Rail Network has had a revolutionary impact on people's lifestyles and Italian Transportation in general: together with the A1 Milan-Naples motorway, it is the nation's most important infrastructural project of the post-war period. It changed the concept of nationwide travel and the duration of the commutes and became a point of excellence that has been used to support major international events, like EXPO 2015.

3.1 The Story of the Italian High Speed Rail Network

The first step of the Italian High Speed Rail Network was the "Direttissima" line between Rome and Florence; the new service was launched in 1992, 17 years after construction began. In December of 2005, the first train ran along the Rome-Naples High speed/High capacity line; this new 204 km long segment was the first commercial line to be managed by the level 2 ERMTS system. Just a few months later, in February of 2006, the Turin-Novara HS line was inaugurated. In March of 2007, the opening of the Padua-Mestre HS line marked the start of the Milan-Venice route; 9 years later, in 2016, another section of this route was added with the Milan-Brescia line. In mean time, the construction of the Turin-Salerno route was in full swing, and in 2008 the Milan-Bologna and Naples-Salerno segments were inaugurated, followed by the Bologna-Florence, the Turin-Milan, and the Rome-Naples HS lines in 2009. These last steps marked the completion of the Turin-Salerno route. Meanwhile, RFI was also engaged in the High Speed Stations project, with the Roma Tiburtina station being restyled and expanded in 2011, and with three new stations for High Speed passengers being opened in 2013. These were the underground stations of Bologna and Torino Porta Susa, and the new Reggio Emilia HS Mediopadana station designed by Calatrava. The Napoli Afragola station, designed by Zaha Hadid and located on the Rome-Naples HS line, was also inaugurated in 2017. However, in addition to the infrastructure, the story of the Italian High Speed Rail Network also regards its management and the evolution of the market. The deregulation of the rail transport sector began in the early 2000s due to the transposition of the European directives into the national legislation. In the meantime, the Ferrovie dello Stato organisation was restructured into various companies with different roles. Trenitalia was that which was dedicated to passenger transport; with regard to the HS market, in 2009 it launched the "Freccia" brand, which consisted of three different tiers and types of trains: the Frecciarossa, the Frecciargento, and the Frecciabianca, each of which ran different route types. In 2006, the first HS competitor appeared on the Italian market; NTV Italo began operating on the Naples-Rome-Milan line in April of 2012 (Fig. 1).

4 Methodology and Data

In accordance with the purpose of this study, the analysis of the rail services was conducted using the results of an assignment model: the model allows for the reproduction of the journey experience, consisting of travel time, number of services, transfer waiting times, and number of transfers. The model's outputs do not only

Fig. 1 HSR development story

Table 1 Supply inserted into the assignment model

Year	Supply	Source
2008	Espresso	Timetable
	Eurostar Italia	
	Eurocity	
	Intercity	
	Euronotte	
2019	Frecciarossa	GTFS[1]
	Frecciargento	
	Frecciabianca	
	Intercity	
	Intercity notte	
	EC	
	ECN	
	Italo NTV	Timetable

consist of the number of services present on a certain route, but those which are actually available to the users, either due to the users' willingness to transfer and to wait for the transfers, or due to the users' preferred times. The simulation was performed using PTV-Visum software. The train services were extrapolated from the timetables for the two simulation years (2008 and 2019), before and after the development of the HSR network. Both in 2008 and in 2019, the only services included in the analysis were the long-haul ones (see Table 1).

[1] Google Inc. "General Transit Feed Specification Reference". Accessed July 19, 2014 from http://developers.google.com/transit/gtfs/reference [71].

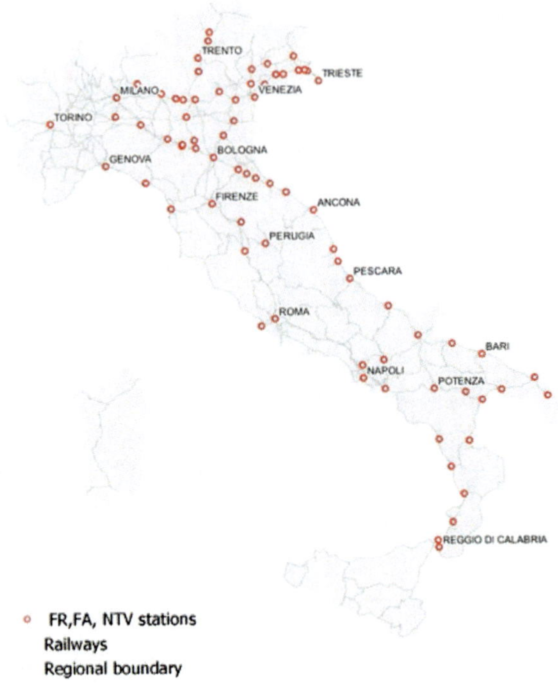

Fig. 2 HS stations used as TAZ

The demand was organised into 75 Transport Analysis Zones; these corresponded to the stations with at least one HS service (Frecciarossa, Frecciargento, Italo), based on the 2019 timetable. For cities with two or more HS stations, a unique traffic zone connected to all of its HS stations was considered; this was the case for Turin, Milan, Genoa, Venice, Florence, Rome, and Naples (Fig. 2).

The OD pairs correspond to all the zone combinations; an artificial demand was associated with every OD pair, and was uniformly allocated to the time period simulation, from 5:00 AM until 10:00 PM. The analysis didn't consider the access time to the HS stations; between 2008 and 2019 there weren't any major infrastructural change besides the HSR network's development, so access time to the stations would be considered unaltered. The assignment model was a schedule-based service; it didn't consider monetary costs, the maximum number of possible transfers was equal to two, and the ideal transfer time was set to 15 min. The model allows multi-operator trips.

Other data used for the study included the train modal share associated with certain Origin/Destination routes with HSR services. These data were collected by Trenitalia with customer interviews on major HS routes and other mobility sector sources in order to determine the typical user profile and to estimate the modal share.

5 Results and Discussion

The analysis consists of two distinct phases, both of which, however, focus on the issue of accessibility; the first phase seeks to determine the attributes of train travel that most influence the choice of this means of transport, while second phase investigates the change in accessibility from 2008 to 2019 linked to the development of high speed rail.

5.1 The Drivers of Travel Choices

The modal share of public transport has sometimes been correlated to the accessibility that it offers [72]. Expanding upon this concept, it can be hypothesised that the average values associated with movement by train on a certain OD route, which have the greatest impact on the modal share of train travel, can also be significant for the purpose of measuring accessibility. In this context, the historical series of the modal shares estimated by Trenitalia, based on targeted surveys conducted on the main HS routes, were utilised. A linear regression model was therefore set up, in which the dependent variable is the observed modal share, while the independent variables are the values associated with movement by train. The modal shares considered are those relating to the two years of simulation (2008 and 2019), for which the travel attributes are used as a model. Since this analysis wasn't aimed at investigating the change in the weight of the attributes within the expression over time, but only their impact on the modal share's value, the two time sets were evaluated as independent observations. The following figure shows the correlation matrix between the modal share and the values associated with the trip (Fig. 3).

It can be noted that the value most correlated to the modal share is the travel speed, followed by the headway of the trips on the specific route. The various outputs of the linear regressions reveal that the travel time on the OD alone cannot explain the users' choices in terms of train type. The headway of the trips offers a better explanation than the choice of train type. The travel speed is a major factor, and the results improve when the headway variable is included in the regression. The value of the determination coefficient remains below 0.5, indicating how other variables come into play in the choice of the vehicle, some of which could be evaluated in the possible extension of the study, such as the travel cost, and others that are more difficult to quantify, such as travel comfort (Tables 2 and 3).

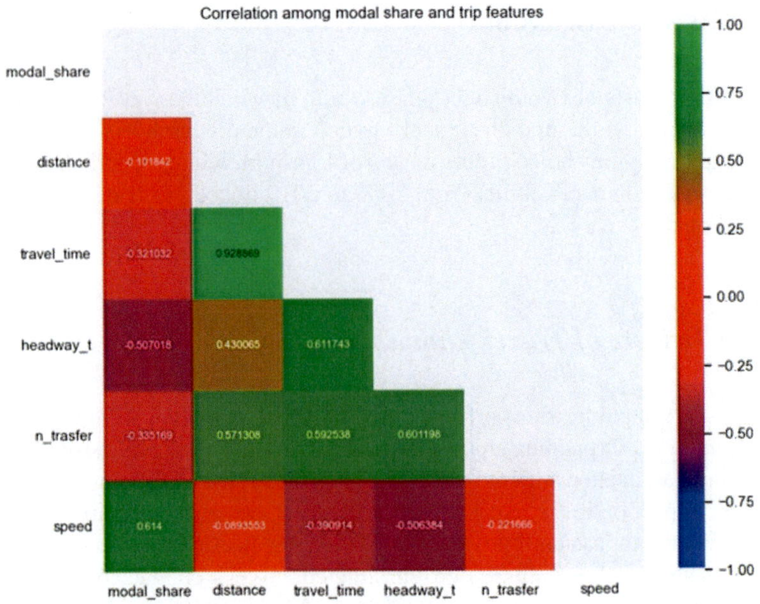

Fig. 3 Correlation among modal share and trip features

Table 2 Model estimation results

Variable test / Model	Constant	Travel time	Headway performance	Speed	No. of transfers	Residual sum of squares	Adj R-squared	F-statitstic
1	56.79 *	−0.048 **				17.794	0.086	0.017
2	56.19 *		−0.22 *			14.740	0.243	7.76E-05
3	50.58 *				−18.13 *	17.511	0.100	0.010
4	−0.98 ***			0.39 *		12.356	0.366	6.08E-07
5	14.45 ****		−0.12 **	0.31 *		11.329	0.407	4.67E-07
6	13.61 ****		−0.08 ****	0.32 *	−5.76 ****	11.183	0.403	1.74E-06

* p-value < 0.01, ** p-value < 0.05, *** p-value < 0.1, *** * p-value > 0.1

Table 3 Model estimation results with Log-transformation variables

Variable test / Model	Constant	Travel time	Headway performance	Speed	No. of transfers	Residual sum of squares	Adj R-squared	F-statitstic
1	2.12 *	−0.22 **				2.28	0.066	0.032
2	2.12 *		−0.33 *			1.56	0.361	7.15E-07
3	1.67 *				−0.69 *	2.19	0.100	0.01
4	−1.10 **			1.32 *		1.5	0.382	2.95E-07
5	0.19 ****		−0.18 **	0.82 *		1.35	0.432	1.51E-07
6	0.003 ****		−0.14 ****	0.89 *	−0.19 ****	1.35	0.427	6.25E-07

* p-value < 0.01, ** p-value < 0.05, *** p-value < 0.1, *** * p-value > 0.1

5.2 *The Effects of HSR to Accessibility in Italy: Analysis of Ten Years Data*

This section only considers the routes present in both simulated scenarios (2008 and 2019). The comparison shows a substantial reduction in travel times (−14% on average), accompanied by a general increase in services (+25%). Figure 4 shows a decline in the travel time distribution curve in 2019, which is consistent with that which is shown with regard to the distribution of the speeds in Fig. 5; the 2019 curve peaks in the 100–110 km/h range, while the 2008 curve peaks in the 90–100 km/h range; most notably, however, there is a broader bell shape in 2019, indicative of the increase in travel speeds. The speed referred to is that obtained from the simulation, dividing the model's real output travel distance by the average travel time, also including the average waiting time for connections (Fig. 6).

The same considerations can be made by grouping the ODs by distance classes; the graphs below show that the average travel times per distance class are lower, and the number of average services is greater, in 2019 with respect to 2008 (Figs. 7 and 8).

All the previous evaluations are summed up in Fig. 9, in which, for each OD, the travel time is shown on the horizontal axis and the number of trips during the simulation period is shown on the vertical axis. On a graphical level, it can be observed that the red pointers (2019) retreat and move upwards compared to the green ones (2008).

Excluding the monetary travel cost component, that which is shown in the above figures represent the real availability of services for users, as the simulation model from which the above outputs are obtained only takes into consideration the routes

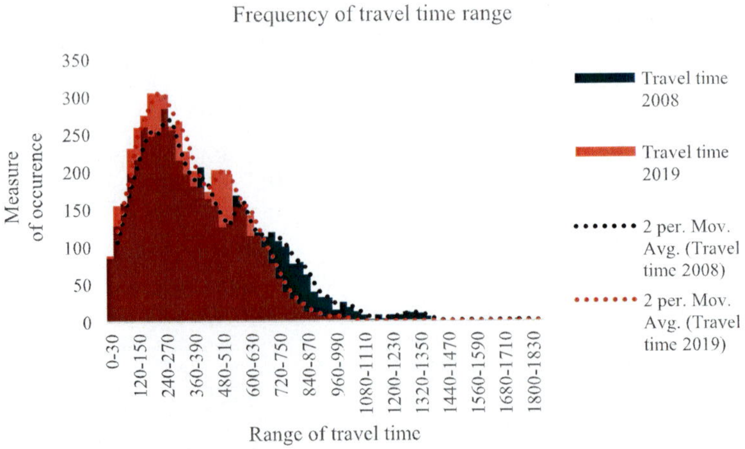

Fig. 4 Frequency distribution of travel times between 2008 and 2019

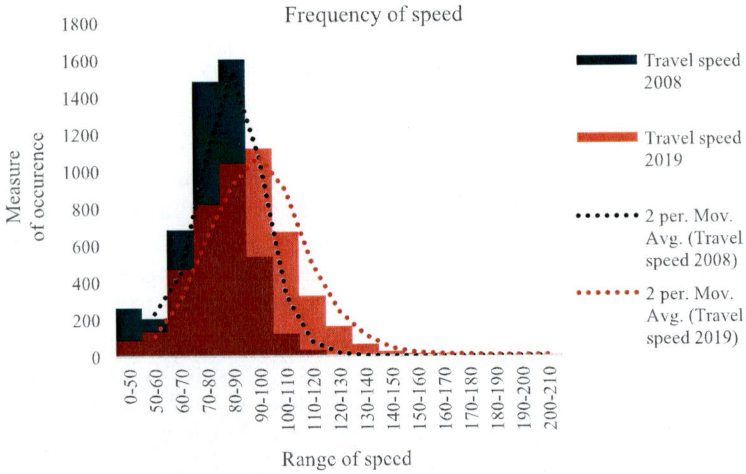

Fig. 5 Frequency distribution of travel speeds per OD between 2008 and 2019

that have similar travel times, and, in the case of connections, the combinations that tend to have the most optimal waiting times. The considerable change in the values associated with long-distance train journeys can be seen in the cartogram below, where the grey area represents the situation in 2008, taken as a baseline reference, and the red area shows how the situation in 2019 has changed in proportion to the variation in travel times.

A comparison of Fig. 10, in which only the travel time has been taken as a distortion parameter, with Fig. 11, in which the distortion parameter is the travel time inclusive of the headway of the trips on the routes to and from Rome, reveals

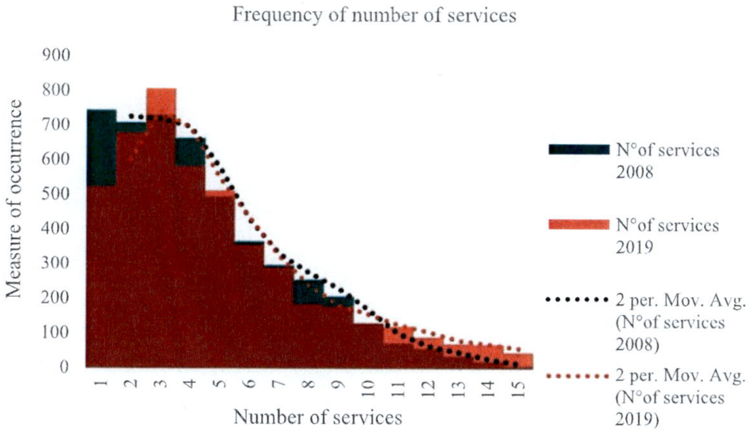

Fig. 6 Frequency distribution of the number of services per OD between 2008 and 2019

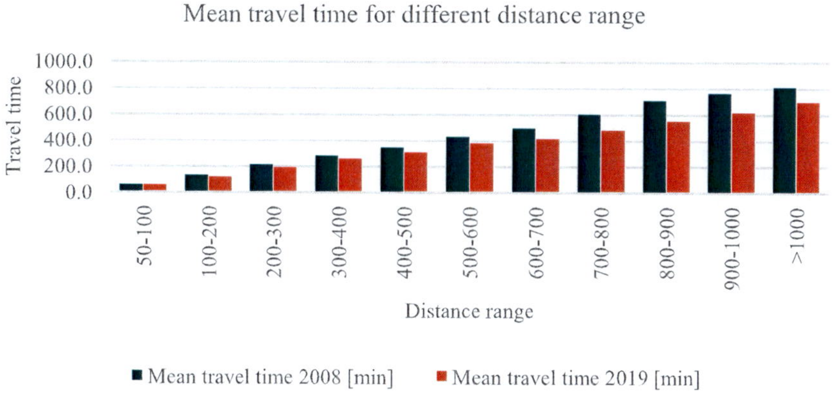

Fig. 7 Average travel time by distance class (2008 versus 2019)

that the evaluation of the changes between the two years must also take into account the number of services present on the day in question.

In addition to the fact that the change is generally more pronounced in the second figure, it should also be noted that the "distancing" of Ancona from Rome in Fig. 10, which would denote an increase in travel times on this route, disappears when the number of connections present on the same route on the day in question is also taken into account, Fig. 11; therefore, despite the shorter travel times in 2008, there was an increase in the number of trains on the route in 2019, which, in fact, results in greater selection for the user (Figs. 12, 13, 14, 15, 16 and 17).

The number of connections available between two cities served by the HS is a decisive factor with respect to the possibility of being able to make the round trip

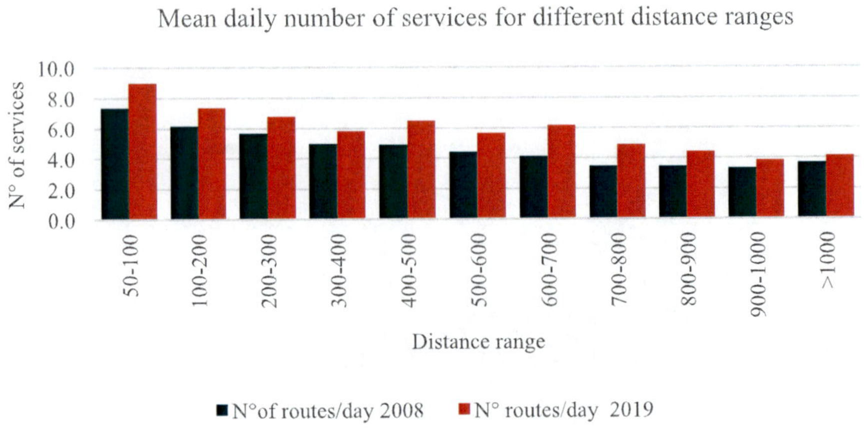

Fig. 8 Average number of services per distance class (2008 versus 2019)

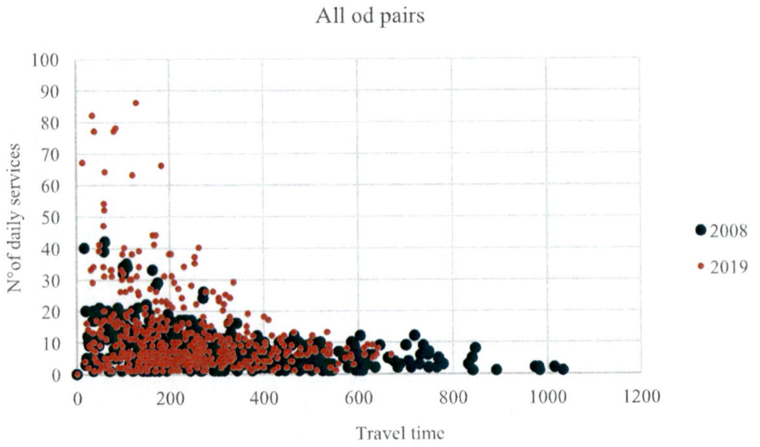

Fig. 9 Comparison between 2008 and 2019, with, for each OD, the travel time on the horizontal axis and the number of trips during the simulation period on the vertical axis

within the same route; in this sense, HS has also changed the magnitude of the journey and the commute. To express this concept in terms of accessibility, a contour measure was used based on the number of cities served by HS that can be reached within a set time; 3 different time thresholds were established (2 h, 3 h, and 4 h), with the latter being hypothesised as the maximum for a return day trip. The number of cities that can be reached within the set time was evaluated first by considering the travel time alone, and then by compounding the travel time with the headway of the services during the simulation period, as a representation of the availability of trips, and therefore the potential for selection on the part of the user. The graphical results of the analysis based on the time limits are shown in (Tables 4, 5 and 6).

Fig. 10 Cartogram based only on travel time

Fig. 11 Cartogram based on travel time and headway performance

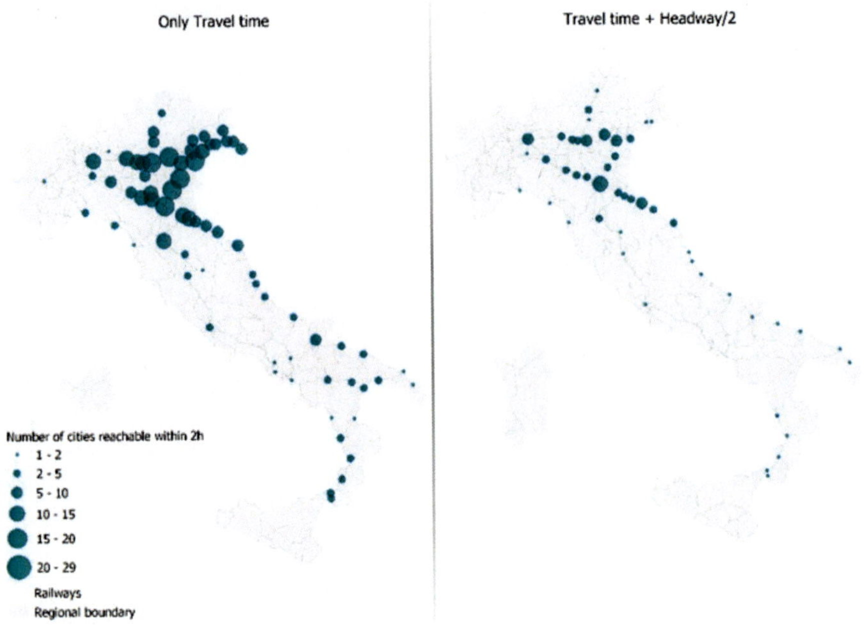

Fig. 12 Number of cities reachable within 2 h, 2008

Fig. 13 Number of cities reachable within 2 h, 2019

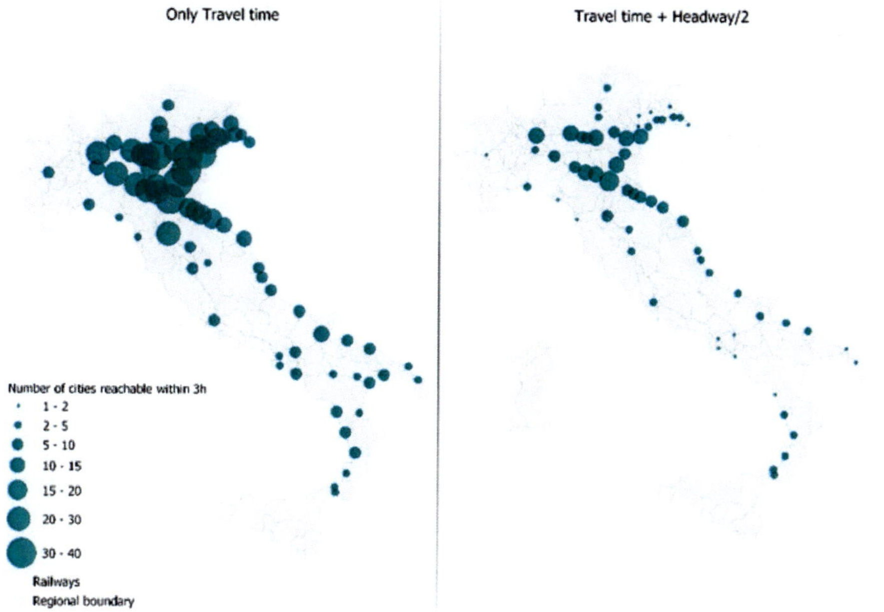

Fig. 14 Number of cities reachable within 3 h, 2008

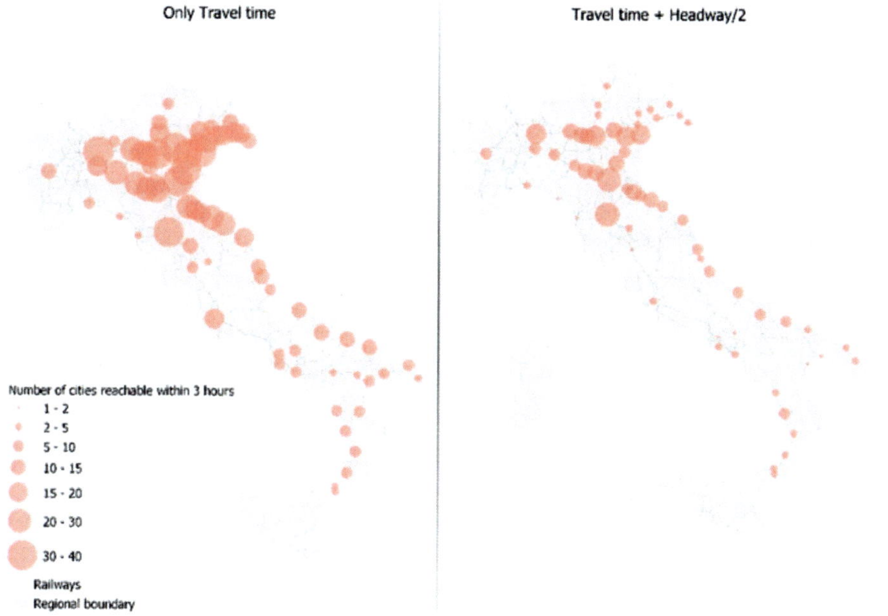

Fig. 15 Number of cities reachable within 3 h, 2019

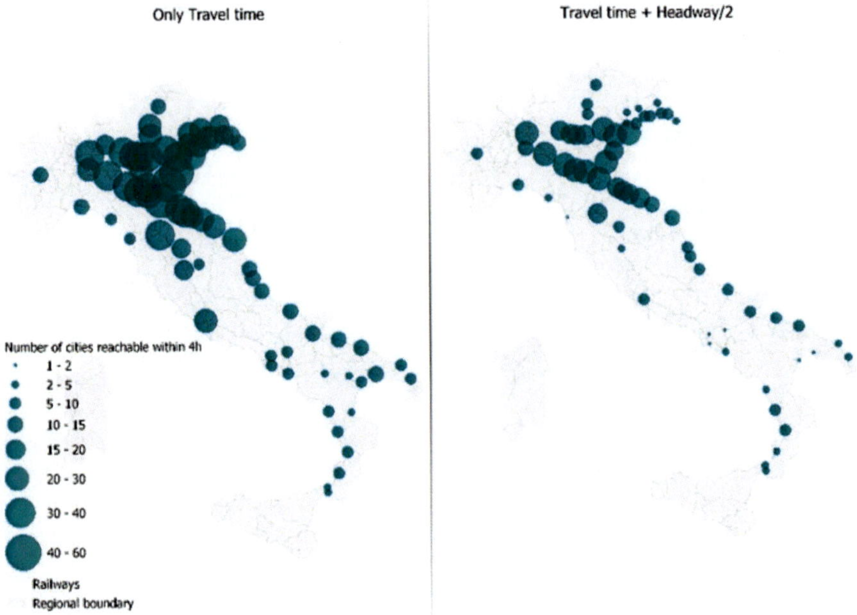

Fig. 16 Number of cities reachable within 4 h, 2008

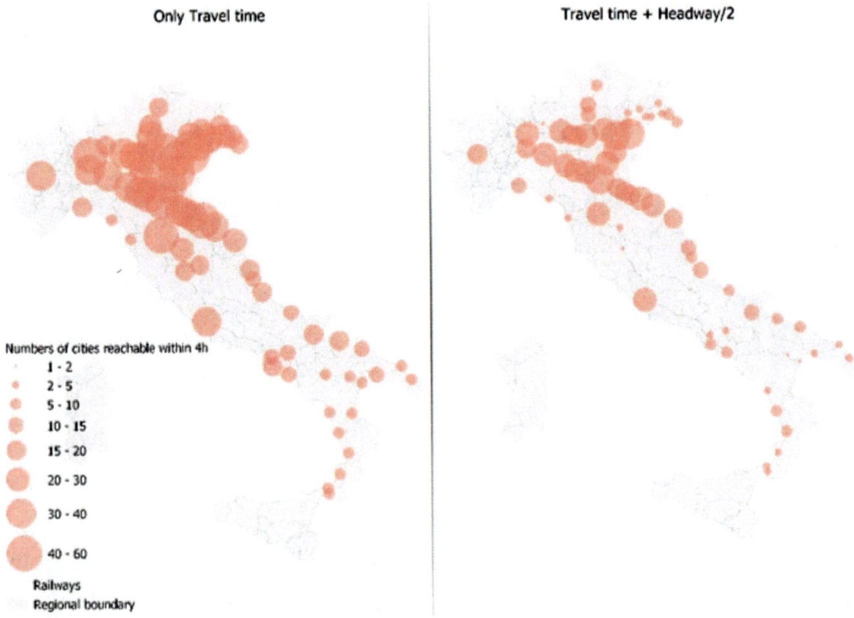

Fig. 17 Number of cities reachable within 4 h, 2019

Table 4 Average number of cities able to be reached in 2 h in the 2 scenarios, based on the different evaluation methods

	A-2008	B-2019	(B-A)/A
1-Only travel time	8.0	8.9	12%
2-travel time + Headway/2	2.4	2.9	23%
Ratio 2/1	30%	33%	

Table 5 Average number of cities able to be reached in 3 h in the 2 scenarios, based on the different evaluation methods

	A-2008	B-2019	(B-A)/A
1-Only travel time	14.1	16.4	17%
2-travel time + Headway/2	5.0	6.4	28%
Ratio 2/1	35%	39%	

Table 6 Average number of cities able to be reached in 4 h in the 2 scenarios, based on the different evaluation methods

	A-2008	B-2019	(B-A)/A
1-Only travel time	21.0	25.0	19%
2-travel time + Headway/2	9.7	11.6	19%
Ratio 2/1	46%	46%	

As can be seen by the representations above and the respective numerical summaries, the evaluation of the cities that can be reached within the various limits varies considerably depending on whether or not the value of the headway of the services is considered within the expression; in fact, the average number of cities that can be reached taking the latter factor into consideration is between 30 and 50% of the average number obtained when only the travel time is taken into account. The improvement between 2008 and 2019 is evident, on the other hand, regardless of the method used for the accessibility measure. For the 2 and 3-h thresholds, the increase in the number of cities that could be reached between 2008 and 2019 is more pronounced if, in addition to the travel time, the headway is also taken into account for the estimate, thus confirming the fact that, in addition to reducing travel times, the HS network's development has also led to an increase in the range of services.

6 Conclusions and Further Perspectives

The analysis carried out places the user at the centre of the accessibility comparison and evaluation between 2008 and 2019 in terms both of the methodological approach utilised to estimate the travel quantities and the attributes used in the accessibility measure. In fact, with regard to the evaluation of the travel attributes, an allocation model for public transport was used; the travel times include the waiting times for any connections, and the number of journeys available on a given OD depends on the number of services with similar travel times and combinations of trips with connection times compatible with the optimal transfer time for the user (15 min). With regard to the accessibility measures, starting with the contour measures type, the number of cities that can be reached within the pre-set time limits was evaluated in terms of travel time alone (again using the model's outputs) and taking into account both travel time and the headway of the services. None of the analyses carried out took into consideration the times for accessing the HS stations; since this is mainly a comparative study between 2008 and 2019, this component can be overlooked considering the fact that the infrastructural networks for accessing the stations involved in the analysis have not undergone any major changes. For the purposes of the study, only the stations present in both scenarios were taken into consideration, and therefore the stations that became operational after 2008 (Reggio Emilia AV, Napoli Afragola) did not fall within the scope of the analysis.

The first part of the study focused on determining the attributes of train travel that had the greatest impact on the choice of this means of transport, expressed by the modal shares resulting from surveys on the main HS routes; as expected, travel speed (as an expression of the efficiency of the service offered) was determined to be the most influential attribute, but greater clarifications were able to be obtained from the model by entering the number of trips offered on the specific route. The picture was then attempted to be completed with the inclusion of travel costs, which would clearly finalise the regression model's set of measurable attributes.

The second part, on the other hand, focused on comparing the services offered in 2008 and 2019 in terms of travel times and number of services, and on evaluating accessibility measures from the user's point of view. A comparison between the two years, before and after the HS infrastructural interventions described in chapter "Comparison of Preliminary, Initial, and Final Construction Costs of Italian High-Speed Railways" and the deregulation of the railway services market, shows a general improvement in the routes served by HS trains. Taking into consideration the routes present in both simulation scenarios, there was a 14% average reduction in travel times and a 25% increase in the number of services, which translates into an average reduction of 60 min of travel per OD, and an average increase of 1.1 trips per OD compared to 2008. On the main HS routes (e.g. those along the Turin—Salerno axis), the reduction in travel times and the increase in services are more pronounced than those described above. On some of the routes covered by the study, due to the number of services offered during the simulation period, extra-urban travel by train nearly equalled urban travel in terms of availability of choice.

The development of the HS rail system has led to a general decrease in travel times and increased flexibility in terms of journey organization, which has resulted in a considerable change in the way the distances and times are perceived by train users. While, with respect to the change in headway, it is possible to attempt a graphic representation of the phenomenon through the distortion of Italy based on the changes in the travel times (Figs. 10 and 11), it is not possible to quantify the changes with respect to other subjective aspects of the train journey, such as, for example, the possibility of using the travel time for recreational and work activities, which are difficult to do with other methods of travel. Finally, by expressing accessibility as the number of cities that can be reached within a certain time limit, it should be noted that, for the three limits evaluated, there was an average increase of 20% from 2008 to 2019. The positive change did not affect accessibility expressed in travel times nor that evaluated in terms of the number of services offered to users; within 3 h of travel, the improvement is even more substantial if we consider accessibility in terms of both travel time and number of services on the ODs combined. The study carried out also revealed a substantial difference between accessibility evaluated in terms of travel time alone and accessibility evaluated in terms of both travel time and headway of the services on the OD; this second measure is intended to be representative of the accessibility experience for the user, for whom the ease of reaching a destination/opportunity also depends on the flexibility of planning the trip.

CRediT authorship contribution statement Conceptualization: Mario Tartaglia;
Methodology: Lorenzo Vannacci;
Resources: Lorenzo Vannacci;
Formal analysis: Martina Farsi;
Visualization: Martina Farsi;
Writing original draft: Abstract Mario Tartaglia, 1. Summary Mario Tartaglia, 2. Introduction Lorenzo Vannacci, 3. Literature review Lorenzo Vannacci, Mario Tartaglia, 4. The case study Martina Farsi; 5. Methodology and data Martina Farsi, 6. Results and discussion Martina Farsi, 7. Conclusions and further perspectives Martina Farsi;
Review: Mario Tartaglia, Lorenzo Vannacci;
Writing, review and editing: Martina Farsi;
Supervision: Mario Tartaglia.

References

1. Curtis, C., Scheurer, J.: Planning for sustainable accessibility: developing tools to aid discussion and decision-making. Prog. Plann. **74**, 53–106 (2010)
2. Litman, T.: A evaluating accessibility for transportation planning, Victoria Transport Policy Institute, www.vtpi.org (2012)
3. Bhat, C., Handy, S., Kockelman, K., Mahmassani, H., Chen, Q., Weston, L.: Development of an Urban Accessibility Index: Literature Review. University of Texas, Austin USA, Center for Transportation Research (2000)
4. Hansen, W.G.: How accessibility shapes land use. J. Am. Instit. Plann. **25**(1) 73–76 (1959)
5. Dalvi, M.Q., Martin, K.M.: The measurement of accessibility: some preliminary results. Transportation **5**, 17–42 (1976)
6. Morris, J.M., Dumble, P.L., Wigan, M.R.: Accessibility indicators for transportation planning. Transp. Res. A **13**, 91–109 (1979)
7. Johnston, R.J., Gregory, D., Pratt, G., Watts, M.: The Dictionary of Human Geography. Blackwell Publishing, Oxford (2000)
8. Bach, L.: The problem of aggregation and distance for analyses of accessibility and access opportunity in location-allocation models. Environ. Plann. A **13**, 955–978 (1981)
9. Ingram, D.R.: The concept of accessibility: a search for an operational form. Reg. Stud. **5**, 101–107 (1971)
10. Handy, S.L., Niemeier, D.A.: Measuring accessibility: an exploration of issues and alternatives. Environ. Plann. A **29**, 1175–1194 (1997)
11. Vickerman, R.W.: Accessibility, attraction, and potential: a review of some concepts and their use in determining mobility. Environ. Plann. A **6**, 675–691 (1974)

12. Ben-Akiva, M., Lerman, S.R.: Disaggregate travel and mobility choice models and measures of accessibility. In: Hensher, D.A., Sopher, P.R. (eds.) Behavioural Travel Modelling. Croom Helm, Andover, Hants, pp. 654–679 (1979)
13. Le Vine, S., Lee-Gosselin, M., Sivakumar, A., Polak, J.: A new concept of accessibility to personal activities: development of theory and application to an empirical study of mobility resource holdings. J. Transp. Geogr. **31**, 1–10 (2013)
14. Geurs, K.T., Ritsema van Eck, J.R.: Accessibility measures: review and applications. RIVM Report 408505 006. National Institute of Public Health and the Environment, Bilthoven (2001)
15. Geurs, K.T., van Wee, B.: Accessibility evaluation of landuse and transport strategies: Review and research directions. J. Transp. Geogr. **12**, 127–140 (2004)
16. Geurs, K.T., van Wee, B.: Discussing equity and social exclusion in accessibility evaluations. EJTIR **11**(4), 350–367 (2011)
17. Curtis, C., Scheurer, J.: Multiple accessibility: developing a tool for evaluating land use-transport integration. GAMUT, Australasian Centre for the Governance and Management of Urban Transport, University of Melbourne, Melbourne (2008)
18. Gutièrrez, J.: Transport and accessibility. In: Kitchin, R., Thrift, N. (eds.) International Encyclopedia of Human Geography, pp. 410–417. Elsevier, Amsterdam (2009)
19. Páez, A., Scott, D.M., Morency, C.: Measuring accessibility: positive and normative implementations of various accessibility indicators. J. Transp. Geogr. **25**, 141–153 (2012)
20. Kwan, M.P.: Space-time and integral measures of individual accessibility: a comparative analysis using a point-based framework. Geogr. Anal. **30**, 191–216 (1998)
21. Rotoli, F., Christidis, P., Vannacci, L., Lopez-Ruiz, H.G., Navajas, C.E., Ibáñez, R.N.: Potential impacts on accessibility and consumer surplus of improvements of the European railway system. In: Euro Working Group on Transportation 2014 Compendium of Papers (2014)
22. Gutierrez, J., Gonzalez, R., Gomez, G.: The European high-speed train network: predicted effects on accessibility patterns. J. Transp. Geogr. **4**(4), 227–238 (1996)
23. Bruinsma, F., Rietveld, P.: The accessibility of European cities; theoretical framework and comparison of approached. Environ. Plann. A **30**, 499–521 (1998)
24. Street, N.: Time Contours: Using Isochrone Visualization to Describe Transport Network Travel Cost. Imperial College London, London (2006)
25. Lightfoot, C., Kelly, F.: Travel-Time Maps–Transforming our View of Transport Asymptopia. University of Cambridge, Cambridge (2006)
26. Gamper, J., Boehlen, M.H., Cometti, W., Innerebner, M.: Defining isochrones in multimodal spatial networks. In: Proceedings of CIKM-11, October 24–28, 2011, Glasgow, Scotland, pp. 2381–2384 (2011)
27. Schürmann, C., Spiekermann, K., Wegener, M.: Accessibility indicators. Berichte aus dem Institut für Raumplanung 39, Institute of Spatial Planning, Universität Dortmund Dortmund (1997)
28. Rodrigue, J.P., Comtois, C., Slack, B.: The Geography of Transport Systems. Routledge, New York (2013)
29. Geurs, K.T., Ritsema van Eck, J.R.: Accessibility evaluation of land-use scenarios: the impact of job competition land-use and infrastructure developments for the Netherlands. Environ. Plann. B **30**(1), 69–87 (2003)
30. Hägerstrand, T.: What about people in regional science? People of the Regional Science Association, 24, pp. 7–21 (1970)
31. Miller, H.J.: Modelling accessibility using space–time prism concepts within geographical information systems. Int. J. Geogr. Syst. **5**(3), 287–301 (1991)
32. Miller, H.J.: Measuring space–time accessibility benefits within transportation networks: basic theory and computational procedures. Geogr. Anal. **31**(2), 187–212 (1999)
33. Recker, W.W., Chen, C., McNally, M.G.: Measuring the impact of efficient household travel decisions on potential travel time savings and accessibility gains. Transp. Res. A **35**(4), 339–369 (2001)
34. Ben-Akiva, M., Lerman, S.R.: Discrete Choice Analysis. MIT Press, Cambridge MA (1985)

35. Williams, H.C.W.L.: Travel demand models, duality relations and user benefit analysis. J. Reg. Sci. **16**(2), 147–166 (1976)
36. Silva, C., Pinho, P.: The Structural Accessibility Layer (SAL): revealing how urban structure constrains travel choice. Environ. Plann. A **42**, 2735–2752 (2010)
37. Charalambous, N., Mavridou, M.: Space syntax: spatial integration accessibility and angular segment analysis by metric distance (ASAMeD). In: Hull, A., Silva, C., Bertolini, L. (eds.) Accessibility Instruments for Planning Practice, pp. 57–62. COST Office, Amsterdam (2012)
38. Vandenbulcke, G., Steenberghen, T., Thomas, I.: Mapping accessibility in Belgium: a tool for land-use and transport planning? J. Transp. Geogr. **17**, 39–53 (2009)
39. Yang, R., Yan, H., Xiong, W., Liu, T.: The study of pedestrian accessibility to rail transit stations based on KLP model. Proc. Social Behav. Sci. **96**, 714–722 (2013)
40. Monzón, A., Ortega, E., López, E.: Efficiency and spatial equity impacts of high-speed rail extensions in urban areas. Cities **30**, 18–30 (2013)
41. Chen, C.L., Hall, P.: The impacts of high-speed trains on British economic geography: a study of the UK's InterCity 125/225 and its effects. J. Transp. Geogr. **19**(4), 689–704 (2011)
42. Martinez Sánchez-Mateos, H.S., Givoni, M.: The accessibility impact of a new High-Speed Rail line in the UK—a preliminary analysis of winners and losers. J. Transp. Geogr. **25**, 105–114 (2012)
43. Chen, C.L., Hall, P.: The wider spatial-economic impacts of high-speed trains: a comparative case study of Manchester and Lille sub-regions. J. Transp. Geogr. **24**, 89–110 (2012)
44. Cao, J., Liu, X.C., Wang, Y., Li, Q.: Accessibility impacts of China's high-speed rail network. J. Transp. Geogr. **28**,12–21 (2013)
45. Chandra, S., Vadali, S.R.: Evaluating accessibility impacts of the proposed America 2050 high-speed rail corridor for the Appalachian Region. J. Transp. Geogr. **37**, 28–46 (2014)
46. Chen, G., de Abreu e Silva, J.: Estimating the provincial economic impacts of high-speed rail in Spain: an application of structural equation modelling. Proc. Soc. Behav. Sci. **111**, 157–65 (2014)
47. Albalate, D., Bel, G.: High speed rail: lessons for policy makers from experiences abroad. In: IREA Working Papers, Barcelona (2010)
48. Adler, N., Pels, E., Nash, C.: High speed rail and air transport competition: game engineering as tool for cost-benefit analysis. Transp. Res. Part B **44**(7), 821–833 (2010)
49. Preston, J.: High speed rail in britain: about time or a waste of time? J. Transp. Geogr. **22**, 308–311 (2012)
50. Chang, J.S., Lee, J.H.: Accessibility analysis of Korean high speed rail: a case study of the Seoul metropolitan area. Transp. Rev. **28**(1), 87–103 (2008)
51. Levinson, D.M.: Accessibility impacts of high-speed rail. J. Transp. Geogr. **22**, 288–291 (2012)
52. Gutiérrez Puebla, J.: El tren de alta velocidad y sus efectos espaciale. Invest. Reg. **5**, 199–221 (2004)
53. Vickerman, R.: High-speed rail and regional development: the case of intermediate stations. J. Transp. Geogr. **42**, 157–165 (2014)
54. Venter, C.J., Cross, C.: Location, mobility, and access to work. Mobility **13** (2011)
55. Willigers, J., van Wee, B.: High-speed rail and office location choices: a stated choice experiment for the Netherlands. J. Transp. Geogr. **19**, 745–754 (2011)
56. Dubé, J., Thériault, M., Des Rosiers, F.: Commuter rail accessibility and house values: the case of the Montreal South Shore, Canada, 1992–2009. Transp. Res. Part A **54**, 49–66 (2013)
57. Spiekermann, K., Wegener, M.: The shrinking continent: accessibility, competitiveness, and cohesion. In: Faludi, A. (ed.) European Spatial Research and Planning, pp. 115–140. Lincoln Institute of Land Policy, Cambridge, MA (2008)
58. Kotavaara, O., Antikainen, H., Rusanen, J.: Population change and accessibility by road and rail networks: GIS and statistical approach to Finland 1970–2007. J. Transp. Geogr. **19**, 926–935 (2011)
59. Ortega, E., López, E., Monzón, A.: Territorial cohesion impacts of high-speed rail at different planning levels. J. Transp. Geogr. **24**, 130–141 (2012)

60. Ortega, E., Lòpez, E., Monzón, A.: Territorial cohesion impacts of high-speed rail under different zoning systems. J. Transp. Geogr. **34**, 16–24 (2014)
61. Monzón, A., Ortega, E., López, E.: Social impacts of high speed rail projects: addressing spatial equity effects. In: Proceedings, 12th WCTR, July 11–15, 2010–Lisbon, Portugal. E.T.S.I. Caminos, Canales y Puertos (UPM), Madrid (2010)
62. Gutiérrez, J.: Location, economic potential and daily accessibility: an analysis of the accessibility impact of the high-speed line Madrid–Barcelona–French border. J. Transp. Geogr. **9**(4), 229–242 (2001)
63. Gutiérrez, J., Condeço-Melhorado, A., López, E., Monzón, A.: Evaluating the European added value of TEN-T projects: a methodological proposal based on spatial spillovers, accessibility and GIS". J. Transp. Geogr. **19**, 840–850 (2011)
64. Martín, J.C., Gutiérrez, J., Román, C.: Data envelopment analysis (DEA) index to measure the accessibility impacts of new infrastructure investments: the case of the high-speed train corridor Madrid-Barcelona-French border. Reg. Stud. **38**(6), 697–712 (2004)
65. Martín, J.C., Reggiani, A.: Recent methodological developments to measure spatial interaction: synthetic accessibility indices applied to high speed train investments. Transp. Rev. **27**(5), 551–571 (2007)
66. Gutièrrez, J., Urbano, P.: Accessibility in the European Union: the impact of the Trans-European road network. J. Transp. Geogr. **4**(1), 15–25 (1996)
67. Hou, Q., Li, S.M.: Transport infrastructure development and changing spatial accessibility in the Greater Pearl River Delta, China, 1990–2020. J. Transp. Geogr. **19**, 1350–1360 (2011)
68. Shaw, S.L., Fang, Z., Lu, S., Tao, R.: Impacts of high speed rail on railroad network accessibility in China. J. Transp. Geogr. (in press) (2014)
69. Vannacci, L., Tartaglia, M., Elena, N., Rotoli, F.: The use of open data for estimating rail accessibility in Europe. In: Ingegneria Ferroviaria LXX (7–8), pp. 611–634 (2015)
70. Vannacci, L.: European accessibility using real timetable data. In: Proceedings of the Nectar Cluster 6 Meeting, Seville 6th February 2014
71. Google Inc.: General transit feed specification reference. http://developers.google.com/transit/gtfs/reference (2014). Accessed 19 July 2014
72. Renne, J.L., Hamidi, S., Ewing, R.: Transit commuting, the network accessibility effect, and the built environment in station areas across the United States. Res. Transp. Econ. **60**, 35–43 (2017)

A Geographically Weighted Poisson Regression Approach for Analyzing the Effect of High-Speed Rail on Tourism in China

Filomena Mauriello, Zhenhua Chen, and Francesca Pagliara

Abstract In the international literature, several studies have analyzed the impact of HSR on tourists' behavior with qualitative and quantitative approaches. However, they have not been able to solve the problem of capturing the spatial and temporal variation by fitting a regression model at a local point. The spatial heterogeneity within local models, such as Geographically Weighted Regression (GWR) models, provides a better platform allowing exploring the different spatial relationships between HSR and tourism. In this chapter, a spatio-temporal analysis has been proposed to evaluate the variables affecting tourists 'choices, specifically the impact of HSR on both Chinese and Foreign tourists. Two advanced methods were adopted: firstly, we used the Weighted Regression with Poisson distribution (GWPR) modelling approach, which considers the problem of the temporal and spatial autocorrelation differently with respect to the Generalized Estimating Equations method. The results of this study support the use of the GWPR as a promising tool for tourism planning, especially because it makes it possible to model non-stationary spatially counting data. As far as the authors know, this methodology has never been applied in the international literature to this context. Secondly, we combined both temporal autocorrelation and spatial autocorrelation by applying models of Geographical and Temporal Weighted Regression (GTWR) types to take into account the local effects from the temporal point of view.

Keywords High-Speed Rail · Tourism · Geographically weighted poisson regression · China

F. Mauriello (✉) · F. Pagliara
Department of Civil, Architectural and Environmental Engineering, University of Naples Federico II, Via Claudio 21, 80125 Naples, Italy
e-mail: filomena.mauriello@unina.it

F. Pagliara
e-mail: fpagliar@unina.it

Z. Chen
City and Regional Planning, Austin E. Knowlton School of Architecture, The Ohio State University, 295 Knowlton Hall, 275 West Woodruff Ave, Columbus, OH 43210, USA
e-mail: chen.7172@osu.edu

F. Pagliara (ed.), *Socioeconomic Impacts of High-Speed Rail Systems*,
Springer Proceedings in Business and Economics,
https://doi.org/10.1007/978-3-031-26340-8_17

1 Introduction

The "second railway age" is brought about by significant advancements in transportation technology and the ongoing building of High-Speed Rail (HSR) [1, 2]. China's HSR network has expanded significantly during the past ten years, affecting the geographical organization of cities within the transportation system.

The longest HSR in the world is, in fact, in China (see Fig. 1). The total operating mileage of the national railroad in 2021 was higher than 150,000 km, including 40,000 km of HSR. Additionally, 2168 km of new mileage was built and put into service in 2021. From 28% in 2012 to 93% in 2021, the HSR network's coverage of cities with a population of more than 500,000 people increased. Except for Lhasa on the mainland, all province capital cities have been connected to the HSR. In August 2020, China National Railway Group Co., Ltd. issued and released the Outline of Railway Development Plan for A Powerful Transportation Country, defining the development blueprint of China Railway in the next 15 years and 30 years [3]. According to the plan, in 2035, China will have built about 200,000 km of the national railway network, including about 70,000 km of HSR. The HSRs mileage will be double-sized w.r.t. the year 2019, and HSR will serve all provincial cities and cities with a population of more than 500,000. The 3-h HSR circle is basically realized between the provincial capitals in adjacent regions (http://www.xinhuanet.com/fortune/2020-08/14/c_1126366741.htm).

The main focus of this study is the investigation of the elements influencing tourists' decisions for the case study of thirty Chinese provinces, where the influence of HSR has been investigated. There are contributions in the literature that primarily focus on the accessibility and mobility impacts of HSR in China [4–6], as well as the influence of HSR on the growth of regional tourism [7] and foreign visitors [8]. Although extensive studies have examed the impact of HSR on tourism in China, the findings are not always consisistent, as they discovered significant effects in some instances [9, 10] and insignificant effects in others [11]. For example, Chen and

Fig. 1 The High-Speed Rail network in China

Haynes [8] demonstrated that the Chinese provinces served by HSR experienced an increase in the number of foreign tourists of 20% and an increase in tourism revenue of almost 25%. In addition, Chen et al. [12] evaluated the spatial impacts of HSR on domestic tourism demand in China using spatial econometric analysis for the period 1999–2016. Their study confirmed that HSR has diverse spatial impacts on tourism output, with a particularly strong effect in the less developed west regions, moderate impact in the central region, and less significant in the developed east regions.

The effect of HSR on tourism in the Yangtze River Delta was analyzed by Taotao et al. [13]. The Yangtze River Delta's development in regional tourism demonstrated a "HSR effect," and the demand and supply of tourism-related goods significantly improved. Yuhua and Jun [14] demonstrated how the HSR's introduction impacted the growth of tourism in the cities it served.

After studying Huangshan City's tourist spatial structure before, immediately after, and two years after the HSR's inauguration, Lei et al. [15] concluded that the HSR's inauguration had little to no effect on Huangshan City's tourism.

An additional investigation revealed that the expansion of the HSR network in the HSR also affected tourist flows and spatial relationships of the two cities, Beijing and Tianjin [16, 17].

Ziyang et al. [18] used Xiamen City as a case study and confirmed a relationship between HSR and tourism. According to Yongze et al. [19], the inauguration of the HSR had a substantial impact on encouraging regional tourism. Still, as the country moved from the east to the west, this influence gradually diminished.

The impact of high-speed rail on tourism development can be also predicted through random utility models (RUM). A study on their predictive capability in terms of market share was conducted by [20]. Travellers' and transport users' preferences can be inferred from different data sources, such as trip diaries or trajectories (e.g. [21]).

In this chapter, a spatio-temporal analysis has been adopted to evaluate the variables affecting tourists 'choices and, specifically, the impact of HSR on both Chinese domestic and foreign tourists.

The two methods were adopted. The GWPR modelling approach was first chosen. The latter considers the problem of temporal and spatial autocorrelation in a different way with respect to the Generalized Estimating Equations method. Specifically, the results of this study support the use of the Geographically Weighted Regression with Poisson distribution (GWPR) as a useful tool for tourism planning, since it makes possible to model non-stationary spatially counting data. This methodology, as far as the authors know, has never been applied in the international literature to this context.

Secondly it takes into account a further analysis which combines both the temporal autocorrelation and spatial-autocorrelation by the application of models of Geographical and Temporal Weighted Regression (GTWR) types in order to take into account also the local effects from the temporal point of view.

The chapter is organized as follows. Section 2 deals with the description of the data set and the methodology. In Sect. 3, the results are reported. Conclusions and further perspectives are reported in Sect. 4.

2 Description of the Data Set and the Methodology

The dataset collected for this study contains information concerning thirty-four Chinese provinces. Hong Kong, Macao, Taiwan and Tibet are excluded from the dataset due to a data limitation. In total, the data covers the period from 2001–2019 (see Fig. 2).

As shown in Table 1, eight variables were adopted in this evaluation.

The impacts of HSR projects on tourism can be quantified in different ways.

In this study, the dependent variables take only non-negative integer values, the statistical treatment differs from that of the normally distributed one, which can assume any real value, positive or negative, integer or fractional. Count data can be modeled using different methods, the most popular is the Poisson distribution, which is applied to a wide range of transportation count data contexts. In a Poisson regression model, the probability of city i having y_{it} number of tourist per year is given by [22–24]:

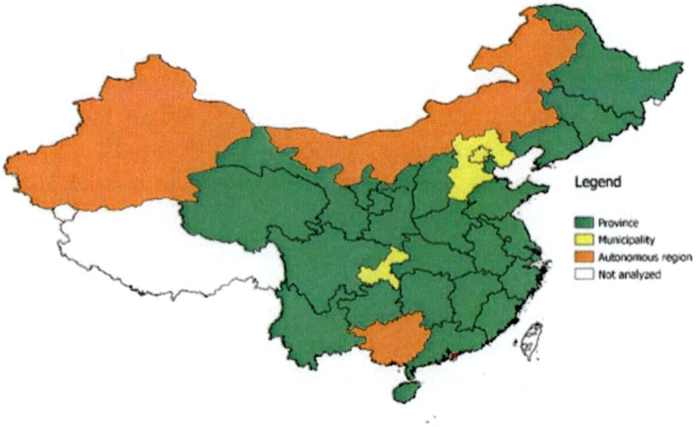

Fig. 2 Provinces and regions under study

Table 1 Variables

Dependent variables	
Domestic	No. of Chinese tourists
Overseas	No. of foreign tourists
Independent variables	
GDP	Gross Domestic Product of the province
Passengers	Total number of passengers
Resorts	No. of resorts in the province
HSR stations	Dummy: 0 if no HSR stations are present in the province, 1 otherwise
IntAirport	Dummy: 0 if no international airports are present in the province, 1 otherwise

$$P(y_i) = \frac{\lambda_i^{(y_i)} \times e^{-\lambda_i}}{y_i!} \tag{1}$$

where $P(y_i)$ is the probability of city i having y_i tourist per year and λ_i is the Poisson parameter for city i, which is equal to the expected number of tourists per year at city i, E[y_i]. The mean and the variance are given by E[y_i]= λ_i and V[y_i] = λ_i. . Generalized Linear Models (GLMs) are considered the most suitable to determine the relationship between count data and the dependent variables. GLMs aim to extend ordinary regression models to non-normal response distributions [25, 26]. Furthermore, the data considered involve measurements over time for the same cities, to avoid the serial correlation seriously affecting the estimated parameters, leading to inappropriate statistical inferences. Therefore, the panel data regression models have been considered. Panel model analysis provides a general, flexible approach in these contexts, since it allows the modeling various correlation patterns. To consider these possible unknown correlations, an extension of GLMs, namely Generalized Estimating Equations (GEEs), has been considered. The relationship between the explanatory variables and the Poisson parameter is given by [27]:

$$E[y_{it}] = e^{(\beta_0+\beta_1 x_1 t+\beta_2 x_2 t+\cdots+\beta_p x_p t+\Phi y_{(it-1)}+u_i t)} \tag{2}$$

where β_0 is the intercept, the β_p, i = 0,1,…,p, are the regression coefficients, ϕ is the parameter for the autoregressive component and u_{it} is the error component model for the disturbances. The main problem is that the u_{it} is auto-correlated with y_{it-1}. In order to fix this, the model is fitted by using population-averaged

Poisson models and by imposing an AR(1) process in the error term. These models are suitable when the random effects and their variances are not of inherent interest, as they allow for the correlation without explaining its origin. The aim is to estimate the average response over the population rather than the regression parameters that would enable the prediction of the effect of changing one or more components of the predictor variable on a given individual.

The parameters of this model are estimated by a backward elimination procedure, which considers all the variables in the model. At each step of the backward process, a variable is removed. The latter is the one assuming the largest p-value. The process ends when all the variables in the model have a p-value less than 0.05 or until there is no variable remaining [25].

The significance of each variable has been tested with the t-student statistic, therefore, a coefficient is significant when t is greater than 1.96.

Then, the Geographically Weighted Generalised Linear model (GWGL) was developed by integrating the GLM and the GWR ones and extending the concept of the Geographical Weighted Regression (GWR) models in the context of the Generalized Linear Models (GLM). Given that the dependent variables are count data with discrete and non-negative integer values, GWR models have been performed using the Poisson distribution error [28].

The Geographically Weighted Poisson Regression (GWPR) approach has been adopted [29] to capture the heterogeneity of the independent variables concerning each province. These models capture the spatial variation by fitting a regression model at each sample point. The result of this process is a set of local spatial parameters, described in Eq. (3).

$$E[y_i] = e^{\left(\sum_p \beta_{jp}(u_j;v_j)x_{jp}\right)} \tag{3}$$

where $(u_j; v_j)$ are the coordinates of the different areas, β_{jp} represents the regression coefficient for the independent variable p and x_{jp} is the independent variable with $p = 1, \ldots, P$. The basic idea of the GWR is that the observed data next to point i has a higher influence on the estimation of $\beta_j(u_i)$'s than the data located further away. A weighting function describes this influence. GWR tries to capture the spatial variation by adjusting a regression model to each point individually and using a distance function denominated kernel spatial function. Models have been estimated yearly to capture the variability over time and space.

Lastly, an extension of geographically weighted regression (GWR), geographical and temporal weighted regression (GTWR), is developed to account for local effects in both space and time [30]. The result of this process is a set of local spatial parameters, described in Eq. (4).

$$E[y_i] = e^{\left(\sum_p \beta_{jp}(u_j;v_j;t)x_{jp}\right)} \tag{4}$$

where similar to the GWPR model, $(u_j; v_j)$ are the coordinates of the different areas with the addition of the term t to indicate the dependence on the dimension time, β_{jp} represents the regression coefficient for the independent variable p and x_{jp} is the variable value.

3 Results

The estimation results of GLM are reported in Tables 2 and 3. The independent variable, that is not statistically significant at the 0.5 level of significance were removed from the models.

The GLM models' results show that both the number of domestic and foreign tourists are influenced by the presence of HSR stations, the presence of IntAirport, GDP, and the Number of Passengers.

The results of the GEE models, reported in Tables 4 and 5, confirm the results obtained by the GLM models. However, GEE models are more conservative than GLMs. A higher standard error of the GLM can be observed, moreover, also the value of the log-likelihood is lower, indicating a greater ability to estimate the models.

Starting from the statistically significant variables obtained from the GLM and GEE models, the GWPR models have been estimated for every year. An example

Table 2 GLM model: domestic tourist

Variables	Coefficient	Odd ratio	Std. Error	P-value
Intercept	7.4150	1660.7091	0.3280 e-02	<0.001
Presence of HSRstations	1.1550	3.1740	0.8674e-03	<0.001
Presence of IntAirport	1.3830	3.9868	3.3270e-03	<0.001
GDP	0.1070e-03	1.0001	2.8300e-08	<0.001
Number of passengers	0. 2190e-03	1.0002	9.2590e-08	<0.001
Log-likelihood	−1,809,529			

Table 3 GLM model: overseas tourist

Variables	Coefficient	Odd ratio	Std. Error	P-value
Intercept	1.1400E+01	89,321.7233	3.8500E-04	<0.001
GDP	4.6800E-06	1.0000047	2.5000E-09	<0.001
Number of hotels	2.550E-03	1.0025533	1.0400E-07	<0.001
Number of passengers	7.0500E-05	1.0000705	7.5100E-09	<0.001
Presence of IntAirport	1.4200	4.1371204	3.8900E-04	<0.001
Presence of HSRstations	2.0700E-01	1.2299826	6.8400E-05	<0.001
Log-likelihood	−341,017,701			

Table 4 GEE model: domestic tourist

Variables	Coefficient	Odd ratio	Std. Error	P-value
Intercept	7.3633	1.5770E+03	0.0949	<0.001
Presence of HSRstations	0.9887	2.6877	0.9640e-03	<0.001
Presence of IntAirport	1.0541	2.8694	0.3690e-02	<0.001
Number of passengers	0.0001	1.0001	9.6000e-08	<0.001
Number of hotels	0.0004	1.0004	3.0800e-06	<0.001
Log-likelihood	−739,015.13			

Table 5 GEE model: overseas tourist

Variables	Coefficient	Odd ratio	Std. Error	P-value
Intercept	13.231	557,378.599	0.1670	<0.001
Presence of HSRstations	0.593	1.8094085	0.6940e-04	<0.001
Presence of IntAirport	0.621	1.8607879	0. 4290e-04	<0.001
Passengers	0. 2150e-04	1.0000215	5.8700e-09	<0.001
Number of hotels	0. 6720e-03	1.0006722	2.1900e-07	<0.001
Log-likelihood	−64,035,671			

of the results obtained, for the years 2001, 2010 and 2019 for Domestic tourists and Foreign tourists, is shown in Tables 1 and 2, respectively, where the minimum, the maximum, the first quartile, the median, the third quartile and the global values are reported (Tables 6 and 7).

In particular, for each year, it is possible to observe a variability of coefficients, indicating that the impact of this variable is not the same for each province and in the time. It generally refers to a diversified mixture of spatial events, which relates to the intensity of a spatial phenomenon.

Table 6 GWPR model: domestic tourist

	Min	1° Qu	Median	3° Qu	Max	Global
2001						
Intercept	1.1362E+01	1.1431E+01	1.1482E+01	1.1511E+01	1.1546E+01	1.1439E+01
GDP	6.1300E-05	1.0300E-04	1.3900E-04	1.8100E-04	2.1700E-04	2.0000E-04
Number of hotels	1.9500E-03	2.5000E-03	3.1200E-03	3.6000E-03	3.9700E-03	3.1000E-03
Number of passengers	−9.0000E-07	1.5600E-05	2.2200E-05	3.1000E-05	5.6800E-05	0.0000E+00
Presence of HSRstations	0.0000E+00	0.0000E+00	0.0000E+00	0.0000E+00	0.0000E+00	0.0000E+00
Presence of IntAirport	5.3400E-01	5.8900E-01	6.0400E-01	6.2400E-01	6.7490E-01	6.1760E-01
2010						
Intercept	1.0171E+01	1.0223E+01	1.0273E+01	1.0343E+01	1.0376E+01	1.028E+01
GDP	6.8000E-06	9.2000E-06	1.2300E-05	1.4900E-05	1.8000E-05	0.0000E+00
Number of hotels	1.7400E-03	1.9200E-03	2.0800E-03	2.3700E-03	2.5900E-03	2.2000E-03
Number of passengers	8.6600E-05	9.6500E-05	1.0400E-04	1.1100E-04	1.2500E-04	1.0000E-04
Presence of HSRstations	−2.6700E-01	−1.9400E-01	−4.4100E-02	1.1400E-01	1.9900E-01	-1.5000E-02
Presence of IntAirport	2.5500E+00	2.6260E+00	2.6600E+00	2.7400E+00	2.8300E+00	2.6210E+00
2019						
Intercept	8.1100E+00	8.1440E+00	8.1520E+00	8.1690E+00	8.2020E+00	8.1209E+00
GDP	9.0000E-07	1.6000E-06	2.7000E-06	3.2000E-06	3.7000E-06	0.0000E+00
Number of hotels	−1.7100E-05	2.7000E-04	4.2000E-04	5.7800E-04	8.3000E-04	7.0000E-04
Number of passengers	−3.0700E-05	−2.5100E-05	−1.8300E-05	−8.2000E-06	−9.0000E-07	0.0000E+00
Presence of HSR.stations	2.1315E+00	2.2179E+00	2.3875E+00	2.4602E+00	2.5561E+00	2.2217E+00
Presence of IntAirport	0.0000E+00	0.0000E+00	0.0000E+00	0.0000E+00	0.0000E+00	0.0000E+00

Table 7 GWPR model: overseas tourist

	Min	1° Qu	Median	3° Qu	Mass	Global
2001						
Intercept	1.1611E+01	1.1651E+01	1.1660E+01	1.1667E+01	1.1672E+01	1.1656E+01
GDP	5.6300E-05	5.9400E-05	6.5300E-05	6.9000E-05	7.3400E-05	6.4800E-05
Number of Hotels	3.1682E-03	3.2202E-03	3.2745E-03	3.3534E-03	3.3987E-03	3.2862E-03
Number of Passengers	0.0000E+00	0.0000E+00	0.0000E+00	0.0000E+00	0.0000E+00	0.0000E+00
Presence of HSR.stations	0.0000E+00	0.0000E+00	0.0000E+00	0.0000E+00	0.0000E+00	0.0000E+00
Presence of IntAirport	7.1896E-01	7.2009E-01	7.2206E-01	7.2591E-01	7.2966E-01	7.2304E-01
2010						
Intercept	1.0222E+01	1.0252E+01	1.0274E+01	1.0304E+01	1.0309E+01	1.0274E+01
GDP	1.1300E-05	1.2600E-05	1.3900E-05	1.4900E-05	1.6500E-05	1.3900E-05
Number of Hotels	2.0250E-03	2.1100E-03	2.1696E-03	2.3218E-03	2.4464E-03	2.2097E-03
Number of Passengers	9.2300E-05	9.7200E-05	1.0130E-04	1.0350E-04	1.0840E-04	1.0040E-04
Presence of HSR.stations	−1.8921E-01	−1.4437E-01	−5.4679E-02	2.1269E-02	6.5728E-02	-5.7078E-02
Presence of IntAirport	2.5742E+00	2.6249E+00	2.6802E+00	2.7241E+00	2.7523E+00	2.6723E+00
2019						
Intercept	1.1611E+01	1.1626E+01	1.1646E+01	1.1654E+01	1.1666E+01	1.1641E+01
GDP	−2.9000E-06	−2.5000E-06	−2.3000E-06	−1.8000E-06	−9.0000E-07	−2.2000E-06
Number of Hotels	1.2032E-03	1.2915E-03	1.4427E-03	1.6169E-03	1.7528E-03	1.4537E-03
Number of Passengers	8.5600E-05	8.7400E-05	8.9400E-05	9.0900E-05	9.2400E-05	8.9300E-05
Presence of HSR.stations	2.0693E+00	2.1332E+00	2.1658E+00	2.1974E+00	2.2267E+00	2.1599E+00
Presence of IntAirport	0.0000E+00	0.0000E+00	0.0000E+00	0.0000E+00	0.0000E+00	0.0000E+00

Looking at the global value of the HSR coefficient, it is interesting to notice, in general, an increase from 2001 to 2019 for both Domestic and Foreign tourists. Observing the values of the HSR coefficient for the provinces of Beijing, Hainan, Hebei, Heilongjiang, Hubei, Qinghai, and Shandong, an increase is observed starting from 2013 (see Figs. 3 and 4).

The results of the three years have been presented, i.e. the 2013, 2015 and 2019 for the Chinese tourists. The results in the years before 2013 are reported since the coefficients are not significant.

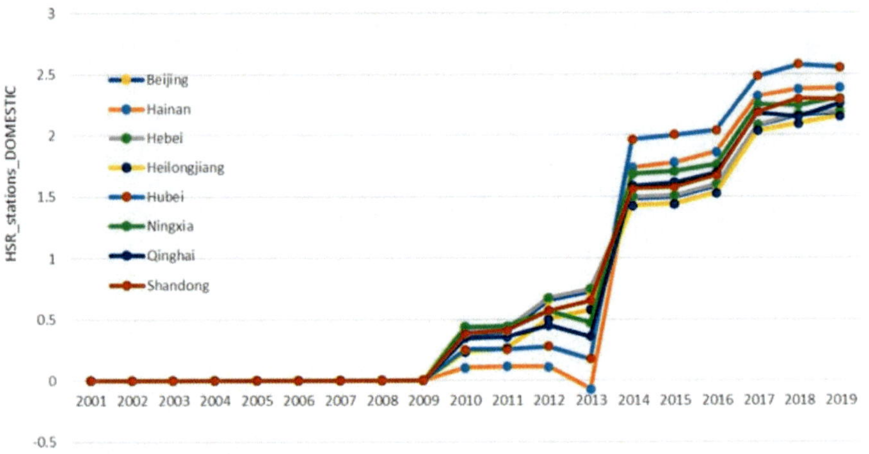

Fig. 3 The impact of HSR on eight provinces—GWPR: domestic tourists

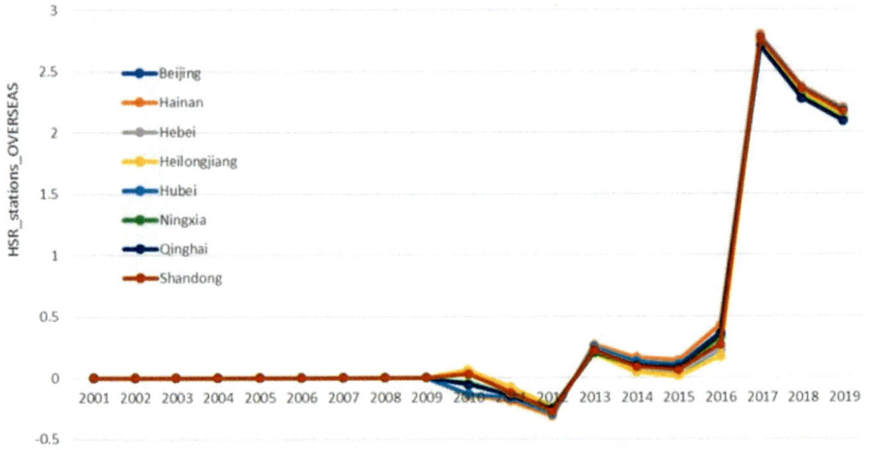

Fig. 4 The impact of HSR on eight provinces—GWPR: overseas tourists

Indeed, in Figs. 5 and 6, the weight that the coefficient of the variable HSR has on each province is reported, i.e. the objective is to demonstrate the effect of HSR of the neighboring provinces on their tourism.

It appears that central provinces, such as Hubei, Chongqing, Jiangxi and Hunan, have experienced more substantial impacts from HSR on domestic tourism, while

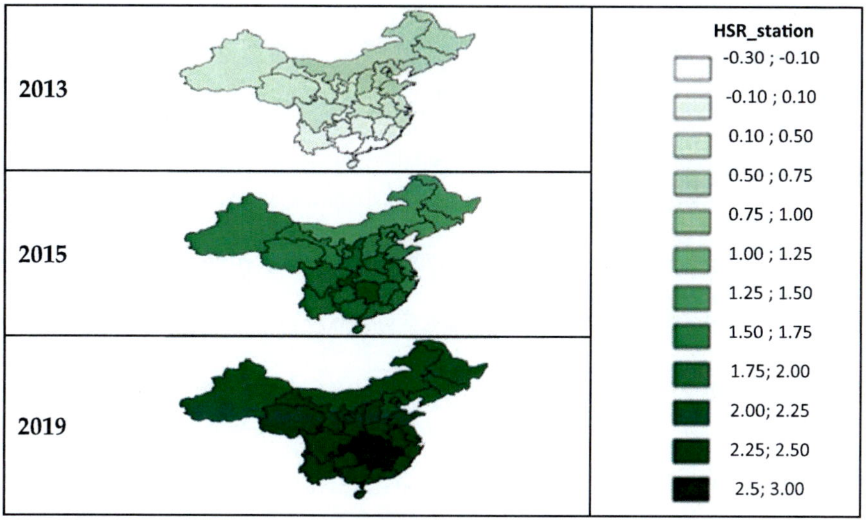

Fig. 5 The coefficients of the variable HSR for each provinces—GWPR: domestic tourists

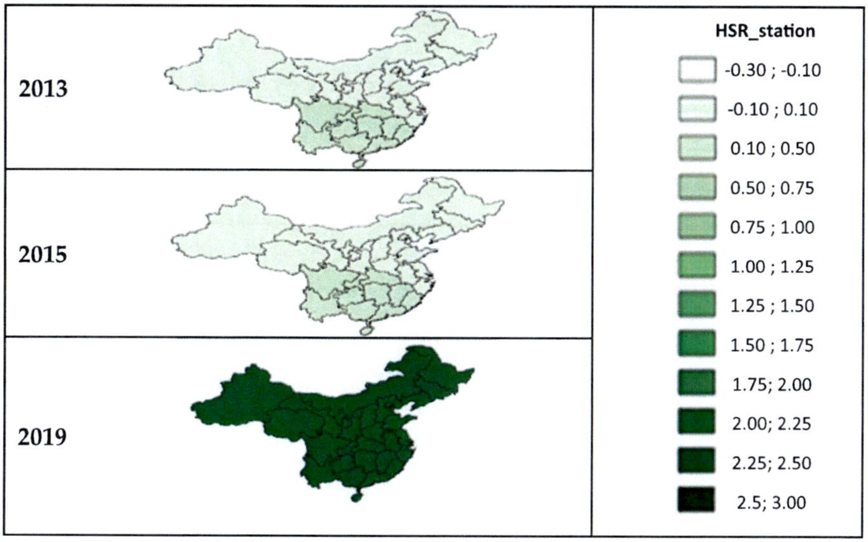

Fig. 6 The coefficients of the variable HSR for each provinces—GWPR: overseas tourists

the HSR's impact on international tourism is relatively evenly distributed among all the provinces.

The results of the GWTR essentially confirm the results of the GWPR models (Tables 8 and 9).

However, being more conservative than the GWPR models, a lower spatial and temporal variability is observed (Figs. 7, 8, 9 and 10).

Table 8 GWTR model: domestic tourist

	Min	1° Qu	Median	3° Qu	Max	Global
2001						
Intercept	6.8706238	6.9250047	6.9374501	6.9469109	6.9594562	6.9344493
Number of hotels	0.0010068	0.0010128	0.0010240	0.0010551	0.0010863	0.0010348
Number of passengers	0.0000315	0.0000458	0.0000522	0.0000592	0.0000806	0.0000534
Presence of HSR_Stations	0.7763535	0.8678959	0.8878655	0.9050140	0.9300897	1.0650239
Presence of Int.Airport	0.0000000	0.0000000	0.0000000	0.0000000	0.0000000	0.0000000
2010						
Intercept	6.9696561	7.0192340	7.0378405	7.0448903	7.0609800	7.0314459
Number of hotels	0.0008254	0.0008308	0.0008387	0.0008436	0.0008635	0.0006783
Number of passengers	0.0000360	0.0000393	0.0000406	0.0000414	0.0000474	0.0000407
Presence of HSR_Stations	0.9515641	0.9577023	0.9665857	0.9781305	1.0066854	0.9692117
Presence of Int.Airport	0.0000000	0.0000000	0.0000000	0.0000000	0.0000000	0.0000000
2019						
Intercept	6.9375215	7.0782946	7.1412861	7.1622839	7.2168691	7.1216771
Number of hotels	0.0005435	0.0005837	0.0006122	0.0006523	0.0007480	0.0006242
Number of passengers	0.0000285	0.0000315	0.0000318	0.0000322	0.0000327	0.0000315
Presence of HSR_Stations	0.9960165	1.0459379	1.0504587	1.0815223	1.2817223	1.0650239
Presence of Int.Airport	0.0000000	0.0000000	0.0000000	0.0000000	0.0000000	0.0000000

Table 9 GWTR model: overseas tourist

	Min	1° Qu	Median	3° Qu	Max	Global
2001						
Intercept	10.8122757	10.8925679	10.9182942	10.9468405	12.5702143	11.0778745
Number of hotels	0.0029198	0.0029751	0.0030080	0.0030441	0.0031375	0.0030168
Number of passengers	0.0000959	0.0000991	0.0001010	0.0001044	0.0001200	0.0001031
Presence of HSR_Stations	0.3073316	0.3455802	0.3649636	0.3681468	0.3731885	0.3540725
Presence of Int.Airport	1.1168181	1.1757089	1.1892301	1.1957628	1.2097021	1.1826228
2010						
Intercept	10.8054438	10.8181583	11.0149457	12.7040881	11.0709957	7.0314459
Number of hotels	0.0028211	0.0028413	0.0028536	0.0028658	0.0029047	0.0028550
Number of passengers	0.0000965	0.0000976	0.0000981	0.0000989	0.0001000	0.0000982
Presence of HSR_Stations	0.3453301	0.3621550	0.3653997	0.3683348	0.3731885	0.3636030
Presence of Int.Airport	1.4131167	1.4430550	1.4750866	1.4906659	1.5594577	1.4713796
2019						
Intercept	10.5316952	10.6866346	11.3139122	12.2241796	10.7766177	7.1216771
Number of hotels	0.0025690	0.0026384	0.0027019	0.0027497	0.0028768	0.0026973
Number of passengers	0.0000950	0.0000966	0.0000972	0.0000983	0.0000997	0.0000973
Presence of HSR_Stations	0.3486482	0.3784870	0.3959278	0.4200536	0.4998510	0.4037856
Presence of Int.Airport	1.4234591	1.4734820	1.5177593	1.5355523	1.6323726	1.5111878

4 Conclusion

In this chapter, we provided a spatio-temporal analysis using two advanced methods, the Weighted Regression with Poisson distribution (GWPR) and the Geographical and Temporal Weighted Regression (GTWR) model, to evaluate the spatial impact of HSR on tourist behavior. Using the Chinese HSR system as an example, our study confirms that the impact of HSR varies both spatially and temporally. The results suggest that the impacts of HSR on tourism in China have increased constantly since 2013. In terms of the spatial impacts, central provinces, such as Hubei, Chongqing, Jiangxi and Hunan, have experienced more substantial impacts

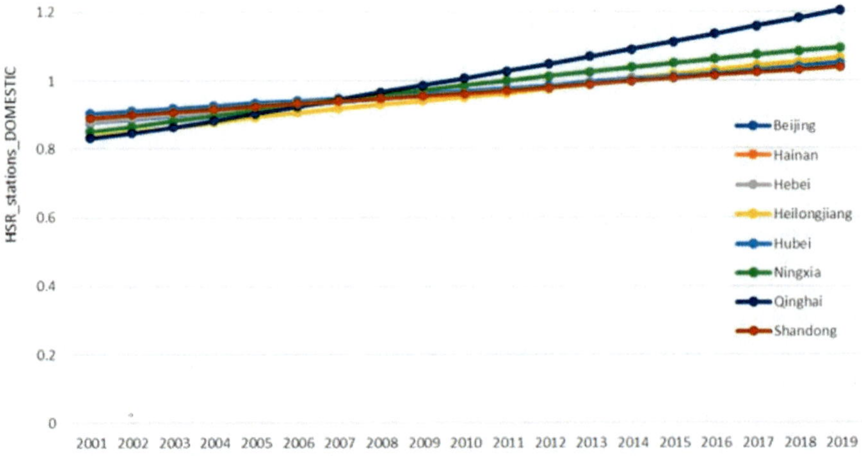

Fig. 7 The impact of HSR on eight provinces—GWTR: domestic tourists

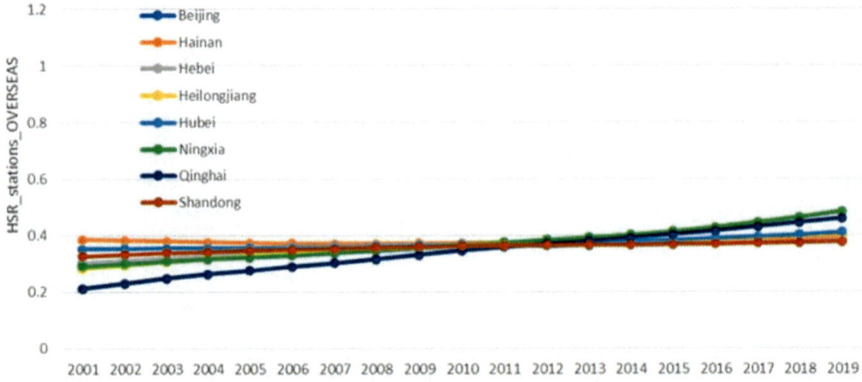

Fig. 8 The impact of HSR on eight provinces—GWTR: overseas tourists

from HSR on domestic tourism, while the HSR's impact on international tourism is quite relatively evenly distributed among all the provinces.

Overall, the study reveals some consistent patterns as Chen et al. [12], which adopted spatial econometric models. For instance, the impact of HSR on domestic tourism is found to be relatively strong in central provinces. Such a result suggests that HSR system tends to enhance the attractiveness of central regions and promote tourism due to improved regional accessibility and connectivity.

Future transport infrastructure project evaluation should consider adopting more advanced spatial modeling techniques, such as spatial weighted regression models and spatial econometric models, to capture both the spatial–temporal variation of

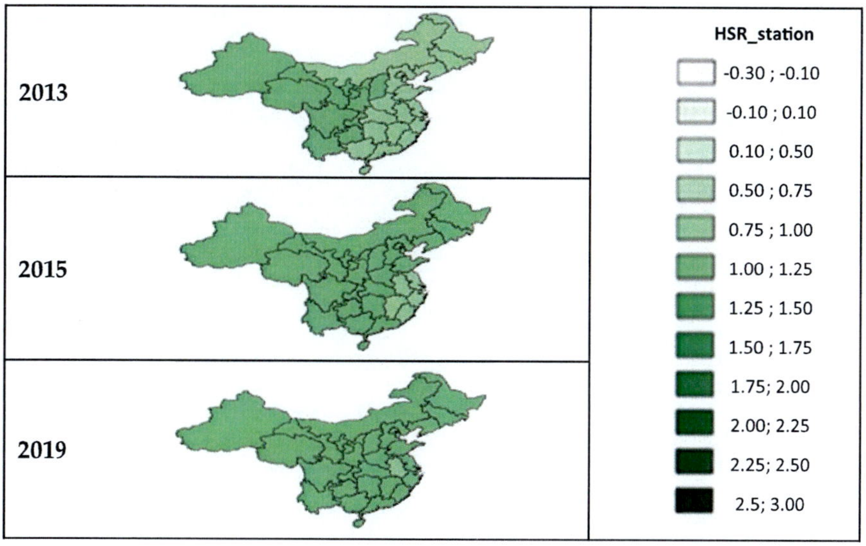

Fig. 9 The coefficients of the variable HSR for each provinces—GWTR: domestic tourists

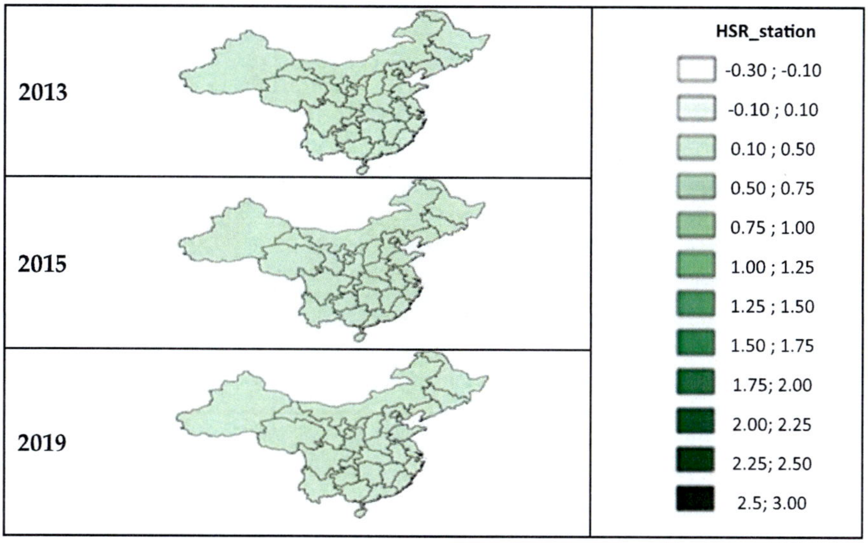

Fig. 10 The coefficients of the variable HSR for each provinces—GWTR: overseas tourists

impacts as well as the spatial dependence of impacts. Only a full understanding of the spatial and temporal impacts of the system may provide sound implications to guide future planning and policy decision-making.

References

1. Pagliara, F., Mauriello, F., Garofalo, A.: Exploring the interdependences between High Speed Rail systems and tourism: some evidence from Italy. Transp. Res. Part A Policy Pract. **106**, 300–308 (2017)
2. Pagliara, F., Mauriello, F.: Modelling the impact of high speed rail on tourists with geographically weighted poisson regression. Transp. Res. Part A Policy Pract. **132**, 780–790 (2020)
3. Yin, P., Pagliara, F., Wilson, A.: How does high-speed rail affect tourism? a case study of the capital region of China. Sustainability **11**(2), 472 (2019)
4. Cao, J., Liu, X.C., Wang, Y., Li, Q.: Accessibility impacts of China's high-speed rail network. J. Transp. Geogr. **28**, 12–21 (2013)
5. Jiao, J., Wang, J., Jin, F.: Impacts of high-speed rail lines on the city network in China. J. Transp. Geogr. **60**, 257–266 (2017)
6. Shaw, S.L., Fang, Z., Lu, S., Tao, R.: Impacts of high speed rail on railroad network accessibility in China. J. Transp. Geogr. **40**, 112–122 (2014)
7. Wang, W.C., Chang, Y.J., Wang, H.C.: An application of the spatial autocorrelation method on the change of real estate prices in Taitung City. ISPRS Int. J. Geo Inf. **8**(6), 249 (2019)
8. Chen, Z., Haynes, K.E.: Impact of high-speed rail on international tourism demand in China. Appl. Econ. Lett. **22**(1), 57–60 (2015)
9. Degen, W., Jia, Q., Yu, N.: Spatial pattern and evolution of urban tourism field strength in China under high speed railway network. Acta Geogr. Sin. **71**(10), 1784–1800 (2016)
10. Xin, W., Tongqiao, Z.: The influence of high speed railway network on the development and layout of China's regional tourism industry. Econ. Geogr. **30**(7), 1189–1194 (2010)
11. Feng, F., Linhao, C.: The opening of high-speed railway and the development of tourism in station cities: engine or corridor? Econ. Manag. **42**(2), 175–191 (2020)
12. Chen, Z., Haynes, K.E., Zhou, Y., Dai, Z.: High speed rail and China's new economic geography: Impact assessment from the regional science perspective. Edward Elgar Publishing, Chapter 6, pp. 148–168 (2019)
13. Taotao, D., Lei, Z., Mulan, M.: Research on the impact of Yangtze River delta high speed railway network on urban tourism development. Econ. Manag. **38**(1), 137–146 (2016)
14. Yuhua, Z., Jun, C.: Heterogeneous impact of high speed rail on tourism development of station cities: a study based on double difference method. Tour. Sci. **32**(6), 79–92 (2018)
15. Lei, L., Lin, L., Chenglin, M., Xiaolong, S.: Evolution of spatial structure of tourism flow in typical tourism cities in the era of high speed railway network: a case study of Huangshan City. Econ. Geogr. **39**(5), 207–216 (2019)
16. Yin, P., Lin, Z., Prideaux, B.: The impact of high-speed railway on tourism spatial structures between two adjoining metropolitan cities in China: Beijing and Tianjin. J. Transp. Geogr. **80**, 102495 (2019)
17. Yin, P., Tonghao, Z., Hanyan, Y.: Study on the change characteristics of urban tourism spatial function of Beijing Tianjin Hebei under the influence of high speed railway. Tour. Guide **4**(1), 30–44 (2020)
18. Ziyang, W., Dan, L., Yuting, Z.: Research on the impact of high speed railway on tourism development based on grey correlation analysis: a case study of Xiamen City. Logist. Eng. Manage. **42**(8), 163–165 (2020)
19. Yongze, Y., Yu, F., Haitao, Z.: The impact of the opening of high speed railway on the development of regional tourism. Res. Financ. **1**, 31–38 (2020)
20. Tinessa, F., Papola, A., Marzano, V.: The importance of choosing appropriate random utility models in complex choice contexts. In: 5th IEEE International Conference on Models and Technologies for Intelligent Transportation Systems (MT-ITS), pp. 884–888, IEEE, Naples, Italy (2017)
21. Tinessa, F., Marzano, V., Papola, A., Montanino, M., Simonelli, F.: CoNL route choice model: numerical assessment on a real dataset of trajectories. In: 6th International Conference on

Models and Technologies for Intelligent Transportation Systems (MT-ITS), pp. 1–10, IEEE, Cracow, Poland (2019)
22. Washington, S.P., Karlaftis, M.G., Mannering, F.: Statistical and Econometric Methods for Transportation Data Analysis. CRC Press, Boca Raton (2010)
23. Montella, A., Marzano, V., Mauriello, F., Vitillo, R., Fasanelli, R., Pernetti, M., Galante, F.: Development of macro-level safety performance functions in the city of Naples. Sustainability **11**(7), 1871 (2019)
24. Cafiso, S., Montella, A., D'Agostino, C., Mauriello, F., Galante, F.: Crash modification functions for pavement surface condition and geometric design indicators. Accid. Anal. Prev. **149**, 105887 (2021)
25. Jobson, J.D.: Applied Multivariate Data Analysis: Regression and Experimental Design. Springer, New York (1991)
26. Agresti, A.: Categorical Data Analysis. John Wiley, New Jersey (2002)
27. Fitzmaurice, G.M., Laird, N.M., Ware, J.H.: Applied Longitudinal Analysis, vol. 998. John Wiley & Sons (2012)
28. Fotheringham, A.S., Brunsdon, C., Charlton, M.: Geographically Weighted Regression: The Analysis of Spatially Varying Relationships. Wiley, Chichester (2002)
29. Nakaya, T., Fotheringham, A.S., Brunsdon, C., Charlton, M.: Geographically weighted Poisson regression for disease association mapping. Stat. Med. **24**, 2695–2717 (2005)
30. Kopczewska, K.: Applied Spatial Statistics and Econometrics Data Analysis in R. Routledge (2021)